Mechanical Wear Fundamentals and Testing

MECHANICAL ENGINEERING

A Series of Textbooks and Reference Books

Founding Editor

L. L. Faulkner

*Columbus Division, Battelle Memorial Institute
and Department of Mechanical Engineering
The Ohio State University
Columbus, Ohio*

Additional Volumes in Preparation

Mechanical Engineering Software

Mechanical Wear Fundamentals and Testing

Second Edition, Revised and Expanded

Raymond G. Bayer

Tribology Consultant
Vestal, New York, U.S.A.

CRC Press
Taylor & Francis Group
Boca Raton London New York

CRC Press is an imprint of the
Taylor & Francis Group, an **informa** business

CRC Press
Taylor & Francis Group
6000 Broken Sound Parkway NW, Suite 300
Boca Raton, FL 33487-2742

First issued in paperback 2019

© 2004 by Taylor & Francis Group, LLC
CRC Press is an imprint of Taylor & Francis Group, an Informa business

No claim to original U.S. Government works

ISBN-13: 978-0-8247-4620-9 (hbk)
ISBN-13: 978-0-367-39437-0 (pbk)

Library of Congress Cataloging-in-Publication Data
A catalog record for this book is available from the Library of Congress.

Visit the Taylor & Francis Web site at
http://www.taylorandfrancis.com

and the CRC Press Web site at
http://www.crcpress.com

To my wife, Barbara

Preface

It has been a decade since the first edition of this book was published. During that period important changes in the field of tribology have occurred. As a consultant I have also gained additional tribological experience in a wide range of industrial applications. It was thus decided to develop a second edition with the goal of incorporating this new information and additional experience into a more useful and current book, as well as clarifying and enhancing the original material. While doing this, the purpose and perspective of the first edition were to be maintained, namely, *"to provide a general understanding ... for the practicing engineer and designer ... engineering perspective ..."*. As rewriting progressed it became clear that the greatly expanded text would develop into a much larger volume that the first. We therefore decided to divide the material into two volumes, while keeping the basic format and style. Essentially the first two parts of the original edition on the fundamentals of wear and wear testing are combined into a single volume, *Mechanical Wear Fundamentals and Testing*. The remaining two parts of the first edition, which focus on design approaches to wear and the resolution of wear problems, are the basis for a second volume, *Engineering Design for Wear: Second Edition, Revised and Expanded*.

While a good deal of background material is the same as in the first edition, significant changes have been made. The most pervasive is the use of a new way of classifying wear mechanisms, which I have found to be useful in formulating approaches to industrial wear situations. As a result, Part A, Fundamentals, has been reorganized and rewritten to accommodate this new classification and to include additional material on wear mechanisms. The treatment of thermal and oxidative wear processes has been expanded, as well as the consideration of galling and fretting. The treatment of frictional heating is also expanded. A section on wear maps has been introduced. Additional wear tests are described in Part B, Testing, which has been expanded to include friction tests. The last two parts of the first edition are discussed in *Engineering Design for Wear*. Additional appendixes have been added, providing further information for use in engineering situations. These new appendixes include tables on threshold stress for galling and sliding wear relationships for different contact situations. A glossary of wear mechanisms has also been added.

These books demonstrates the feasibility of designing for wear and using analytical approaches to describe wear in engineering situations, which has been my experience over the last 40 years.

Raymond G. Bayer

v

Contents

1
Terminology and Classifications

1.1. WEAR, FRICTION, AND LUBRICATION

A number of different definitions, which have varying degrees of completeness, rigor, and formalism, can be found for wear (1–6). However, for engineering purposes, the following definition is adequate and contains the essential elements. Wear is progressive damage to a surface caused by relative motion with respect to another substance. It is significant to consider what is implied and excluded by this. One key point is that wear is damage and it is not limited to loss of material from a surface. However, loss of material is definitely one way in which a part can experience wear. Another way included in this definition is by movement of material without loss of mass. An example of this would be the change in the geometry or dimension of a part as the result of plastic deformation, such as from repeated hammering. There is also a third mode implied, which is damage to a surface that does not result in mass loss or in dimensional changes. An example of this might be the development of a network of cracks in a surface. This type of damage might be of significance in applications where maintaining optical transparency is a prime engineering concern. Lens and aircraft windows are examples where this is an appropriate definition for wear. As will be shown in subsequent sections, the significant point is that wear is not simply limited to loss of material, which is often implied in some, particularly older, definitions of wear. While wear is not limited to loss of material, wear damage, if allowed to progress without limit, will result in material loss. The newer and more inclusive definitions of wear are very natural to the design or device engineer, who thinks of wear in terms of a progressive change to a part that adversely affects its performance. The focus is on adverse change, which simply may be translated as damage, not necessarily loss of material. The implications of this generalization will be further explored in the discussion of wear measures.

Older definitions of wear and application oriented definitions often define wear in terms of limited contact situations, such as sliding or rolling contact between solid bodies. However, the definition of wear given does not have such limitations. It includes contact situations involving sliding, rolling, and impact between solid bodies, as well as contact situations between a solid surface and a moving fluid or a stream of liquid or solid particles. The wear in these latter situations is normally referred to as some form of erosion, such as cavitation, slurry, or solid particle erosion.

At least in the context of engineering application and design, these considerations essentially indicate what wear is. A brief consideration as to what it is not is of importance as well. Engineers, designers, and the users frequently use the phrase "it's worn out." Basically, this means that as a result of use, it no longer works the way it should or it is broken. In this context, the part or device may no longer function because it has experienced severe corrosion or because a part is broken into two pieces. In terms of the definitions for wear, these two failures would not be considered wear failures nor would the two mechanisms, that is, corrosion and fracture, be considered wear. Corrosion is not a form of wear because it is not caused by relative motion. Brittle and fatigue fracture in the sense referred to above are not considered forms of wear because they are more a body phenomenon rather than a surface phenomenon and relative motion and contact are not required for these mechanisms to occur.

While corrosion and fracture, per se, are not forms of wear, corrosion and fracture phenomena are definitely elements in wear. This is because in a wearing situation, there can be corrosive and fracture elements contributing to the damage that results from the relative motion. An illustration of this point is sliding, rolling, and impact situations in which material is lost as a result of the formation and propagation of cracks near the surface. In such situations, fatigue and brittle fracture mechanisms are generally involved in the wear. In addition to be involved in the wear, corrosion and fracture, per se, can be influenced by wear. An example of wear being a factor in fracture of a part is a situation where the wear scar might act as a stress concentration location to initiate fracture or where fracture results from the propagation of a crack formed in the wearing process. An example of a situation where both types of relationships can occur is wear situations involving the pumping of slurries. In such situations, wear behavior involves both chemical and mechanical factors and the severity of the corrosion can be influenced by the wear. These interactions and involvement of fracture and corrosion phenomena in wear will be further discussed and illustrated in subsequent chapters.

While illustrated by corrosion and fracture, the important point is that all failures of devices or life-limiting aspects associated with use or exposure are not the result of wear and wear processes. To be considered wear failures, there generally has to be some surface, mechanical, and relative motion aspects involved. However, as will be shown, wear mechanisms involve a very large number of physical and chemical phenomena including those involved in fracture and corrosion.

In view of these considerations, another way of defining wear for engineering use is that wear is damage to a surface resulting from mechanical interaction with another surface, body or fluid, which moves relative to it. Generally, the concern with wear is that ultimately this damage will become so large that it will interfere with the proper functioning of the device. While not the subject of this book, it is interesting to note that machining and polishing are forms of wear. As such, there is a positive side to wear and wear phenomena.

In situations involving sliding or rolling contact, a companion term with wear is friction. Friction is the force that occurs at the interface between two contacting bodies and opposes relative motion between those bodies. It is tangential to the interface and its direction is opposite to the motion or the incipient motion. Generally, the magnitude of the friction force is described in terms of a coefficient of friction, μ, which is the ratio of the friction force, F, to the normal force, N, pressing the two bodies together

$$\mu = \frac{F}{N} \tag{1.1}$$

Distinction is frequently made between the friction force that must be overcome to initiate sliding and that which must be overcome to maintain a constant relative speed.

The coefficient associated with the former is usually designated the static coefficient of friction, μ_s, and the latter the dynamic or kinetic coefficient of friction, μ_k. A frequently encountered impression is that the two terms, wear and friction, are almost synonymous in the sense that high friction equates to a high wear rate or poor wear behavior. The complimentary point of view is that low friction equates to a low wear rate or good wear behavior. As a generality, this is an erroneous concept. While there are common elements in wear and friction phenomena, as well as interrelationships between the two, that simple type of correlation is frequently violated. This point will become clear as the mechanisms for wear and friction, as well as design relationships, are presented and discussed. However, the point can be illustrated by the following observation. Teflon is noted for its ability to provide a low coefficient of friction at a sliding interface, for example, a dry steel/Teflon system typically has a value of $\mu \leq 0.1$. However, the wear of the system is generally higher than can be achieved with a lubricated hardened steel pair, where $\mu \approx 0.2$.

Another element that can be considered in differentiating between friction and wear is energy dissipation. Friction is associated with the total energy loss in a sliding system. The principal form of that energy loss is heat, which accounts for almost all of the energy loss (7–9). The energy associated with the movement or damage of the material at the surface, which is wear, is normally negligibly small in comparison to the heat energy.

Often in rolling situations, an additional term, related to friction, is used. This is traction. Traction is defined as a physical process in which a tangential force is transmitted across the interface between two bodies through dry friction or an intervening fluid film, resulting in motion, stoppage, or the transmission of power. The ratio of the tangential force transmitted, T, and the normal force, N, is called the coefficient of traction, μ_T

$$\mu_T = T/N \tag{1.2}$$

The coefficient of traction is equal to or less than the coefficient of friction. In rolling situations, the amount of traction occurring can often be a significant factor in wear behavior. In sliding situations, the coefficient of traction equals the coefficient of friction.

There are two other terms, lubrication and lubricant, which are related to friction and wear behavior and that need to be defined. One is lubrication. Lubrication may be defined as any technique for: (a) lowering friction, (b) lowering wear, or (c) lowering both. A lubricant is a material that, when introduced to the interface, performs one of those functions. Understood in this manner, any substance, solid, liquid, or gas, may be a lubricant; lubricants are not just liquid petroleum-based materials. It should be recognized that some materials may act as a friction reducer and a wear riser in some situations, as well as the converse. Different types of lubrication and lubricants are discussed in later sections and reasons for this apparent anomaly are pointed out. This is also a further illustration of the distinction between friction and wear.

1.2. WEAR CLASSIFICATIONS

There are three apparent ways in which wear may be classified. One is in terms of the appearance of the wear scar. A second way is in terms of the physical mechanism that removes the material or causes the damage. The third is in terms of the conditions surrounding the wear situation. Examples of terms in the first category are pitted, spalled, scratched, polished, crazed, fretted, gouged, and scuffed. Terms like adhesion, abrasion, delamination, oxidative are examples of the second type of classification. Phrases are commonly used for the third method of classification. Examples of this are lubricated wear, unlubricated wear, metal-to-metal sliding wear, rolling wear, high stress sliding wear, and high temperature metallic wear. All three methods of classification are useful to the engineer but in different ways. Classification in terms of appearance aids in the comparison of one wear situation with another. In this manner, it helps the engineer extrapolate experience gained in one wear situation to a newer one. It also aids in recognizing changes in the wear situation, such as differences in the wear situation at different locations on a part or at different portions of the operation cycle of a device. It is reasonable that if the wear looks different, different ways of controlling it or predicting it are required; if similar in appearance, the approaches used should also be similar.

Some of these aspects can be illustrated by considering the wear of gears. Scuffing is a term used to characterize the appearance of a wear scar produced as a result of sliding with poor or no lubrication in metal-to-metal systems. With gears, different portions of the tooth experience different types of relative motion. If designed and fabricated properly, near the pitch line it should be pure rolling. As you move further out, sliding occurs. If scuffing features are observed at the pitch line, it can be inferred that sliding is occurring, pointing to a possible contour or alignment problem. In a lubricated situation, there may be little evidence of wear near the tip. However, if evidence of scuffing wear is found to occur with time or with different operating conditions, it suggests a possible lubrication problem. Increased scuffing in such a case could be the result of lubricant degradation or loss, or the use of the wrong lubricant for the different condition. These observations would guide engineering action to resolve the problems.

The usefulness of classification by physical mechanism would be in guiding the engineer in using the correct models to project or predict wear life and to identify the significance of design parameters that can be controlled or modified. Given that the mechanism for wear is known, the engineer can then identify the dependency of such parameters as load, geometry, speed, and environment.

From a designer's viewpoint, the third type of classification is the most desirable and potentially the most useful. It describes a wear situation in terms of the macroscopic conditions that are dealt with in design. The implication is that given such a description, a very specific set of design rules, recommendations, equations, etc., can be identified and used.

While wear is generally described in terms of these three classifications, there is no uniform system in place at the present time. In addition, the same term might be used in the context of more than one classification concept. For example, the term scuffing is used in several ways. One author may use this term simply to describe the physical appearance. Another author may use this to indicate that the wear mechanism is adhesive wear. A third may use it to indicate wear under sliding conditions. This leads to another point that needs to be recognized with respect to these classifications.

While relationships exist between these classifications, the classifications are not equivalent nor are the interrelationships necessarily simple, direct, unique, or complete. A common error is to assume that a category in one is uniquely associated with ones in the other two, such as unlubricated metal-to-metal sliding is always associated with a scuffing appearance and adhesive wear. Basically, this is because there are numerous ways by which materials can wear and the way it wears is influenced by a wide number of factors. With the present state of knowledge within tribology, complete correlation between operating conditions, wear mechanisms, and appearance generally are not possible, particularly in relationship to practical engineering situations. Because of the complex nature of wear behavior, it has even been argued that it may never be possible or even practical to establish complete relations of this type (10,11). While this is the case, analytical relationships of more limited scope can be used effectively in engineering (12,13).

All three of types of classifications are used in this book, since individually they are of use to the designer and any one classification method is not sufficient to provide an adequate description in engineering situations.

REFERENCES

1. F Bowden, D Tabor. The Friction and Lubrication of Solids, Part I. New York: Oxford U. Press, 1964, pp 3, 285.
2. E Rabinowicz. Types of wear. In: Friction and Wear of Materials. New York: John Wiley and Sons, 1965, p 109.
3. M Peterson, W Winer, eds. Introduction to wear control. In: Wear Control Handbook. ASME, 1980, p 1.
4. M Neale, ed. Mechanisms of wear. In: Tribology Handbook. New York: John Wiley and Sons, 1973, p F6.
5. Glossary of Terms and Definitions in the Field of Friction, Wear and Lubrication (Tribology). Paris: Research Group on Wear of Engineering Materials Organization for Economic Cooperation and Development, 1969.
6. Standard Terminology Relating to Wear and Erosion, G40. Annual Book of ASTM Standards. ASTM International.
7. H Uetz, J Fohl. Wear as an energy transformation process. Wear 49:253–264, 1978.
8. D Rigney, J Hirth. Plastic deformation and sliding friction of metals. Wear 53:345–370, 1979.
9. F Kennedy. Thermomechanical phenomena in high speed rubbing. In: R Burton, ed. Thermal Deformation in Frictionally Heated Systems. Elsevier, 1980, pp 149–164.
10. D Rigney, W Glaeser. Wear resistance. In: ASM Handbook. Vol. 1, 9th ed. Materials Park, OH: ASM, 1978, pp 597–606.
11. K Ludema. Selecting Materials for Wear Resistance. Proceedings of International Conference on Wear of Materials. ASME, 1981, pp 1–6.
12. R Bayer. Wear Analysis for Engineers. HNB Publishing, 2002.
13. R Bayer. Comments on engineering needs and wear models. In: K Ludema, R Bayer, eds. Tribological Modeling for Mechanical Designers, STP 1105. ASTM International, 1991, pp 3–11.

2
Wear Measures

Previously wear was defined as damage to a surface. The most common form of that damage is loss or displacement of material and volume of material removed or displaced can be used as a measure of wear. For scientific purposes, this is generally the measure used to quantify wear. In many studies, particularly material investigations, mass loss is frequently the measure used for wear instead of volume. This is done because of the relative ease of performing a weight loss measurement. While mass and volume are often interchangeable, there are three problems associated with the use of mass as the measure of wear. One is that direct comparison of materials can only be done if their densities are the same. For bulk materials, this is not a major obstacle, since the density is either known or easily determined. In the case of coatings, however, this can be a major problem, since their densities may not be known or as easily determined. The other two problems are more intrinsic ones. A mass measurement does not measure displaced material, only material removed. In addition, the measured mass loss can be reduced by wear debris and transferred material that becomes attached to the surface and cannot be removed. It is not an uncommon experience in wear tests, utilizing mass or weight loss technique, to have a specimen "grow", that is, have a mass increase, as a result of transfer or debris accumulation.

From the above, it can be seen that volume is the fundamental measure for wear when wear is equated with loss or displacement of material. This is the case most frequently encountered in engineering applications. However, in engineering applications, the concern is generally not with volume loss, per se. The concern is generally with the loss of a dimension, an increase in clearance or a change in a contour. These types of changes and the volume loss are related to each other through the geometry of the wear scar and therefore can be correlated in a given situation. As a result, they are essentially the same measure. The important aspect to recognize is that the relationship between wear volume and a wear dimension, such as depth or width, is not necessarily a linear one. This is an important aspect to keep in mind when dealing with engineering situations, since many models for wear mechanisms are formulated in terms of volume. A practical consequence of this is illustrated by the following example.

Consider the situation where there is some wear experience with a pair of materials in a situation similar to one currently under study. Both applications are sensitive to wear depth. In the prior situation, it was observed that a reduction of load by a factor of 2 increased wear life by a factor of 2 and by implication reduced wear rate by the same factor. In the current situation the wear is too large and there is the possibility to reduce the load by a factor of 2. Because of the prior experience, it is assumed that this decrease in load should result in reducing wear by a 50%; however, when tried, only a 30% improvement

7

is found. The difference in the results of load reduction in the two situations can be explained if the primary relation is between wear volume and load, not depth and load. In the first situation, the part had a uniform cross-section and, as a result, the volume wear rate and the depth rate of wear rate would both have been proportional to load. In the second situation, the geometry of the wearing part was such that the wear volume was proportional to the square of the wear depth. This results in the depth rate of wear being proportional to square root of the load. In which case a factor of 2 reduction in load would result in only a 30% improvement. This is not a very profound point but is one that is frequently overlooked or not initially recognized in design work.

Volume, mass, and dimension are not the only measures for wear that are used in engineering. There is a wide variety of what may be termed operational measures of wear that are used. Life, vibration level, roughness, appearance, friction level, and degree of surface crack or crazing are some of the measures that are encountered. Generally, these measures are parameters that are related to performance, correlatable to wear, and typically monitored or can be monitored. There are two other practical reasons why such measures are used or needed. One is that the volume, dimension, or mass change associated with the wear feature of significance is so small that it is impractical to measure it, so another type of measurement relatable to wear is required. Since mass or volume loss is typically negligible in the wear of lenses, using the amount of scratches or haziness on a lens surface is an example of this. The other reason is that volume, dimension, or mass, while significant and measurable, cannot be conveniently measured, while the appearance of the part or response of the mechanism to that wear can be. For example, in a high-speed printer, the degradation of print quality can easily be monitored, but it may take several hours or days to disassemble the printer to obtain a wear depth measurement on the part.

While the utilization of these types of wear measures is often a practical necessity, they do add one more complication to engineering considerations of wear. It must be recognized that these operational measures of wear are generally indirectly related to primary wear behavior. As a result, additional factors have to be considered when extrapolating from one situation to another or relating to fundamental wear theory. One example of this would be the need to consider aspects, which are similar to the one discussed previously regarding the dependency on load. Another example is the need to consider the possibility that other elements, not related to the wear, could produce similar operational changes. For example, poor print quality in a high-speed printer could be the result of timing problems in the electronic controls, rather than excessive wear. Another example would be in the use of vibration levels to monitoring roller bearing performance. The noise level tends to increase or change with wear, but it could also change as a result of contamination of the bearing or loss of lubricant. Generally in such cases additional measurements or observations are needed to eliminate the effects of these other elements on the operational parameter.

3
Wear Mechanisms

3.1. OVERVIEW

This chapter focuses on the classification of wear in terms of the manner in which material is lost, displaced, or damaged as a result of a wearing action. As a starting point, it has been observed that wear, when it involves loss of material, generally occurs through the formation of particles rather than by loss of individual atoms (1). A similar statement can be made when wear is considered as displacement or even as damage. These aspects are generally the precursors to the formation of particles and their initial manifestations are generally on a much larger scale than atomic dimensions and involve more than individual atoms. A corollary of these observations is that wear mechanisms can generally be thought of as typical material failure mechanisms, occurring at or near the surface, which produce particles, rather than as atomic processes. As a consequence, most wear mechanisms are built around concepts such as brittle fracture, plastic deformation, fatigue, and cohesive and adhesive failures in bonded structures. In the case of wear, the complexities associated with each of these types of mechanisms are compounded by the fact that more than one body is involved as well as the unique properties and features of surfaces and the effect of wear on these. While this is the case, wear in some situation can result from atom removal processes. For example, as a result of high temperatures developed during machining, a contributing mechanism in tool wear can be by diffusion of atoms into the work piece (2–8). An additional example of a situation in which wear can occur by loss of atoms is an electrical contact situation that involves sparking. In this case, material can be lost as a result of the arcing process, which is usually described in terms of atom removal [(9,10), Sec. 7.6 of *Engineering Design for Wear: Second Edition, Revised and Expanded* (EDW 2E)]. Except for a few unique situations, such as these, atomic loss processes are not important in current engineering. However, atomic processes may become more important with the evolution of micro-electro-mechanical devices (MEMS) (11), where they may become life-limiting factors as a result of their sensitivity to very small amounts of wear.

A cursory review of tribological literature would tend to indicate that there is an extremely large number of wear mechanisms. For example, the glossary of this book contains over 80 terms for wear mechanisms. While extensive, this is not an all-inclusive list; others mechanisms or terms for them can be found in the literature, as well. While this is the case, it is possible to group wear mechanisms into a few generic categories. In the 1950s wear mechanisms were broken down into the following categories: adhesion, abrasion, corrosion, surface fatigue, and minor categories (12). While increased knowledge has made these categories somewhat simplistic and incomplete, wear behavior is still often categorized in these terms (13–18). However, a more refined and extensive set of categories is more appropriate and useful (19). These categories are given in Table 3.1. In general, wear

Table 3.1   Generic Wear Mechanisms

1. Adhesion
2. Single-cycle deformation
3. Repeated-cycle deformation
4. Oxidation (corrosion, chemical)
5. Thermal
6. Tribofilm
7. Atomic
8. Abrasion

behavior can be described either by specific mechanisms in these eight categories or by some combination of mechanisms in these categories.

The generic type of mechanism that is included in the adhesion or adhesive wear category may be described in the following manner. The basic concept is that, when two surfaces come into contact, they adhere to one another at localized sites. As the two surfaces move relative to one another, wear occurs by one surface pulling the material out of the other surface at these sites. For single-cycle deformation wear the concept is that of mechanical damage that can be caused during a single contact, such as plastic deformation, brittle fracture, or cutting. Repeated-cycle deformation wear mechanisms are also mechanical processes but ones were repeated contact or exposure is required for the damage to result. Examples are fatigue or ratcheting mechanisms (20–27). These three types of mechanisms are illustrated in Fig. 3.1.

Oxidation wear mechanisms, also referred to as corrosive or chemical wear mechanisms, are those in which chemical reactions are the controlling factor. With this type of

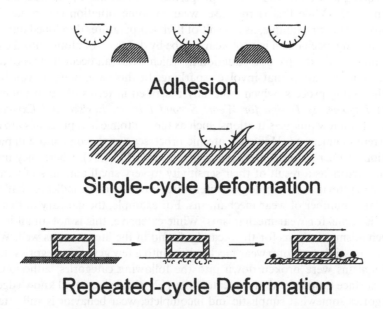

Figure 3.1   Conceptual illustration of adhesive and deformation wear mechanisms. A generic form of adhesive wear is shown. Single-cycle deformation wear is illustrated as a cutting process, while a fatigue process is used to illustrate repeated-cycle deformation wear.

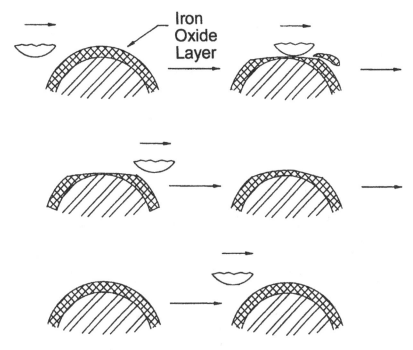

Figure 3.2 Illustration of the basic concept of chemical or oxidative wear, showing the removal, reformation, and removal of the reacted layer.

mechanism, the growth of the reacted or oxide layer controls wear rate. However, material removal occurs by one of the deformation modes, either in the layer or at the interface between the layer and the un-reacted material. An illustration of this process on an asperity on a steel surface is shown in Fig. 3.2.

Thermal wear mechanisms are those in which surface temperature or local heating of the surface controls the wear. Such mechanisms involve both a temperature increase, which is the driving factor, and material removal mechanisms, such as atomic, adhesion, and single or repeated cycle deformation mechanisms. Friction and hysteretic heating in wear situations can result in the formation of thermally soften layers or regions, melting, thermal cycling of the surface, regions of localized expansion, such as thermal mounding (28), and evaporation. The degree to which any of these phenomena occur can influence wear and, in some cases, they can be the primary and controlling factor in the wear. Figure 3.3 illustrates some thermal wear processes.

Films or layers composed of wear debris can form on or between surfaces. The existence of these films, which are called tribofilms, results in another type of wear mechanism or process. With the tribofilm type of mechanism, wear is controlled by the loss of material from the tribofilm. The basic concept is that the tribofilm is in a state of flux. The majority of the material circulates within the film and between the film and the surfaces, with a small amount being displaced from the interface. Under stable conditions, the amount of fresh wear debris that can enter the tribofilm is determined by the amount of material displaced from the interface. This process is illustrated in Fig. 3.4.

Atomic wear mechanisms are mechanisms that are based on the removal of individual of atoms or molecules from a surface. Mechanisms, such as electrical discharge, diffusion, and evaporation, are examples (6–8).

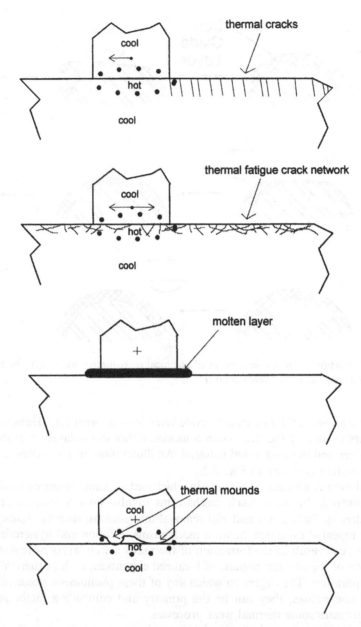

Figure 3.3 Illustration of several thermal wear mechanisms, cracking, fatigue, melting, and thermoelastic instability (TEI). With the thermoelastic instability mechanism, the real area of contact is reduced to a few isolated regions or mounds, as illustrated.

Abrasion or abrasive wear mechanisms are deformation mechanisms caused by hard particles or hard protuberances. This category is different from the others; it is primarily a classification based on wear situation, not a type of physical mechanism. However, it is a worthwhile classification because of the unique nature of wear by hard particles and the dominance and importance of this type of wear in many situations. With the older, simpler classification the term abrasion was used somewhat differently. It referred to wear mechanisms associated with hard protuberances or particles that resulted in grooves,

film composed of primary wear debris from surfaces

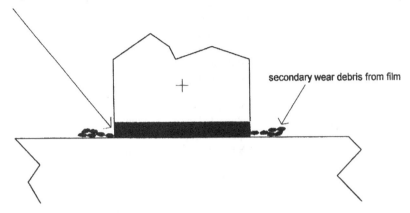

secondary wear debris from film

Figure 3.4 Conceptual illustration of a tribofilm wear process, showing the tribofilm composed of wear debris separating the surfaces and the loss of material from this layer.

scratches, or indentations on a surface. Using the newer, more extensive classification, this would correspond to the single-cycle deformation category. However, this is not the only way hard particles can cause wear. Such mechanisms are the typical type associated with hard particles when the wearing surface is softer than the particles. When the wearing surface is harder, the type of mechanism changes to the repeated-cycle deformation type. Abrasion in the current classification includes both types of deformation mechanisms.

A fifth category, called minor mechanisms, was also identified in the older classification (29). This was used for what was considered to be unique wear mechanisms, which were only encountered under special situations. However, knowledge gained over the past 50 years has shown that many of these unique mechanisms are variations of the more general types or particular combination of these. An illustration of this is delamination wear (25) and lamination wear (30). Each tends to explain the physical wear process in somewhat different manners and emphasize different aspects. Both involve the idea of crack nucleation and the eventual formation of a particle or fragment as a result of repeated engagement. Hence both can be viewed as subcategories of repeated-cycle deformation.

Another illustration is fretting and fretting corrosion. These were considered as unique wear mechanisms associated with small amplitude sliding. It is now generally recognized that these two modes of wear can be described in a two-step sequence. First wear debris is produced by either adhesion, single-cycle or repeated-cycle deformation. The wear debris, as a result of work hardening or oxidation, then acts as an abrasive, accelerating and controlling the wear from that point. By including oxidation of the surface, this sequence is used to explain fretting corrosion as well as fretting (31).

While this generic classification applies to all wear situations, the relevance and importance of the individual type of mechanism tends to vary with the nature of the situation. For example, Table 3.2 lists the more significant types typically associated with different types of motion. In addition, abrasion can be important in all situations when there are significant amounts of abrasives, that is hard particles, present. It is generally not important otherwise. Also the importance of chemical and thermal mechanisms tends to vary with the type of material involved and lubrication conditions. Repeated-cycle deformation, oxidation, and tribofilm mechanisms tend to become the more dominant mechanisms in long-term sliding wear behavior.

Table 3.2 Significant Wear Mechanisms

Sliding motion
 Adhesion
 Single-cycle deformation
 Repeated-cycle deformation
 Oxidation
 Tribofilm
Rolling motion
 Repeated-cycle deformation
Impact motion
 Single-cycle deformation
 Repeated-cycle deformation

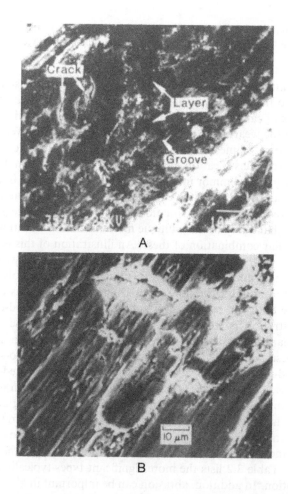

Figure 3.5 Examples of the simultaneous occurrence of several wear mechanisms during sliding. Cracks and severe plastic deformation indicate repeated-cycle deformation mechanisms. Grooves are indicative of single-cycle deformation. Adhered and deformed material is an indicator of adhesive wear. Debris layers and layers of compacted material are characteristics of tribofilm mechanisms. ("A" from Ref. 174, "B" from Ref. 87 reprinted with permission from ASME.)

More than one type of mechanism can be present in a wear situation. Typically, one can find in the examination of a wear scar features indicative of different mechanisms, as illustrated by the micrographs shown in Fig. 3.5. When more than one mechanism is present, they can interact serially to form a more complex wear process, as illustrated by the fretting situation discussed previously in this section. They can also act in a parallel or simultaneous fashion, with each contributing to the total wear. While this is the case, most situations can usually be characterized in terms of one controlling or dominant mechanism. There are some situations, however, where this cannot be done and it is necessary to consider the contributions of each (10,32–34).

There is another approach to classifying wear mechanism that can also be useful (35). In this classification, wear mechanisms are divided into cohesive wear and interfacial wear categories. Under cohesive wear are those wear mechanisms which occur primarily in the relatively large volumes adjacent to the interface. Interfacial wear, on the other hand, includes those mechanisms related to the interface alone. Both types of deformation mechanisms would be included in the former, while adhesion, tribofilm, and oxidation mechanisms in the latter. Thermal could be of either type, depending on the depth of the heat-affected zone. This alternate classification focuses on the significance of the energy densities involved in the two regions, that is, in the thin layers at the interface and in the larger regions adjacent to it. A corollary to this classification is that bulk properties and responses are generally major aspects in the mechanisms included in the cohesive category, while surface properties and phenomena are key in the interfacial category. While this classification is not particularly useful in grouping physical mechanisms, it is useful for identifying aspects that must be considered in the treatment of wear and offers the opportunity for some insight into what are controlling factors in certain wear situations, that is surface vs. bulk phenomena.

The classification of basic wear into the eight categories shown in Table 3.1 is not necessarily a complete or rigorous classification. However, it does provide a useful basis for an effective engineering understanding of wear, particularly as it relates to design.

3.2. ADHESIVE MECHANISMS

Before adhesive mechanisms are discussed some general concepts regarding the nature of the contact between two surfaces and the behavior of inter-atomic forces need to be considered. The first aspect that will be considered is the area of contact.

In engineering, the macro-geometry or contour of the bodies in contact is often used to determine contact area. This is usually done by geometrical projection or by models, which are based on the elastic or plastic deformation. For example, the Hertz contact theory is frequently used not only to determine stress levels in the contact but the size of the contact region as well (10,36). In these approaches, the surfaces are generally assumed to be smooth. Actual surfaces, on the other hand, always exhibit some degree of roughness and as a result the actual contact situation is different from that implied by these macro-methods. Figure 3.6 illustrates the actual situation. What this illustrates is that actual physical contact occurs at localized spots within the area that is defined by the macro-geometry. These points at which the actual contact occurs are referred to as junctions. The sum of the individual contact areas of these junctions is called the real area of contact. The area of contact that is determined through the macro-considerations is called the apparent area of contact. As will be seen, fundamental physical models regarding wear generally are

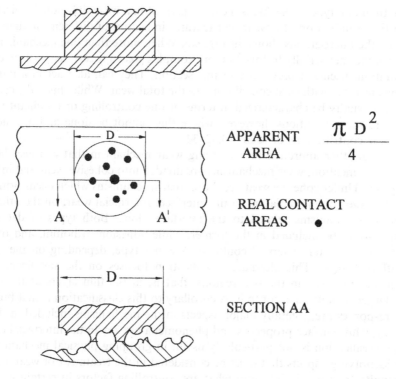

APPARENT $$\dfrac{\pi D^2}{4}$$
AREA

REAL CONTACT
AREAS   •

SECTION AA

Figure 3.6 Apparent and real area of contact. Contact occurs are discrete locations, called junctions.

developed in terms of real area considerations, while engineering formulations and models generally are related to the apparent area of contact.

The roughness characteristics of the surface have a significant influence on the number of junctions formed, as well as on the ratio of the real area of contact to the apparent area of contact. The degree to which one surface penetrates the other, which is a function of the normal force pressing the bodies together, can also influence both these aspects. Figure 3.7 shows how the real area of contact changes, assuming one surface to be flat and smooth, as load and penetration is increased. The real area in this illustration increases not only because the cross-sectional area of an asperity increases with penetration but also because the number of asperities encountered increases with penetration.

The size and number of these junctions and their relationship to the apparent area of contact have been investigated by both theoretical and experimental means (37–40). Because of the potential range of the parameters involved, a wide variety of contact conditions is possible; however, some generalization may be made. One is that the real area of contact is generally much less than the apparent area. The ratio might be as small as 10^{-4} in practical situations (41). A similar generalization can be made regarding individual junctions. It has been estimated that the diameter of typical junctions is in the range of 1–100 μm (42). The larger value would most likely occur for a very rough surface and high loads. Diameters of the order of 10 μm would be more likely in more typical contact situations. While on the basis of stability, it can be argued that there must be at least three junctions involved, the number generally is larger. Estimates based on the yield point of materials and junction size generally indicate that the number ranges from the order of

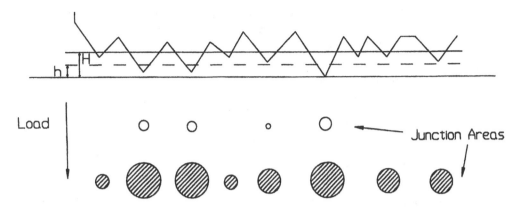

Figure 3.7 The effect of increased load on the real area of contact. h is the penetration of a rough, hard, surface into a smooth, flat, soft surface at a low load; H, at a higher load. Both the number and size of the junctions increase with increasing penetration.

10 to the order of 10^3, with 10 to 10^2 being more likely (43). The deformation properties of the materials involved and the loading conditions on the junctions also influence the real area of contact. Junction growth as a result of applied shear, that is, friction, also occurs.

The deformation at the junction can be plastic as well as elastic. Just how much of each occurs depends on the number of junctions and their size, as well as the properties of the materials involved. While it is not impossible to have only elastic deformation on all the junctions, this is generally not the case. Models, assuming typical surface profiles, indicate that some junctions would generally be plastically deformed (37,38). This tends to be confirmed by the topography found in the initial stages of wear. Some evidence of local plastic deformation can usually be found on these.

The contact between rough surfaces and the effect of shear have been modeled and equations for the real area of contact have been developed (37,38,44,45). A summary of the equations for the real area of contact obtained from these models is given in Table 3.3. If the plasticity index, a measure of the state of stress at the junctions, is less than 0.6, all the junctions are elastically deformed; if greater than 1, all are plastically deformed. For intermediate values, there are some junctions in both states. It can be seen from the equation for the plasticity index that increased hardness, lower modulus, more uniform and rounder asperities reduce the degree of plastic deformation in the real area of contact. For typical unworn surfaces, the plasticity index is generally closer to 1 than 0.6. However, roughness features tend to change with wear and as a result the index changes with wear, often becoming smaller (37). Because of these changes in roughness and the change in apparent area of contact that is often the result of wear, the ratio of real to apparent areas of contact, the number of junctions formed, and the size of the junctions typically change with wear as well.

In summary, the most significant points to be recognized about the contact between two bodies is that actual contact occurs at individual sites within an apparent area of contact and that the real area is generally only a fraction of the apparent area. The features observed in most micrographs of wear scars produced under sliding conditions support this view of the contact between two surfaces, as well as the generalizations regarding the ratio of real and apparent areas, junction size and number, and the plasticity index.

It is important to understand the nature of the interaction that occurs at these junctions on both an asperity and an atomic level. At the asperity level, the focus is on the type

Table 3.3 Equations for the Real Area of Contact

Plasticity index

$$\psi = \frac{E'}{p}\left(\frac{\delta'}{r'}\right)$$

For elastic contact, $\psi < 0.6$
For plastic contact, $\psi > 1.0$
Mixed, $0.6 < \psi < 1.0$

Elastic contact

$$A_R \approx \frac{3.2P}{E'(\delta'/r')^{1/2}}$$

Plastic contact

$$A_R' = \frac{P}{p}$$

Effect of shear

$$A_R' = A_R\left(1 + \alpha\mu^2\right)^{1/2}$$

Symbols
  ψ, plasticity index
  A_R, real area of contact without friction
  A'_R, real area of contact with friction
  p, hardness of softer material
  P, load
  E', composite elastic modulus

$$\left[\frac{(1-\nu_1^2)}{E_1} + \frac{(1-\nu_2^2)}{E_2}\right]^{-1}$$

  E, elastic modulus
  ν, Poisson's ratio
  δ', composite standard deviation of asperity peak heights

$$\left(\sigma_1^2 + \sigma_2^2\right)^{1/2}$$

  σ, standard deviation of asperity peak heights
  r', composite mean radius of curvature of asperity tips

$$\left(\frac{1}{r_1} + \frac{1}{r_2}\right)^{-1}$$

  r, radius of curvature of asperity tips
  μ, coefficient of friction
  α, empirical constant (approximately 12)

of deformation that occurs at these junctions. To understand the interactions on an atomic level, it is necessary to consider the nature of inter-atomic forces. The behavior of the force between two atoms is illustrated in Fig. 3.8. For large separations between the atoms there is a weak attractive force. At separations comparable to inter-atomic spacing the attractive force increases rapidly. With still smaller separations, the attractive force begins to decrease and ultimately the force changes to a repulsive one. Arrays of atoms also exhibit the same general behavior, which is shown in Fig. 3.9 for the case of an Al crystal and a Zn crystal (46). In this figure, A shows the variation in the potential energy of such a contact as a function of the separation of the two crystals. This is the more common way of describing the interactions. A negative potential energy indicates bonding. The slope of the curve is force; a negative slope indicates a repulsive force and a positive slope indicates an attractive force. B shows the corresponding variation in force with separation.

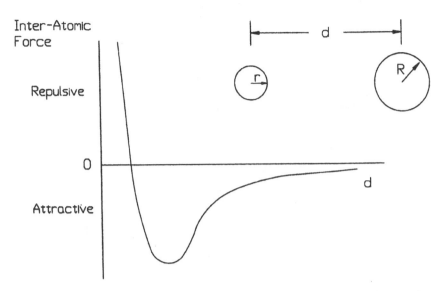

Figure 3.8 General nature of the force between atoms as a function of separation.

Since junctions form as a result of two surfaces being pressed together, the nature of inter-atomic forces indicates that bonding occurs at these junctions. It also means that over some portion of the real area of contact the atoms of the two surfaces must have gone past the point of maximum bonding. This is the only way the forces can be balanced. This implies that some adhesive forces or bonds must be overcome to separate the two surfaces at these sites. This atomic view of the contact situation at the junctions provides the foundation for the concept of adhesive wear.

Consider the diagram shown in Fig. 3.10. This depicts the situation at a junction at which bonding has occurred. As the two surfaces move relative to one another, rupture of the junction will eventually occur. If the rupture occurs along Path 2, which is the original interface, no material will be lost from either surface, though some plastic deformation may have occurred. If, on the other hand, the rupture occurs along some other path, illustrated by Path 1 in the figure, the upper surface would have lost material. The removal of material from a surface in this manner is called adhesive wear.

Scanning electron microscope (SEM) and electron dispersive x-ray (EDX) micrographs, illustrating the adhesive wear process, are shown in Fig. 3.11. These micrographs show the sequence of events associated with a simulated asperity moving across a smooth flat surface. In this case, a small rounded iron stylus simulates the asperity. In the lower right of A, the results of asperity engagement and junction formation are evident. Initially the junction formed by this asperity appears to rupture at the original interface, leaving only a plastically deformed groove in the wake of its motion. Some deformation of the asperity is likely as well during this period. At some point, failure no longer occurs at the original interface but at some depth within the asperity, leaving a portion of the asperity adhering to the flat surface. This is the event indicated in the middle by the adhered wear fragment. B shows that as sliding continued, the same series of events repeated (upper left in the EDX) but with the asperity now modified both by plastic deformation and adhesive wear.

There is a mathematical model for adhesive wear that has been found to be in good agreement with experimental observations (47). It has been used extensively in describing

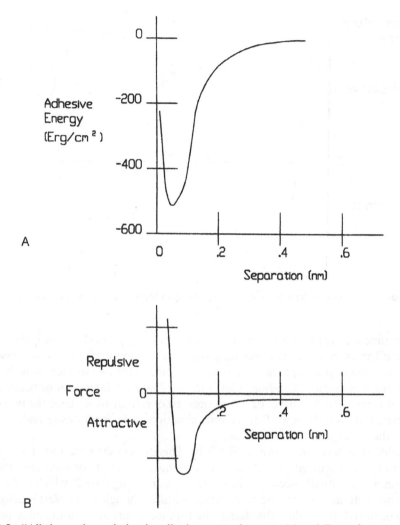

Figure 3.9 "A" shows the variation in adhesive energy between A1 and Zn surfaces as a function of separation (from Ref. 46). "B" illustrates the corresponding variation in the force between the two surfaces.

adhesive wear behavior. This formulation can be developed as follows. Assume that the real area of contact is composed of n circular junctions of diameter d. Further, assume that if an adhesive wear fragment is formed, it will be hemispherical shaped with a diameter d. The total real area of contact, A_r, is then

$$A_r = \frac{n\pi d^2}{4} \qquad (3.1)$$

An assumption frequently used in tribology is that all the junctions are plastically deformed. In which case A_r is also given by the following equation from Table 3.3

$$A_r = \frac{P}{p} \qquad (3.2)$$

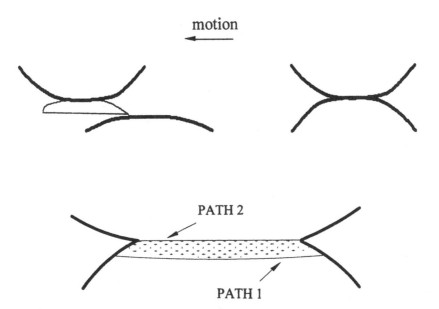

Figure 3.10 The lower diagram illustrates possible separation paths at a junction. Separation along Path 1 does not result in loss of material. Separation along Path 2 results in rupture and loss of material, as indicated in the upper diagrams.

where P is the normal force pressing the two surface together and p is the penetration hardness of the softer material. Combining these equations, the following is obtained for n:

$$n = \frac{4P}{\pi p d^2} \tag{3.3}$$

Now, if the distance of sliding over which any given junction is operative is approximately d, then in a unit distance of sliding each junction must be replaced by another $(1/d)$ times. Therefore, the total number of junctions occurring in a unit distance, N, is

$$N = \frac{n}{d} = \frac{4P}{\pi p d^3} \tag{3.4}$$

If K is the probability that the rupture of any given junction will result in adhesive wear, the number of junctions producing adhesive wear in a unit sliding distance, M, is given by

$$M = KN = \frac{4P}{\pi p d^3} \tag{3.5}$$

Since the volume of an adhesive wear fragment is $\pi d^3/12$, the volumetric wear rate, dV/dx, where V is the volume of wear and x the distance of sliding, is

$$\frac{dV}{dx} = \frac{\pi d^3}{12} \tag{3.6}$$

Integrating and combining the following relationship is obtained for adhesive wear:

$$V = \frac{K}{3p} Px \tag{3.7}$$

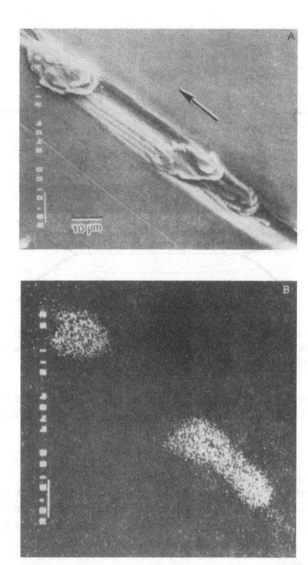

Figure 3.11 Example of adhesive wear process. "A" shows the wear scar produced by an iron pin sliding across the flat surface of a nickel disk. "B" is an EDX map for iron on the disk surface, confirming transfer of material from the pin to the disk. (From Ref. 175, reprinted with permission from ASME.)

This equation was first developed by Archard (47) and because of that it is frequently referred to as Archard's equation.

A key point in the development of this equation is that K is a probability and therefore it cannot be greater than unity. Experimental data are consistent with that. This can be seen in Table 3.4, which gives values for K inferred from sliding wear data for a range of conditions. Such data generally provide an upper-bound estimate of K, since in many cases other wear mechanisms are present, contributing to the wear and possibly even dominating the situation. However, K values in the range of 10^{-4} or higher have been documented for wear situations in which adhesion has dominated (48). This being the case, these data also do indicate that in most situations K is likely quite small, particularly in practical

Table 3.4 K Values for Adhesive Wear

Combination	K
Selfmated metals	
Dry	$2 \times 10^{-4} - 0.2$
Lubricated	$9 \times 10^{-7} - 9 \times 10^{-4}$
Non-self-mated metals	
Dry	$6 \times 10^{-4} - 2 \times 10^{-3}$
Lubricated	$9 \times 10^{-8} - 3 \times 10^{-4}$
Plastics on metals	
Dry	$3 \times 10^{-7} - 8 \times 10^{-5}$
Lubricated	$1 \times 10^{-6} - 5 \times 10^{-6}$

situations. It also indicates that the range of K values possible is very large. The values indicate that the probability of a junction wearing by adhesion can range from one in ten to less than one in a million. As a point of reference, a value of K of 10^{-5} and often much less is required for acceptable wear behavior in applications. This being the case, the large range possible for K makes it an extremely important parameter in controlling this type of wear mechanism. From an engineering point of view, this means selecting design parameters so that high values of K are avoided, i.e., the probability of adhesion is low.

One factor that can affect K is the relative strength of the junction interfaces to the strength of the asperities that make up the junctions. The weaker the adhesion at the interfaces is, the less likely adhesive wear will occur. Consequently, choosing conditions, which inhibit adhesion over those, that promote adhesion reduce K and adhesive wear.

Interface adhesion can be affected by the similarity of the two materials in contact. The more similar they are, the stronger the adhesion. As a result, dissimilar pairs should have lower values of K than more similar or identical pairs. In addition to composition, such aspects as lattice parameters and mutual solubility characteristics, can also be factors in determining the degree of similarity (49–52). Another factor affecting interface adhesion is surface energy. Lower surface energies result in lower adhesion. Therefore, since polymers and ceramics generally have lower surface energies than metals, K values would generally be lower for situations involving these materials than between metals. Also, the presence of oxides, lubricants and contaminates on metal surfaces reduce surface energy and result in lowering of K values.

The data in Table 3.4 illustrate some of these trends. Clean, unlubricated, and similar metal pairs generally have high values for K. Lubricated conditions give the lowest values and conditions involving ceramics and polymers have intermediate values associated with them.

As is shown in the following, there is a minimum asperity load required for transfer to take place and a minimum asperity load for the transferred fragment to remain attached. K, which is the average probability for transfer, will be changed as the percentage of junctions with loads below the critical value for transfer changes, since the probability for failure at these junctions is 0. Consequently, K can also be affected by the distribution of load across the junctions. The percentage of these junctions would tend to be higher as load is decreased, since the average pressure on the junctions decreases with decreasing load (38). A simple model can illustrate the requirement for a minimum asperity load for transfer (53).

Consider a circular junction of diameter d and the formation of hemispherical wear fragment of diameter d, as illustrated in Fig. 3.12. For such an ideal situation, the adhesive wear process can be reduced to the following criteria. For adhesive wear to take place, the elastic energy stored in the volume of the potential fragment, E_v, must be equal to or greater than the energy associated with the new surfaces, E_s. Mathematically,

$$E_v \geq E_s \tag{3.8}$$

Assuming that the tip of the asperity has been plastically deformed, the stored elastic energy per unit volume is

$$e_v = \frac{\sigma_y^2}{2Y} \tag{3.9}$$

where σ_y is the yield point and Y is Young's modulus. Since the volume of the hemispherical region is $\pi d^3/12$,

$$E_v = \frac{\pi d^3 \sigma_y^2}{24Y} \tag{3.10}$$

Noting that two hemispherical surfaces are formed and assuming that the material on both sides is the same,

$$E_s = \pi d^2 \Gamma \tag{3.11}$$

where Γ is the surface energy. Combining, the minimum junction diameter, d', required for an adhesive wear fragment to be formed is

$$d' = \frac{24\Gamma Y}{\sigma_y^2} \tag{3.12}$$

If P_a is the load supported by that asperity,

$$P_a = \frac{\pi p d^2}{4} \tag{3.13}$$

Combining these two equations and utilizing the following the empirical relationships (53):

$$\sigma_y = 3 \times 10^{-3} Y \tag{3.14}$$

$$\sigma_y = \frac{p}{3} \tag{3.15}$$

$$\Gamma = \frac{p^{1/3}}{\beta} \tag{3.16}$$

where β is a constant for different classes of materials, it can then be shown that the minimum asperity load for adhesive wear to take place is inversely proportional to the surface energy, namely,

$$P_a' = \frac{4.5 \times 10^7 \beta^3}{\Gamma} \tag{3.17}$$

Since the sum of the asperity loads must equal the applied macro-load, P, a similar relationship should also exist at the macro-level. For the simple case of a uniform asperity

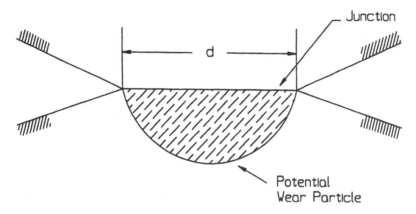

Figure 3.12 Model for the formation of a hemispherical adhesive wear fragment.

distribution, the minimum macro-load, P', would be N times Eq. (3.17), where N is the number of asperities in contact.

A similar approach can be used to develop expressions regarding the formation of loose adhesive wear fragments (53). The expression for the minimum junction size for a loose fragment is

$$d'' = \frac{2 \times 10^4 W_{ab}}{p} \tag{3.18}$$

The expression for the minimum asperity load is

$$P''_a = \frac{\pi \times 10^8 W_{ab}}{p} \tag{3.19}$$

W_{ab} is the interfacial energy between the two surfaces and p is the hardness of the softer of the two materials. The concept behind these relationships is that when junction separation occurs the stored elastic energy in a bonded fragment will cause that fragment to break off, if the elastic energy is greater than the interface energy.

The combined concepts of a minimum load for transfer to occur and K being affected by load are illustrated and supported by the data shown in Figs. 3.13 and 3.14 for unlubricated sliding between noble metal specimens. Figure 3.13 shows how K varies with load for unlubricated sliding between gold specimens. There are two stable regions for K, one below 5 g and the other above 30 g. These two transition points are close to the minimum loads for transfer and loose fragments to occur, respectively. Values of 1 and 25 g are estimated for these, using the simple models described (54). The data suggest that adhesive wear can only be characterized by a stable K value after the mean junction load exceeds the junction load required for the formation of a loose particle. At the same time the data support the concept that K becomes 0 at sufficiently low loads. The fact that in the graph K does not go to 0 and appears to stabilize below 5 g is attributed to the existence of other wear mechanisms. The appearance of wear scars for load under the minimum load for transfer indicates that the wear mechanism is some form of deformation. An example of such a wear scar is shown in Fig. 3.14. Note the absence of features suggestive of transfer.

Since a higher plasticity index implies a higher mean junction load, a corollary to the requirement for a minimum load for transfer is that the probability for transfer would tend

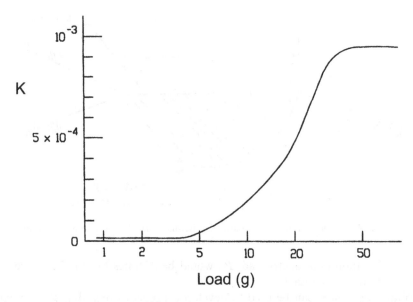

Figure 3.13 Variation in K of Eq. (3.7) as a function of load for unlubricated, Au–Au sliding. (From Ref. 53.)

to be higher with higher values of the index. Consequently, roughness conditions that increase the plasticity index would also tend to increase K.

Transfer is rarely observed in nominal rolling and impact situations. When it is observed in such situations, traction and slip at the interface are generally present (55,56). Also, a greater degree of transfer and adhesion is observed when interfaces are pulled apart after they have been subjected to shear than when they have not (55–57). It is generally concluded from these observations that the probability for transfer

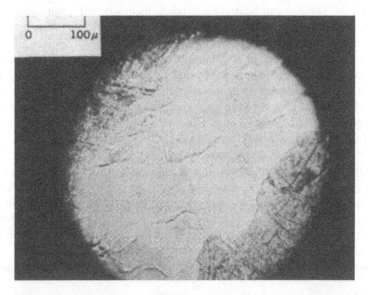

Figure 3.14 Wear scar on silver, below the minimum load required for transfer. (From Ref. 53, reprinted with permission from John Wiley and Sons, Inc.)

is much lower when junctions are pulled apart than when sheared. As a consequence, adhesive wear is primarily considered to be a sliding wear mechanism. K is typically assume to be 0 for pure rolling and normal impact. In rolling and impact situations, it is generally assumed that any adhesive wear is the result of tangential motion, which may be present.

In summary, the model for adhesive wear indicates that the major factor involved is the probability factor, K, which varies over several orders of magnitude. This factor is influenced by a wide variety of parameters, that may be grouped as follows: material pair compatibility, surface energies, lubrication, as well as the nature of asperity contact and load distributions. Also, adhesive wear is most probable in sliding wear situations but may occur in rolling and impact situations when there is slip and traction between the surfaces.

3.3. SINGLE-CYCLE DEFORMATION MECHANISMS

Single-cycle deformation mechanisms are deformation mechanisms that produce plastic deformation, permanent displacement, or removal of material in a single engagement. These processes result from the penetration of a softer body by a harder body. Common forms of this type of mechanism for sliding are plowing, wedge formation, cutting, and microcracking, which are illustrated in Fig. 3.15. Physical examples of three of these mechanisms, plowing, wedge formation and cutting, are shown in Fig. 3.16. An example of microcracking would be the fracturing that occurs when an ice pick is dragged across a piece of ice. The same generic mechanisms of plastic deformation and cracking are also possible for pure rolling and normal impact. In rolling and impact situations, cutting is also possible when there is some sliding or tangential motion involved. The micrographs shown in Fig. 3.17 provide additional examples of the damage resulting from these types of single-cycle deformation mechanisms. Single-cycle deformation mechanisms are the dominant mechanisms in abrasive and erosive wear situations, when the particles are harder than the wearing surface.

It has been found that for sliding situations in which single-cycle deformation mechanisms dominate, the wear can be described by the following equation, which is the same form as that for adhesion [see Eq. (3.7)]. However, while the forms are the same, the coefficients are affected by different parameters

$$V = \frac{KP_X}{p} \tag{3.20}$$

For single-cycle deformation mechanism, the following model provides a basis for this empirical relationship.

The model assumes that the junctions between the two surfaces can be represented by an array of hard conical indenters of different sharpness plastically indenting and penetrating a softer surface. As relative sliding occurs the cones produce wear grooves in the softer surface, whose individual volumes are the cross-sectional area of the indentation times the distance of sliding. The situation for a single cone is shown in Fig. 3.18.

Since penetration hardness, p, is load divided by projected contact area, the load on an individual cone, P_i, is pa_i, where a_i is contact area for the individual cone. Only the leading surface of the cone is in contact. As a result, the projected contact area of a cone

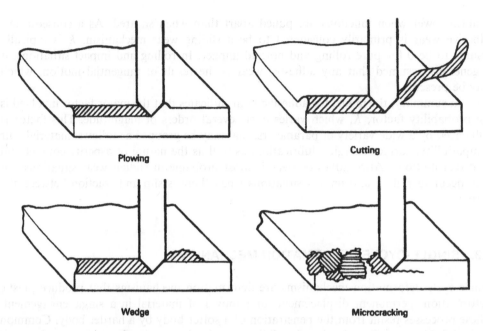

Figure 3.15 Four single-cycle deformation mechanisms. (From Ref. 142, reprinted with permission from ASM International.)

is $\pi r_i^2/2$. This results in the following for P_i:

$$P_i = p\frac{\pi r_i^2}{2} \tag{3.21}$$

By consideration of the geometry of the contact, it can be shown that the cross-sectional area of the indentation is $r_i^2 \tan \theta_i$. The wear volume, dV_i, produced in a sliding distance, dx, by a single cone, is therefore given by

$$dV_i = r_i^2 \tan \theta_i \, dx \tag{3.22}$$

Combining with Eq. (3.21), integrating and summing over the array, the following expression is obtained total amount of wear, V, occurring for a sliding distance of x

$$V = \frac{2}{\pi p}\left\{\sum_{i=1}^{n} P_i \tan \theta_i\right\}x \tag{3.23}$$

It is possible to convert this form to one using the total load, P, by using an effective value for $\tan \theta_i$, $\tan \Theta$. This is defined by the following equation:

$$P \tan \Theta = \sum_{i=1}^{n} P_i \tan \theta_i \tag{3.24}$$

Using this effective value, the equation for V can be written as

$$V = \frac{2 \tan \Theta}{\pi p} Px \tag{3.25}$$

Combing the constants with $\tan \Theta$, Eq. (3.20) is obtained.

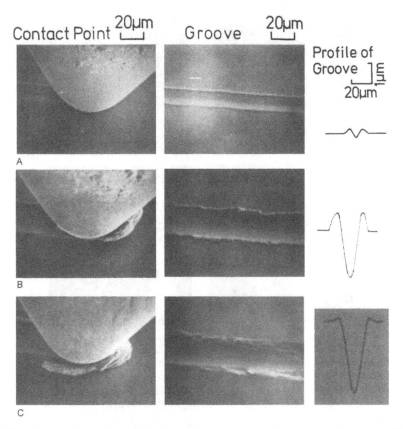

Figure 3.16 Changes in single-cycle deformation wear morphology as a function of increasing load. "A", plowing, "B", wedge formation; "C", cutting. Load increases from A to C. (From Ref. 62, reprinted with permission from ASME.)

In Table 3.5, values of K for different values of Θ are given. The equivalent included cone angle and surface roughness condition, as well as the range for some silicon carbide abrasive papers, are also given in this table (58). These suggest that the nominal range for K in typical situations is between 10^{-2} and 1. Empirical data from situations where this form of wear is known to dominate indicate a narrower but similar range, 2×10^{-2} to 2×10^{-1} (59).

Empirical data also show that K is also affected by material properties and wear mechanism. Conceptually, these effects can be taken into account by modifying the relationship between indentation size and groove size. To account for these affects, it is assumed that a proportionality relationship exists between the area of the groove and the area of the indention, rather than being equal. This modifies the expression for $\tan \Theta$ to the following:

$$P \tan \Theta = \varepsilon \sum_{i=1}^{n} P_i \tan \theta_i \qquad (3.26)$$

In this expression, ε is the ratio of the groove area to the indentation area, which can be affected by material properties, such as ductility, toughness, and elasticity, and vary with the mechanism. As a result, K, which is proportional to ε, would also be influenced by these.

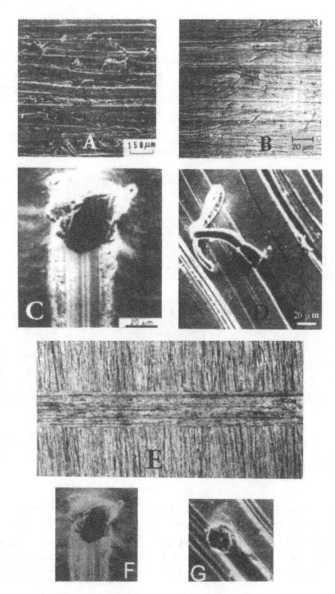

Figure 3.17 Examples of wear scar morphology on metal surfaces, resulting from various single-cycle deformation mechanisms during sliding contact. "A"–"D", "F", and "G" are from three-body abrasive wear situations. "E" is a wear scar resulting from a single sliding stroke between a hard ball and a softer flat. ("A" from Ref. 152, "B" from Ref. 148, "C" & "D" from Ref. 65, "E" from Ref. 67, "F" & "G" from Ref. 64. "A", "B", "C", "D", "F", and "G" reprinted with permission from ASME. "E" reprinted with permission from Elsevier Sequoia S.A.)

K values tend to be higher for microcracking and cutting then for plowing and wedge formation as a result of this effect. With microcracking cracks propagate beyond the indention, which results in material being removed beyond the indentation. Studies have indicated that the cracked area may be up to 10× the indentation area (60). Consequently, for microcracking, ε would be greater than 1, and could be as large as 10. For plowing, wedge formation, and cutting, ε would be less than or equal to 1, because of elastic

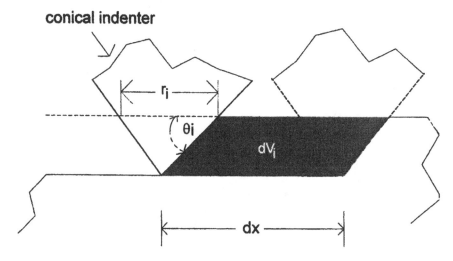

Figure 3.18 Model for single-cycle deformation wear.

recovery. Any elastic recovery, which occurs, will result in the groove area being smaller than the indentation area. As a result, ε would be less than 1 and would become smaller with larger degrees of recovery. With cutting, there is less recovery than with wedge formation and plowing. Therefore, ε would tend to be larger for cutting than either of the two plastic deformation mechanisms. The difference between plowing and cutting is illustrated in Fig. 3.19. In this figure, the area of the groove produced by a stylus in a softer material is plotted as a function of attack angle. A transition from plowing and wedge formation to cutting occurs at a critical attack angle, a_c. In the cutting region, the groove areas are larger and more sensitive to angle than in the plowing region.

Since deformation characteristic, such as ductility, toughness, and elasticity, tend to be different with different classes of materials, K can be different for different classes of materials. K tends to be higher for brittle materials and lower for tougher and ductile

Table 3.5 K Values for Different Cone Angles

Θ (°)	Cone angle (°)	K	R_a (μm)	Abrasive papers
0	180	0		
0.1	179.8	0.001		
0.7	179	0.008	0.01	
1	178	0.01	0.1	
5	170	0.06	1	
10	160	0.1		# 600
20	140	0.2		
30	120	0.4	10	
45	90	0.6		# 100
60	60	1.1	100	
75	30	2.4		
80	20	3.6		
85	10	7.3		

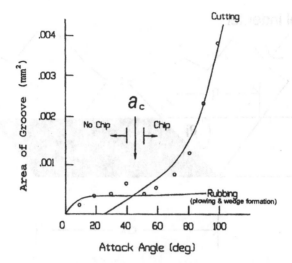

Figure 3.19 The effect of attack angle on chip formation for a hard stylus sliding against a softer metal surface. (From Ref. 176, reprinted with permission from Elsevier Sequoia S.A.)

materials. Such an effect on K is shown in Fig. 3.20. In this figure, wear rate for three classes of materials is plotted as a function of hardness. For each class, the behavior with hardness is that given by Eq. (3.20). However, the value for K is different for each class. The large difference between carbon and the other two classes of materials is primarily related to the poorer ductility of carbon. With carbons microcracking typically occurs, resulting in additional material loss, while it does not with the metals or plastics. The smaller difference between plastics and metals is a result of the difference between cutting and plowing. It has

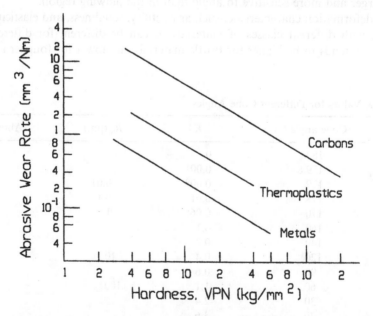

Figure 3.20 The effect of hardness on the abrasive wear rate of different classes of materials. The data are for two-body abrasion, using 100 μm SiC paper. (From Ref. 58.)

been shown that the following relationship exists between the coefficient of friction, μ, and the angle at which the transition from plowing to cutting takes place, a_c (61–65):

$$\tan(90° - a_c) \approx \frac{1 - \mu^2}{2\mu} \tag{3.27}$$

Examination of this equation shows that lowering friction reduces the critical angle. Since the coefficient of friction with plastics is generally lower than with metals, cutting, which is more severe, is a more likely mechanism for plastics than for metals.

Because of this relationship between the critical angle for cutting and friction, K can also be affected by lubrication. Values of K tend to be a factor of two to five times higher when lubrication is involved [(66); see Table 3.8]. There is also another possible explanation or contributing factor for the increase in K with lubrication. In addition to its effect on the critical attack angle, lubrication can also increase single-cycle deformation wear by its effect on debris accumulation. When wear debris is trapped between or coat surfaces, it tends to provide separation, reducing the amount of contact with and penetration by particles or asperities, as is illustrated in Fig. 3.21. Lubrication tends to remove debris and prevent the buildup, which would result in more contact with the abrasives or asperities.

Single-cycle deformation mechanisms are not limited to asperities and particle contacts. These mechanisms can also occur on a macro-scale and be associated with the gross geometry of the contacting bodies (57,67–69). A necessary condition for these mechanisms to occur is that the asperity, particle, or counterface be harder than the wearing surface. This is illustrated by the sharp decrease in abrasive wear that occurs when the surface becomes harder than the abrasive, as shown in Fig. 3.22. Consequently, making the surface harder than the counterface or abrasives can eliminate these mechanisms.

As stated initially in this section, single-cycle deformation mechanisms can occur in rolling and impact situations, as well as in sliding situations. These mechanisms follow the same general trend as for sliding, as illustrated by the following equation for solid particle erosion: [(70); see Sec. 3.8]

$$V = K' \frac{Mv^2}{p} \tag{3.28}$$

In this equation, M is the total mass of the particles producing the wear and v is particle speed. K' is similar to K. It is a function of particle profile, that is, sharpness, and material properties and mechanism affect its value in the same manner as with K. It is also affected by incident angle, because of changes in mechanisms (see Sec. 3.8).

Some major trends for single-cycle deformation wear mechanisms are: (1) they only occur when the surface is softer than the counterface or particle, (2) wear volume is inversely proportional to hardness, and (3) plastic deformation or ductile mechanisms are milder than cracking or cutting. There is a fourth trend related to elasticity. Except for the difference in elasticity between elastomers and other classes of materials, K is generally not affected by differences in elasticity. K values with elastomers tend to be 0 or much lower than with other materials for plowing and wedge formation as a result of their ability to recovery from very large strains. This difference in elasticity is usually not a significant factor with cutting and K values are unaffected for this mechanism (58).

Except in abrasive situations and some sliding situations involving soft materials and very rough surfaces, that is, file-like surfaces, single-cycle wear mechanisms tend to become less significant as wear progresses. This is generally attributed to changes that take place as a result of wear and the emergence of other mechanisms. Typical changes that contribute

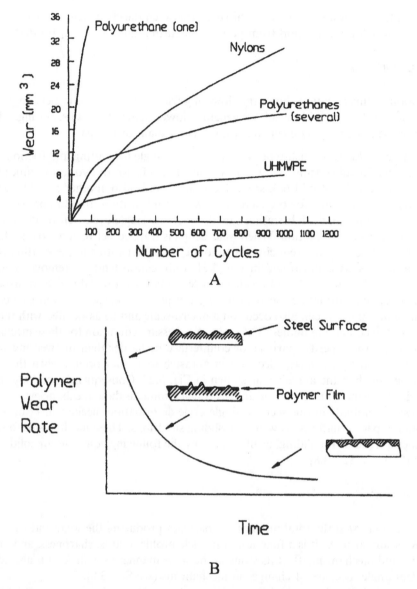

Figure 3.21 "A" shows the wear behavior of several polymers sliding against SiC-coated abrasive paper. The decrease in wear rate with number of cycles is attributed to the accumulation of polymer wear debris on the surface of the paper. The effect of this on the contact situation is illustrated in "B". The polymer film tends to protect the surface from contact with the abrasive particles. (From Ref. 156, reprinted with permission from Elsevier Science Publishers.)

to the reduction of single-cycle wear are reduction in the average junction stress and associated penetration, as described in Sec. 3.2 on adhesion, increased conformity of surfaces, and as well as strain-hardening with some materials. An example of this reduction in significance is shown in Fig. 3.23 for the case of lubricated sliding wear of Cu. As can be seen in the figure, striations, indicative of plowing, are the dominant feature initially. As sliding continues, these features become less pronounced and features indicative of repeated-cycle deformation mechanism appear and become the dominant ones.

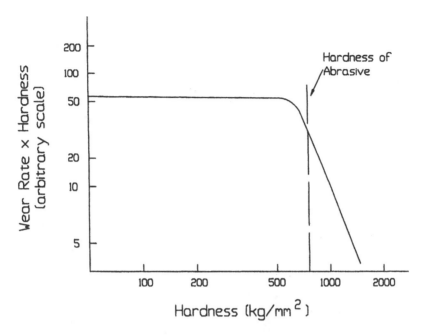

Figure 3.22 Transition in wear behavior when the wearing material becomes harder than the abrasive. (From Ref. 177.)

3.4. REPEATED-CYCLE DEFORMATION MECHANISMS

Repeated-cycle deformation mechanisms are wear mechanisms that require repeated cycles of deformation. There are a number of these mechanisms. Some of these mechanisms involve progressive deformation processes, like creep, compression set, and subsurface flow. However, these are usually limited to particular types of materials in specific wear situations. The more general ones, surface fatigue, delamination, and ratcheting involve fatigue-like or fatigue processes. Such processes involve the accumulation of plastic strain, which ultimately leads to the nucleation and propagation of cracks or fracture, which is similar to conventional fatigue. Micrographs of wear scars associated with these common forms are shown in Figs. 3.24, 3.25, 3.27 and 3.30–3.32. Examples of creep and subsurface flow are shown in Fig. 3.26. In general, the severity of these mechanisms is proportional to some power, often high, of the ratio of an operating stress to a strength property of the material, such as contact pressure to compressive yield stress. The exact form depends on material and wear mechanism. As a class, repeated-cycle deformation mechanisms are not limited to a particular type of motion. They can occur as a result of sliding, rolling, or impact. They are also not limited to contact between two bodies but can occur as a result of contact between surface and abrasive particles. However, they only are important in the latter case when the surface is harder than the particle.

Surface fatigue is a generic term used for repeated-cycle deformation wear mechanisms that result from fatigue processes, which occur on and below the surface of contact. These processes result in the formation of cracks and crack networks on and below the surface and in deformed material. Such processes can also result in the formation of pits. Examples of these features are shown in Figs. 3.24, 3.25, 3.27 and 3.30–3.32. Delamination is a particular form of surface fatigue, which is related to the accumulation of

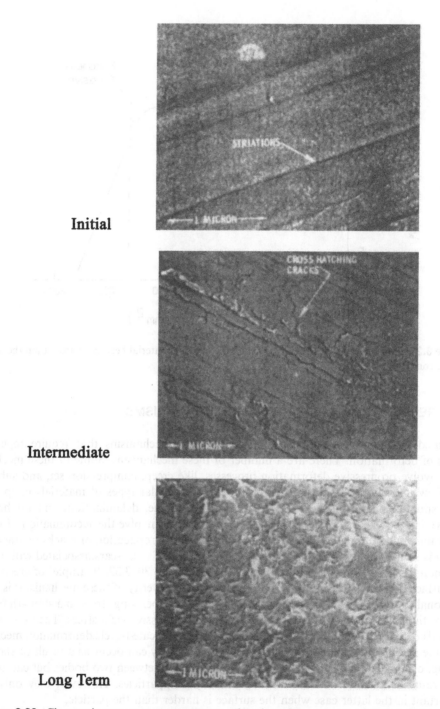

Initial

Intermediate

Long Term

Figure 3.23 Changes in wear scar appearance as a function of the amount of sliding. Data are for lubricated sliding between a steel sphere and a single crystal copper flat. (From Ref. 20.)

dislocations in a narrow band below the surface. This type of wear is illustrated in Fig. 3.27. Ratcheting is another particular form of repeated-cycle deformation wear that is based on incremental plastic flow, the accumulation of plastic strain, and mechanical

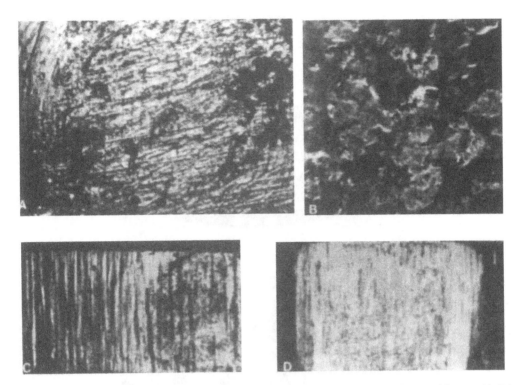

Figure 3.24 Examples of surface fatigue wear in metals under conditions of normal impact (A–D) and rolling (E–H). ("A"–"D" from Ref. 75; "E"–"G" from Ref. 73; "H" from Ref. 71; ("A"–"D" reprinted with permission from Elsevier Science Publishers; "E", original source The Torrington Co., and "F" and "G" reprinted with permission from Texaco's magazine *Lubrication*: "H" reprinted with permission from ASME.)

shakedown. This is illustrated in Fig. 3.28. Again fracture ultimately results from crack formation and propagation, that is, fatigue.

The common concept associated with the typical forms of repeated-cycle deformation wear is fatigue or, more appropriately, fatigue wear. The basic concept of fatigue wear is that with repeated sliding, rolling, or impacting, material in the vicinity of the surface experiences cyclic stress. As a result of this, stress cycling, plastic strain accumulates and cracks are ultimately formed. With further cycling, the cracks propagate, eventually intersecting with the surface and themselves. These intersections then produce free particles, which are easily removed from the surface by a subsequent motion. This worn surface also experiences stress cycling and the process continues, resulting in progressive loss of material from the surface. This concept is illustrated in Fig. 3.29.

This type of wear mechanism is most evident in rolling and impact wear situations, where it is generally recognized as the principal mechanism (71–75). Figs. 3.24 and 3.30 show examples of fatigue wear under such conditions. Fatigue wear is also possible with sliding (21,23,25,76–78). Examples are shown in Figs. 3.25, 3.27, 3.31 and 3.32. In the case of rolling and to a lesser degree with impact, the topological features of the wear scar are often quite suggestive of crack initiation and propagation. Under sliding conditions, the topological features are generally not as suggestive. There are several reasons for this. Features associated with adhesive and abrasive mechanisms frequently confound the appearance in sliding situations. Smearing on the surface also tends to hide surface cracks

Figure 3.24 (*continued*)

with sliding. In addition, the crack network under rolling and impact tends to be more macroscopic or coarser than those often encountered under sliding conditions and frequently result in larger particles or pits being formed. This tends to make fatigue features

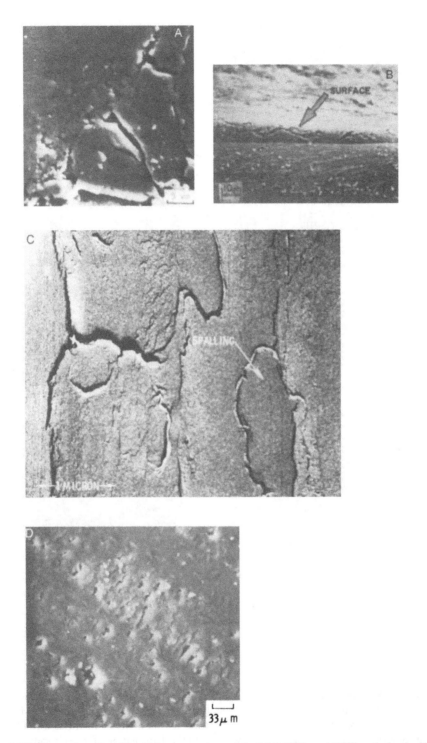

Figure 3.25 Examples of surface fatigue wear in metals ("A", "B", and "C") and plastics ("D") as a result of sliding. ("A" from Ref. 9; "B" from Ref. 24; "C" from Ref. 20; "D" from Ref. 21; A" and "D" reprinted with permission from ASME; "B" and "C" reprinted with permission from Elsevier Sequoia S.A.)

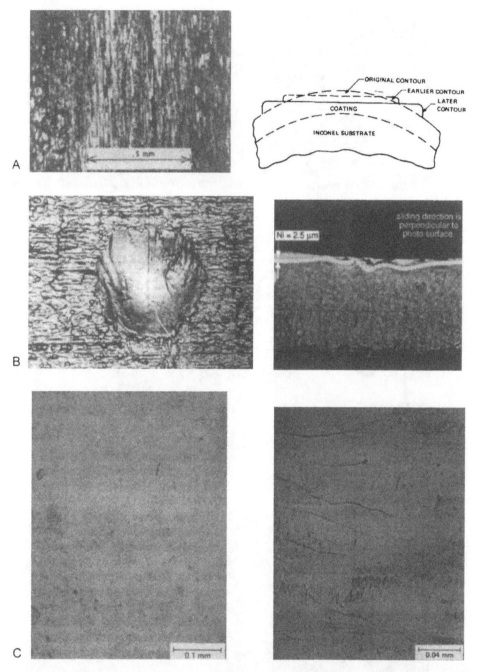

Figure 3.26 "A" shows the wear of a lead coated c-ring as a result of small amplitude oscillations, which results from creep. An example of wear resulting from progressive subsurface flow is shown in "B". These micrographs show the wear of an electrical tab on a circuit board as a result of small amplitude oscillations. The left-hand micrograph shows the deformation of the substrate. "C" shows the worn surface of an elastomer slab subjected to repeated impact. Two modes are shown. In the left-hand micrograph, there is no material loss but the material is permanently deformed, which results from a compression set type of behavior. In the right-hand, there is material loss resulting from fatigue. ("A" is from Ref. 178, "B" is from Ref. 179, "C" is from Ref. 180.)

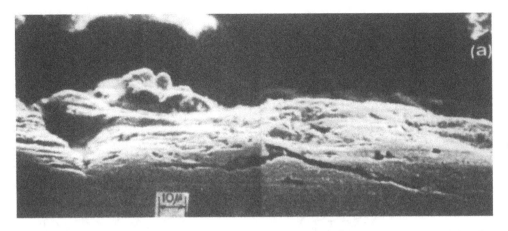

Figure 3.27 Crack structure in delamination wear. (From Ref. 24, reprinted with permission from Elsevier Sequoia S.A.)

more easily detected in the case of rolling and impact. Because of these aspects, often the only way to determine the existence of cracks under sliding conditions is by means of microscopic examination of cross-sections through the worn surface, such as those shown in Figs. 3.25b, 3.27, 3.30, 3.31b and 3.32d. Magnifications of several hundred times or more are generally required for this.

While fatigue wear and fatigue, that is structural fatigue, share a common basic concept, namely the formation and propagation of cracks, they have different characteristics. While both have an incubation period, the periods are not the same. With fatigue, the incubation period is the period of crack formation. With fatigue wear the incubation period extends beyond this. For fatigue wear, the incubation period involves the propagation of the cracks to the surface and generally the formation of loose particles. Some topological changes might be evident during this initial period of fatigue wear, including some evidence of plastic deformation. However, there is no loss of material from the surface or formation of free particles. There are also further distinctions between fatigue wear and fatigue. With fatigue, the process simply involves the formation and propagation of cracks. With fatigue wear the process is a continuous cycle of crack formation,

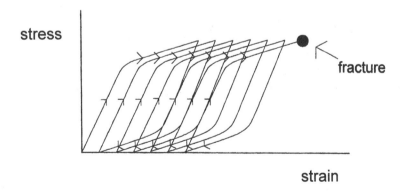

Figure 3.28 Conceptual illustration of the ratcheting wear mechanism. The diagram shows the accumulation of strain as a result of repeated stress cycling, which leads to fracture.

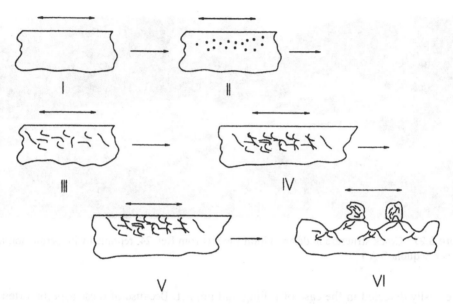

Figure 3.29 General model for surface fatigue wear. Stage I, stress cycling of surface; Stage II, nucleation of cracks in near-surface regions; Stage III, crack growth; Stage IV, crack coalescence; Stage V, crack intersection with surface; Stage VI, formation of loose particles.

propagation, and removal. For fatigue, most materials exhibit an endurance limit, that is, a stress level below which fracture will not occur. In the case of fatigue wear, there does not appear to be such a limit at least in terms of macroscopic loads and stresses. For practical load conditions, no matter how small the load or stress, sufficient rolling, sliding or impact results in the generation of fatigue wear. A further difference is that with fatigue a distinction is often made between low cycle fatigue and high cycle fatigue. A similar distinction is not made with fatigue wear.

For rolling situations, there is a generally accepted empirical relationship between load and the number of revolution defining the incubation period (77,79–81). The general form of the relationship for both point and line contact situations is

$$N_1 P_1^n = N_2 P_2^n \tag{3.29}$$

where N_1 is the number of revolutions required for a load of P_1 and N_2 the number of revolutions required for a load of P_2. For point contact situations, such as in a ball bearing, n is 3; for line contact, such as in a roller bearing, n is 10/3. Frequently this relationship is referred to as Palmgren's equation (81,82). A more fundamental form of this equation relates stress to number of revolutions. Since according to elastic contact theory (83), the maximum stress in a point contact situation, S_m, is proportional to $P^{1/3}$, the stress form of Eq. (3.29) becomes

$$N_1 S_{m_1}^9 = N_2 = S_{m_2}^9 \tag{3.30}$$

Similar relationships exist for sliding and impact, as described later in this section (21,84).

The progression of wear scar morphology for fatigue wear under sliding conditions was studied in Cu. (21) The sliding system consisted of a hardened steel sphere sliding back and forth across the flat surface of Cu single crystals. Boundary lubrication was used and stress levels were maintained well under the yield point of the Cu. Three stages were

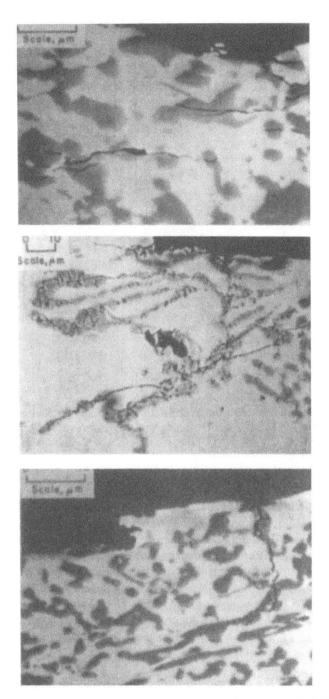

Figure 3.30 Examples of cracks formed under impact conditions. (From Ref. 181, reprinted with permission from ASME.)

identified and are shown in Fig. 3.32. In the first stage, grooves and striations in the direction of sliding were the predominate feature. There was no material loss and the topography would suggest single-cycle deformation. During this stage, as sliding increased, the

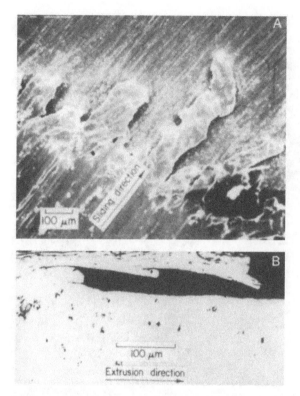

Figure 3.31 Crack structure in extrusion wear. In "A", the surface morphology of the wear scar is shown. In "B", a view of the cross-section through the wear scar. (From Ref. 87, reprinted with permission from ASME.)

density or number of these grooves increased. In the second stage, damage features perpendicular to the sliding direction appeared. Again, there was no loss of material. This feature, termed crosshatching, implied something other than a single-cycle deformation mode was occurring. As sliding continues in this stage, the crosshatching became more pronounced until ultimately spalling and flaking occurred. This is the start of the third and final stage. In this stage, material loss occurs and, with continued sliding, a wear groove of increasing depth is formed. The start of the third stage was considered to be the end of the incubation period.

The striations of the first stage are the result of local stress systems associated with individual asperity contact. However, the crosshatching features occur over many striations and are therefore probably associated with the overall stress system associated with the macro-geometry of the contact. This feature is also considered to be associated with the initiation and growth of subsurface cracks. Micrographs of cross-sections through the wear scar confirmed the existence of sub-surface cracks in this situation, as shown in Fig. 3.32d.

In the same study, it was found that the number of cycles required to initiate the third stage could be correlated to the maximum shear stress associated with the macro-geometry. In fact, a relationship identical to Eq. (3.30) was found. This correlation is shown in Fig. 3.33. It is significant to note that the same type of correlation with stress is found in impact wear situations when the macro-stresses are within the elastic limit of the materials (84). As stated earlier, a similar correlation is found with rolling.

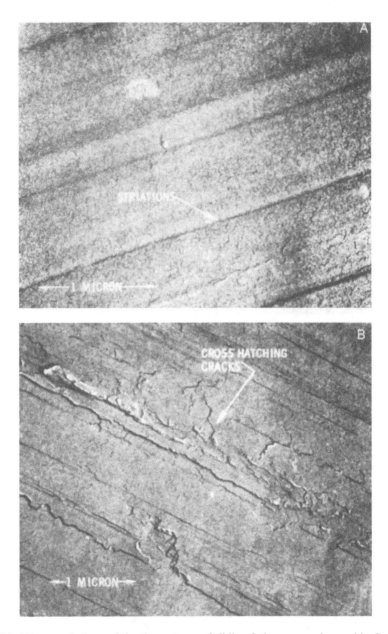

Figure 3.32 The morphology of the three stages of sliding fatigue wear observed in Cu. The initial stage is shown in "A". The intermediate stage is shown in "B" and the final, in "C"."D", which is a TEM of a region below the surface of the wear scar, shows the subsurface cracks found in the final stage. (From Ref. 20, reprinted with permission from Elsevier Sequoia S.A.)

Wear scar morphology, similar to the stage three morphology observed with the Cu single crystal, and cracks have been observed in many sliding systems (23,25,76,77,85–87). While this is the case, the nature of the crack systems is frequently different. The micrographs in Figs. 3.27 and 3.31 serve to illustrate these points. Many of the topological features of the wear scars shown in these two figures are similar to those associated with

Figure 3.32 (*continued*)

stages three Cu wear. However, it is apparent that the crack systems in each of these three cases are different. In Fig. 3.27, the cracks are near and parallel to the surface. This mode was termed delamination wear and was described in terms of dislocation behavior (25,77). In Fig. 3.31, cracks form at the base of extruded wedges or lips. This mode is sometimes referred to as extrusion wear (87). In Fig. 3.32d, it can be seen that in low stress sliding wear of Cu, the cracks had a more random orientation and extended well below the surface.

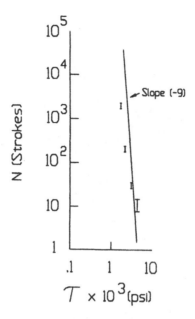

Figure 3.33 The number of cycles required to produce the second stage in the fatigue wear of Cu as a function of shear stress. (From Ref. 20.)

The crack systems found in sliding are generally different from those found under rolling and impact conditions. Figure 3.30 shows some examples of the crack systems for impact. The wear scar topography also varies with the situation, as can be seen by comparing the micrographs in Figs. 3.24, 3.27, 3.31 and 3.32. For impact and rolling, features suggestive of sliding are not evident. Also, in the case of rolling, the features tend to be coarser or larger than typically found in sliding situations. The variation in crack systems and patterns can be related to the response of materials to different stress systems.

Because of this strong influence of stress on fatigue wear, it is worthwhile to consider the nature of the stress systems associated with different contact situations, prior to discussing formulations for fatigue wear. Conceptually, the stress system occurring in a wear contact can be separated into two parts. One part may be termed the macro-stress system and is related to the overall geometry or shape of the contacting members, that is, the features that relate to the apparent area of contact. The second part is the micro-stress system and this is governed by local geometry associated with the asperities. This concept is illustrated graphically in Fig. 3.34.

For the macro-system, there are two general types of contact situations which are illustrated in Fig. 3.35. One is a conforming situation, such as a flat against a flat or a sphere in a socket of the same radius. The second is a nonconforming contact situation, such as a sphere against a plane, two cylinders in contact, or sphere in a socket of a larger radius. For the conforming situation, the pressure distribution across the surface is uniform and the stress level is highest on the surface, decreasing with distance from the surface. For the nonconforming case, the situation is quite different. Hertz contact theory shows that in this case the pressure is greatest in the middle of the contact and that the maximum shear stress is below the surface at a distance of approximately a third of the radius of the apparent contact area (83). It also has a value of approximately one-third of the maximum contact pressure. In this case, significant stress can occur well below

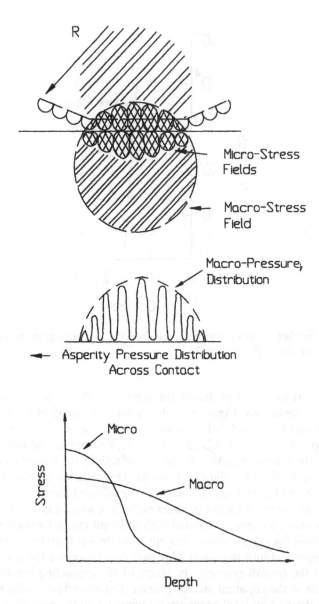

Figure 3.34 General nature of the stress field in contact situation, illustrating the relative effects of contact geometry and asperities on stress.

the surface, up to depths comparable to contact dimensions. The comparative nature of these two contact situations is illustrated in Fig. 3.36.

When shear or traction is applied to the interface, the macro-stress system is modified. The modification is most significant at or near the surface since the shear decays rapidly as a function of depth (88). With μ as the coefficient of friction and $q(x)$ as the pressure distribution, $\mu q(x)$ is the traction across the contact. For the case of a conforming contact, the maximum shear stress is on the surface and can be shown to be approximately $q_o(0.25 + \mu^2)^{1/2}$, where q_o is the contact pressure. For the nonconforming case, the maximum shear stress can occur either on the surface or beneath the surface, depending on the

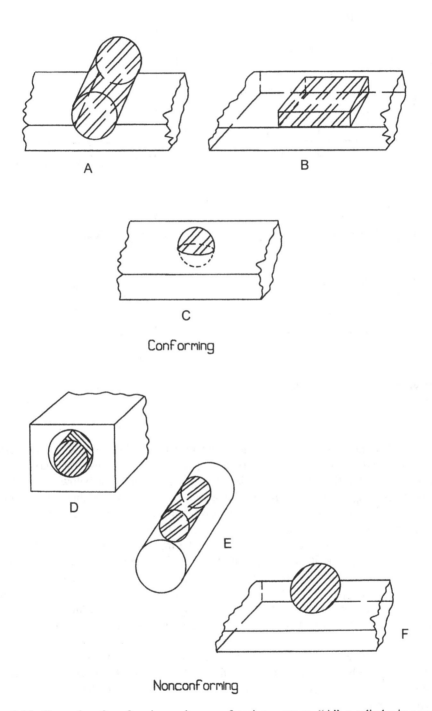

Figure 3.35 Examples of conforming and nonconforming contacts. "A", a cylinder in a groove of matching radius, "B", a flat on a flat, and "C", a sphere in a spherical seat of matching radius, are examples of conforming contacts. "D", a cylinder in a hole of larger radius, "E", parallel cylinders, and "F", a sphere on a flat, are examples of nonconforming contacts.

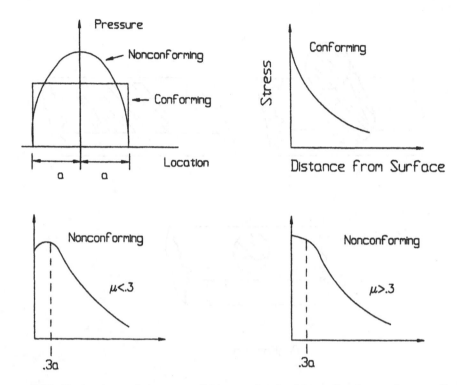

Figure 3.36 Comparison of the stress fields associated with conforming and nonconforming contacts.

value of μ. For nonconforming contacts, the maximum shear stress on the surface is approximately μq_o, where q_o is the maximum pressure. This is also the maximum shear stress if μ is greater than 0.3. If μ is less than 0.3, it would be below the surface and approximately $0.3q_o$. A consequence of this is that for a nonconforming contact situation, lubrication cannot only modify the stress level but also the stress distribution, as illustrated by the change in the location of maximum stress.

The pressure distribution associated with the macro-system can effect the load distribution across the asperities, as illustrated in Fig. 3.37. Since the pressure distributions are different for conforming and nonconforming contact, the micro-stress systems for these two general types of contact will also be different. For nonconforming contacts, asperities in the center of the contact region will tend to be loaded higher than asperities near the edges of the contact region. For conforming contacts, the loading will be more uniform.

While asperities have curvature, the micro-stress fields can be of two types. If the asperity is plastically deformed, the stress field will have the characteristics of a conforming contact. If elastically deformed, the stress field will have the characteristics of a nonconforming contact.

When considering stress in wearing contacts, a further aspect has to be recognized. This is that wear generally changes the micro- and macro-geometrical features of the surfaces in contact. As a result, there can be changes in the two stress systems associated with the contact. The magnitude of the stresses can change as a result of changes in the real and apparent areas of contact, as well as the stress distribution. As discussed in Sec. 3.2, wear

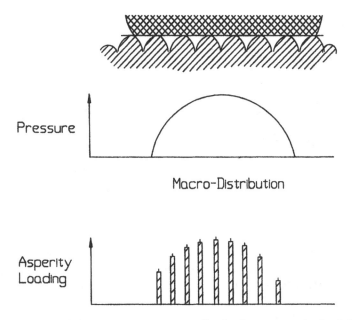

Figure 3.37 Effects of the macro-contact stress distribution on asperity load distribution.

also tends to reduce the plasticity index, implying that asperity deformations become more elastic with wear (37). Wear will also cause an initial nonconforming contact to become a conforming contact, changing the nature of the macro-stress system and increasing the apparent area of contact, as illustrated in Fig. 3.38.

Different features of the wear can be related to these two stress systems. For example, grooving and striations in the direction of sliding can be related to the micro-stress. Also, the general nature of the cracks and crack system can be related to these stress systems. The effect of the macro-stress system on crack formation, illustrated in Fig. 3.39, is an example of this. Also, differences between the wear scars, which are shown in Figs. 3.24, 3.27 and 3.30–3.32, can be related to the stress conditions of the tests. In the examples of sliding that are shown in Figs. 3.27 and 3.31 the contacts were conforming, that is, flat-against-flat. They were also unlubricated and as a result the coefficient of friction, μ, was high. In these cases, the significant stress would be confined to a small region near the surface, essentially at the asperity level and the micro-stress system would be the predominate system. In these cases near-surface cracking is found, as well as surface features related to asperity contact. In the rolling contacts, the macro-geometry was nonconforming and there was negligible friction and traction. The initial geometry in the experiments with Cu was also nonconforming and the tests were performed with lubrication, which resulted in a low value for μ. In this case, the nonconforming nature of the contact would remain until the end of the incubation period. At this point, material loss would result in a change to a conforming contact. In these two situations, significant stresses would occur well beyond that near-surface region and the macro-stress would be significant. For the sliding wear of Cu, as shown in those figures, the micro-stress system would become more significant beyond the incubation period, since the geometry would then become conforming. Also, the average stress level would decrease as a result of increasing contact area. For impact the contacts were initially nonconforming and approach conformity with wear. In these situations, wear behavior is related to the

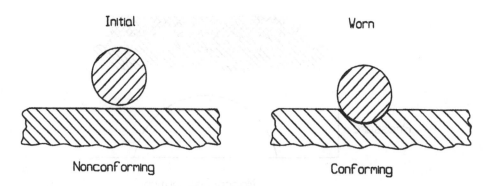

Initial   Worn

Nonconforming   Conforming

Figure 3.38 The changing nature of the contact situation that occurs in the case of a hard steel sphere sliding against a soft copper flat.

macro-stress system and significant stresses occur well below the surface (75). In these three cases, rolling, impact and low-stress sliding, sub-surface cracking is found and damage related to asperity contact is not evident, except for the initial stages of sliding wear.

The main features of several of the models proposed for fatigue wear after the incubation period can be illustrated by the consideration of an idealized and simple model. Assume that the sliding system can be approximated by a smooth, flat surface of area, A_a, sliding against a flat, rough surface, which has by an exponential distribution of asperities of different heights and a tip radius of β. Further, assume that the wear is confined to the smooth flat surface. This situation is shown in Fig. 3.40. The key assumption in these models is that the formation of a fatigue wear particle can be described by Wohler's equation for fatigue (89), namely,

$$N_f = \left(\frac{S_0}{S}\right)^t \tag{3.31}$$

In this equation, N_f is the number of cycles to failure at a stress level of S and S_0 is the stress level required to produce failure in a single stress cycle. Both t and S_0 are material dependent.

On the macro-scale, the contact situation considered is a conforming one. Hence, the principal stress system will be associated with the asperity contact conditions. In a fatigue wear situation, any initial plastic loading conditions would modify asperity geometry so that the material would respond elastically in subsequent load cycles. Consequently, it is generally assumed for fatigue wear that the asperity contacts can be described by elastic contact theory. In the assumed situation, the asperity contacts can be approximated by a sphere pressed against a flat surface, a situation that is covered by Hertzian contact theory (83). The principal equations governing this situation are:

$$q_0 = 0.58 P'^{1/3} E^{*2/3} \beta^{-2/3} \tag{3.32}$$

$$a' = 0.91 P'^{1/3} \beta^{1/3} E^{*-1/3} \tag{3.33}$$

$$E^* = \left(\frac{1-\nu_1}{E_1}\right) + \left(\frac{1-\nu_2}{E_2}\right) \tag{3.34}$$

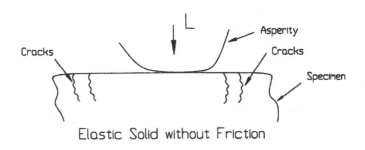

Elastic Solid without Friction

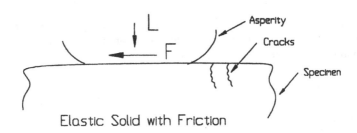

Elastic Solid with Friction

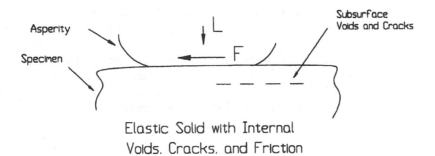

Elastic Solid with Internal
Voids, Cracks, and Friction

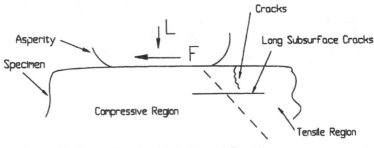

Elastic/Plastic Solid with Friction

Figure 3.39 Examples of the influence of the nature of the stress system on crack formation. (From Ref. 63.)

where q_o is the maximum contact pressure at the asperity contact, a', the radius of the contact spot, P', the load on the asperity, and E's and ν's, the Young's modulus and Poisson's ratio for the two materials in contact. Assuming that the load is uniformily distributed

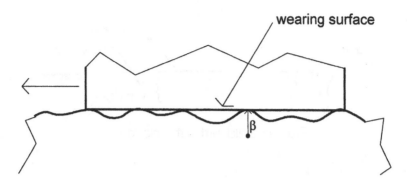

Figure 3.40 Contact situation between a smooth surface and a rough surface, used in the development of a fatigue wear model. It is assumed that the rough surface can be characterized by an asperity distribution of different heights but the same tip radius.

over the asperities,

$$P' = \frac{P}{\Phi A_a} \tag{3.35}$$

where P is the load between the two surfaces and Φ is the number of asperities per unit area in contact at load P.

While shear stress is frequently related to fatigue behavior, some studies have indicated that, in the case of wear, it can be correlated with the maximum tensile stress, which occurs at the leading edge of the contact area (90). This is not significant in the development of the model since both are proportional to the maximum contact pressure. The maximum shear stress is either μq_0 or $0.31q_0$ with the former occurring at the interface and the latter at a distance of approximately $0.3a'$ below the asperity tip. The maximum tensile stress is approximately $0.5\mu q_o$. All of these cases can be covered by the following relationship:

$$S = \Gamma q_0 \tag{3.36}$$

In this form, Γ can be viewed as an empirically determined coefficient, which is material dependent. Ultimately, a zone of the surface will experience enough loading cycles so that a free particle will form. Assuming that the dimensions of this particle can be approximated by the dimensions of the region of significant stress under the contact, the volume of the fragment may be estimated. This may be approximated as a spherical shell of diameter $2a'$ and depth $0.3a'$. In that case it can be shown that the volume of the wear fragment, v', is given by

$$v' = \frac{0.36P'\beta}{E^*} \tag{3.37}$$

For a sliding distance of L the number of stress cycles the surface will experience is given by $L\Phi^{1/2}$. The number of times a wear fragment will form during this amount of sliding is, therefore, $L\Phi^{1/2}/N_f$ and the total volume, V, that is lost is given by

$$V = \frac{v'\Phi^{3/2}A_aL}{N_f} \tag{3.38}$$

Using the relationships developed for flat rough surfaces, it can be shown that for an exponential distribution of asperity heights with a standard deviation of σ, Φ is given by the following equation (37):

$$\Phi = \frac{3}{4}\left(\frac{P}{E^*\beta^{1/2}\sigma^{3/2}}\right) \tag{3.39}$$

Substituting in Eq. (3.38) and reducing, the following equation is obtained:

$$V = \Omega M G \Gamma^t P^{1.5} L \tag{3.40}$$

where

$$\Omega = 0.42 \times 0.64^t \tag{3.41}$$

$$M = \frac{E^{*(t-1.5)/3}}{S_0^t} \tag{3.42}$$

$$G = \frac{\sigma^{(t-1.5)/2}}{A_a^{t/3}\beta^{(t-1.5)/2}} \tag{3.43}$$

This simple model provides a general identification of the typical parameters, which influence fatigue wear. Fatigue parameters of the material are one type of parameter which affect fatigue wear, as illustrated by t and S_0 in the wear equation. In addition several of what might be termed mechanical parameters of the system are also involved. These are the geometrical features of the surfaces (roughness and apparent area of contact), load, elastic constants of the materials (Young's modulus and Poisson's ratio), and the coefficient of friction of the material pair. The significance of individual parameters is influenced by the overall fatigue behavior of the material, as illustrated by the effect of t, the exponent of Wohler's equation, on exponents associated with these parameters. This overriding influence of the fatigue behavior can be illustrated with the present model by noting that in fatigue studies values of t as low as 2 and in excess of 20 have been found for different materials (78,90).

While the general nature of this equation for fatigue wear does not change with the nature of the asperity distribution, the exponents can change. For example, if a uniform distribution is assumed, the following is obtained:

$$V = \Omega M G \Gamma^t P^{(1+t/3)} L \tag{3.44}$$

$$\Omega = 0.36 \times 0.58^t \tag{3.45}$$

$$M = \frac{E^{*(2t/3-1)}}{S_0^t} \tag{3.46}$$

$$G = \frac{1}{\Phi^{(t/3-0.5)}A_a^{t/3}\beta^{(2t/3-1)}} \tag{3.47}$$

As illustrated by these results, different fatigue relationships and assumptions regarding asperity loading and distributions can affect the dependency on load. Other models for fatigue wear and experimental data indicate that the load dependency can generally be represented by a power relationship, P^n. While some models for fatigue wear result in values of n near 1 for specific conditions, significantly larger values, for example, 3 or larger are also possible (26,27,78,85,86,91,92). Equations (3.40) and (3.44) illustrate this.

This range of n values is consistent with empirical observations. For example, the wear test data for bainitic steels, shown in Fig. 3.41, indicate a range of 2–6 for n. In studies by the author values in the range of >1 to <4 have been observed, as well (93). On the other hand, a near-linear relationship was found in some studies of polymer wear (94).

Theoretical models and empirical observations suggest that the general form for fatigue-like repeated-cycle deformation wear is

$$V = KP^nS, \quad n \geq 1 \tag{3.48}$$

where S is the distance of sliding. For rolling and impact, S can be replaced by number of impacts or revolutions. As can be inferred by comparison with Eqs. (3.40) and (3.44), K depends on a range of material and contact parameters, as well as the type of fatigue process, but not directly on hardness. As a result, there are two significant differences between this equation and the ones for adhesive, Eq. (3.7), and single-cycle deformation wear, Eq. (3.25). One is that the relationship for repeated-cycle deformation does not contain an explicit dependency on hardness as the ones for adhesive and single-cycle deformation wear. The other is that the dependency on load is different. For adhesion and single-cycle deformation, there is a linear relationship, while for repeated-cycle deformation, it is generally non-linear. Models and empirical information indicate that n is a function of materials, wear process, and asperity distribution.

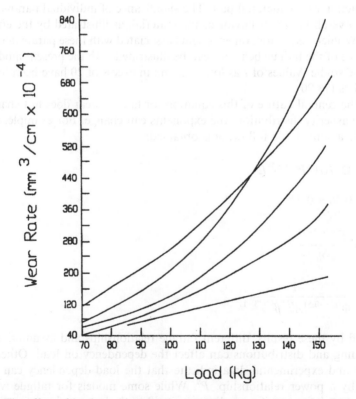

Figure 3.41 The effect of load on the unlubricated sliding wear of several bainitic steels. (From Ref. 96.)

The dependency of fatigue wear on the radius of the asperity tip, β, was investigated for a variety of materials (95). In general, a high-order dependency on β was found. Some of the data are shown in Fig. 3.42. These data suggest that wear rate is proportional to β^{-6} or β^{-5} for several of the system investigated. In terms of the Wohler-based models, this implies a value of the order of 10 for t, which is similar to the exponents relating stress and incubation cycles, as illustrated by Eq. (3.30).

More fundamental approaches to fatigue wear have also been proposed, such as dislocation theory (26,27,96), and fracture mechanics (85,86). These models, while indicating some of the underlying features and concerns in fatigue wear, have not been as useful in practice, as the models or concepts based on more simple engineering concepts for fatigue or using Eq. (3.44) as an empirical relationship. However, such concepts can provide some insight into the relative behavior of different materials with respect to this type of wear.

Because of the incubation period of fatigue-like repeated-cycle deformation mechanisms, these mechanisms tend not to be significant in the early stages of wear or early life of a component. Adhesive and single-cycle deformation mechanisms tend to be more significant in these. Fatigue-like mechanisms become more significant and often are the dominant mechanisms in later stages of wear associated with long-term behavior. The severity of the wear resulting from repeated-cycle deformation mechanisms tend to be proportional to (stress/strength parameter) ^{n}. The exponent is typically greater than 1 and can be high, for example, in the range of 10. However, the strength parameter is generally something other than hardness.

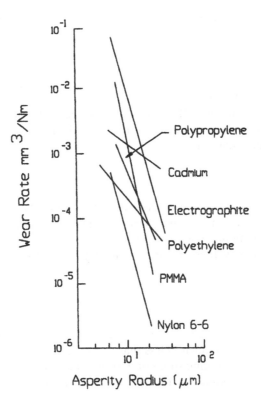

Figure 3.42 Effect of asperity radius on initial wear rate of several materials sliding against an unlubricated mild steel surface. (From Ref. 22.)

3.5. OXIDATIVE WEAR PROCESSES

The basic concept for these processes is that wear occurs by the continuous removal of oxide layers as a result of sliding contact between asperities. In between contacts, the oxide regrows on these denuded areas of the surface and is again removed with subsequent asperity engagement. Characteristic of such processes is the formation of a glassy-like layer on the surface and subsequent appearance of fractures and denuded regions in the layer (97,98). Examples of this are shown in Fig. 3.43. Under these conditions, the wear rate is generally low and fine wear particles of oxides are observed.

A simple model for metals can be used to describe the basic elements of oxidative wear (99,100). The implicit assumption of the model is that the weakest point is at the interface between the substrate and the oxide and that as the result of sliding engagement the oxide layer flakes off at the interface, much like a coating or plating with poor adhesion. The overall sequence is shown in Fig. 3.44.

It is assumed that the real area of contact can be represented as a uniform array of circular junctions as shown in Fig. 3.45. The wear rate associated with a junction, w_i, is given by

$$w_i = \frac{\pi a^2 d}{2a} \tag{3.49}$$

$$w_i = \frac{\pi a d}{2} \tag{3.50}$$

where $2a$ is the diameter of the circular junction and d is the thickness of the oxide film. The wear rate of the surface would then be

$$w = \frac{\pi n a d}{2} \tag{3.51}$$

where n is the number of junctions.

Assume that the growth of the oxide follows a logarithmic law, which is generally true for the initial growth on clean metal surfaces (99). In this case, the thickness, d, is given by the following equation:

$$d = \beta \ln\left(\frac{t}{\tau} + 1\right) \tag{3.52}$$

where t is time and β and τ are parameters associated with the kinetics of the oxidation process. β is a constant dependent on material and temperature and τ is a constant dependent on material. Assuming that each time a junction is formed, the oxide layer is removed, t would be the average time it takes for a junction to reform. If S is the average spacing between junctions,

$$t = \frac{S}{v} \tag{3.53}$$

where v is the sliding velocity. For many sliding situations, this relationship may be simplified by noting that t/τ is frequently less than 1. For example, for the case of iron, τ is in the range of seconds. For a sliding speed of 0.01 in./s and asperity spacing of 0.002 in., t is less than 1 s. Hence, for sliding Eq. (3.52) can be written as

$$d \approx \frac{\beta t}{\tau} \tag{3.54}$$

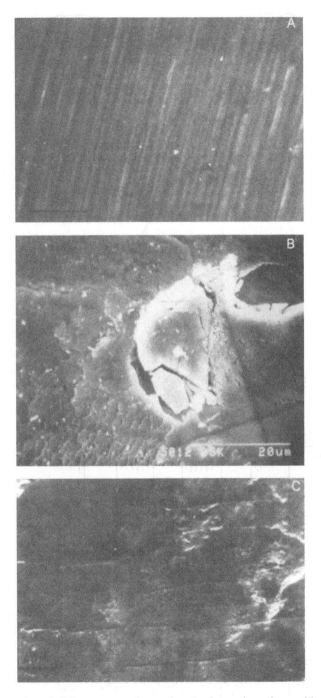

Figure 3.43 Examples of sliding wear surfaces after the formation of an oxide layer. In "A", the layer appears continuous and uniform. In "B" and "C", the layers are cracked and fractured. "A" and "C" are for self-mated unlubricated fretting between Inconel specimens at elevated temperature, 540°C and 700°C, respectively. "B" shows the appearance of the wear scar on a steel pin after sliding on an unlubricated molybdenum disk. ("A" and "C", from Ref. 182; "B", from Ref. A107; reprinted with permission from ASME.)

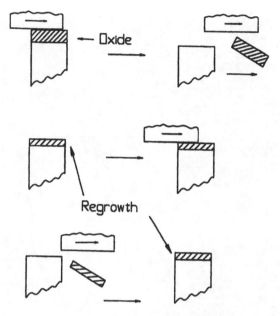

Figure 3.44 Model for oxidative wear.

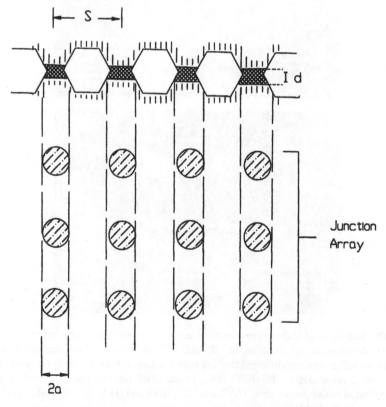

Figure 3.45 Junction array used with model for oxidative wear.

or

$$d = \frac{\beta S}{v\tau} \tag{3.55}$$

The average spacing of junctions, S, is given by

$$S = \left(\frac{A_a}{n}\right)^{1/2} \tag{3.56}$$

where A_a is the apparent area of contact. Assume that the oxide layer is too thin to significantly affect the mechanical properties of the surface and consequently the contact situation. Assuming that the asperities are plastically deformed, the real area of contact is equal to the load, P, divided by the penetration hardness of the softer material, p. Consequently,

$$n = \frac{P}{\pi a^2 p} \tag{3.57}$$

Utilizing these relationships, it can be shown that

$$w = \left(\frac{\beta}{2\tau v}\right) \left(\frac{\pi A_a P}{p}\right)^{1/2} \tag{3.58}$$

or

$$W = \left(\frac{\beta}{2\tau v}\right) \left(\frac{\pi A_a P}{p}\right)^{1/2} L \tag{3.59}$$

where W is the volume of wear and L is the distance of sliding.

The same equation would result if one did not assume that the oxide is always removed at each junction formation. If K is the probability that the rupture of a junction would result in the formation of a wear particle, the average time for oxide growth would be $K^{-1}(S/v)$ and the K's would cancel in the final expression. Simply, this means that frequent removal of a thin oxide layer is equivalent to infrequent removal of thick oxide layer.

Other assumptions regarding the real area of contact can modify the dependencies on mechanical parameters. For example, a refined version of this model, which assumes that the surface topography is described as a Gaussian distribution of conical asperities, results in the following equation for W (101):

$$W = \frac{\pi}{4} \left(\frac{\beta A_a}{\tau v}\right) \left(1 - \frac{\phi(\xi)}{2\xi \int_{\xi_0}^{\xi_0} \phi(\xi) d\xi}\right) \tag{3.60}$$

where

$$\phi(\xi) = \frac{1}{(2\pi)^{1/2}} e^{-\xi^2/2} \tag{3.61}$$

$$\xi = \frac{\psi}{\sigma} \tag{3.62}$$

Ψ is the separation of the center lines of the surfaces and σ is the composite surface roughness for the surfaces. ξ_o is the value of ξ corresponding to the separation when there is initial contact. As can be seen by comparison of the two equations for W, the dependencies on the reaction parameters and speed remained the same but the dependencies on apparent area of contact, load, and hardness changed. Analysis of the term in the bracket shows that while ξ depends on both load and roughness, its value is almost independent of load, hardness, and roughness (101). This does not mean, however, that the wear is independent of these parameters. As was stated previously, β is a function of temperature. This implies that β is also a function of load, hardness, and sliding speed.

In general, β is related to temperature by means of an Arrhenius type of relationship, namely,

$$\beta = \beta_o e^{-Q_o/RT} \tag{3.63}$$

where β_o is the Arrhenius constant for the reaction, Q_o is the activation energy associated with the oxide, R is the gas constant, and T is the temperature of the surface. On the basis of a simple model for asperity temperature (102), T can be related to P, p and v by the following:

$$T = T_0 + \frac{\mu P^* v}{4J(k_1 + k_2)a} \tag{3.64}$$

where T_0 is the nominal temperature of the surface; μ, the coefficient of friction; P^*, the load on the junction; a, the radius of the junction; J, Joule's constant; the k's are the thermal conductivities of the two bodies. P^* and a are functions of P, p, and the asperity distribution as illustrated by Eq. (3.57). (See Sec. 3.6 for a discussion of frictional heating and alternate equations for T.)

This simple model for oxidative wear indicates the various factors or parameters of a tribosystem that can influence these types of mechanisms. These processes are dependent on the chemical nature of the surface, reaction kinetics, mechanical and thermal properties of the materials, micro- and macro-geometrical features of the two surfaces, and operating conditions, that is, load, speed, and environment.

It has been shown that a similar model can be used to describe some of the general trends observed for cases of dry, sliding wear of steel surfaces (98,103,104). In this model, it is assumed that there is a thin layer of oxide on the surface at all times. Since the growth rate on clean surfaces and on oxidized surfaces tend to be different, this model uses a different relationship for oxide growth. Growth on oxide layers tends to follow a parabolic relationship rather than a logarithmic one. As a result, this model uses the following equation rather than Eq. (3.52):

$$m^2 = \beta t \tag{3.65}$$

where m is the amount oxygen a unit area of surface has taken up in time t. m is related to oxide thickness by the following equation:

$$m = f \rho_0 d \tag{3.66}$$

ρ_0 is the density of the oxide and f is the fraction of the oxide that is oxygen. β again is described in terms of an Arrhenius relationship.

This model allows the possibility of multiple engagements before a wear particle is formed by assuming that a critical oxide thickness, d_c, is required for fracture to occur.

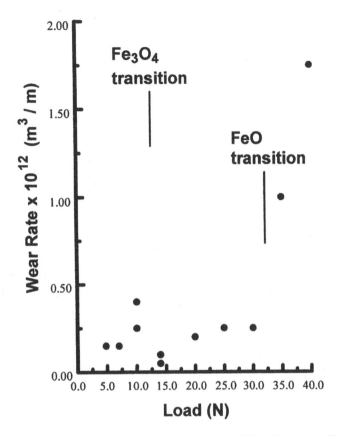

Figure 3.46 Wear rate as a function of load for unlubricated sliding between self-mated steel. The transitions in oxide formation are also shown. (From Ref. 104, reprinted with permission from ASME.)

This model resulted in the following equation for wear rate:

$$w = \frac{2P^{3/2}\beta_0 e^{-Q_0/RT}}{\pi^{1/2}\nu p^{3/2}f^2 d_c^2 \rho_0 n^{1/2}} \tag{3.67}$$

In using this model to explain the behavior of wear rates observed in dry sliding experiments with steels, it is necessary to make additional assumptions, primarily regarding n and d_c. Dry sliding data for EN8 steel are shown in Fig. 3.46. As can be seen in the figure, transitions in wear rate were found to correlate with the occurrence of different oxides. Oxidation studies have shown that there are three distinct regions of oxide growth with different activation energies (105). These regions are described in Table 3.6. Regression analysis of that data using this model indicated that it was necessary to assume that n and d_c were functions of load (104). This is shown in Fig. 3.47 for one sliding speed. It can be seen that speed and the region of oxidation affect the relationships between these parameters and load. A similar regression analysis of dry sliding data for EN31 was also done. These data are shown in Fig. 3.48. In this case, it was found that a correlation existed between these parameters, T, and the state of oxidation of the surface, that is, the mixture of Fe_2O_3 and Fe_3O_4. These correlations are shown in Fig. 3.49 (98). These results imply that n, d_c, T and w are interrelated and characteristic of a state of oxidation.

Table 3.6   Oxidation Kinetics of Steel Surfaces

Temperature (°C)	Oxide	β_0 (kg^2/m^{4}s)[a]	Q_0 (kJ/mole)
$T < 45$	Fe_2O_3	10^{16}	208
$45 < T < 600$	Fe_2O_3	10^3	0.96
	Fe_3O_4		
$600 < T$	Fe_2O_3	10^8	210
	Fe_3O_4		
	FeO		

$\Delta m^2 = \beta_0 e^{-\frac{Q_0}{RT}}$

Symbols: Δm, mass oxygen taken up per unit time; R, gas constant.
[a]Determined by regression analysis of wear data.
Source: Ref. 104.

The state of oxidation is determined by the operating conditions and the heat flow characteristics of the interface which is affected by the properties of the oxide. The regression analysis used in these studies involved the simultaneous satisfaction of wear and heat flow conditions.

Oxidative wear is primarily a sliding wear mechanism. It generally does not occur with lubrication. Since this mechanism is related to the chemical reactivity, it is more significant with metals than other materials. However, oxidative wear processes have been found to occur with ceramics, as well (106). It is important to recognize that not all unlubricated sliding situations with metals involve oxidative wear processes. For example, in

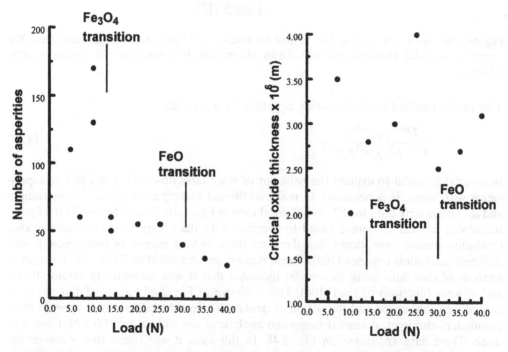

Figure 3.47 Variation in the number of junctions and critical oxide thickness as a function of load for unlubricated sliding between self-mated steel. The transitions in oxide formation are also shown. (From Ref. 104, reprinted with permission from ASME.)

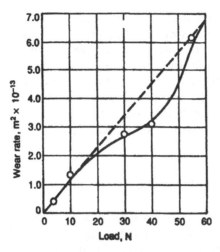

Figure 3.48 Example of the variation in wear rate with load for unlubricated sliding between self-mated steel. (From Ref. 98 reprinted with permission from ASM International.)

the wear study using EN31, discussed previously, oxidative wear did not occur for loads under 4 N. It is also possible that under some loading conditions oxidative wear processes may not be significant, even though oxidation occurs, because the dominant wear process involves failure underneath the oxide layer.

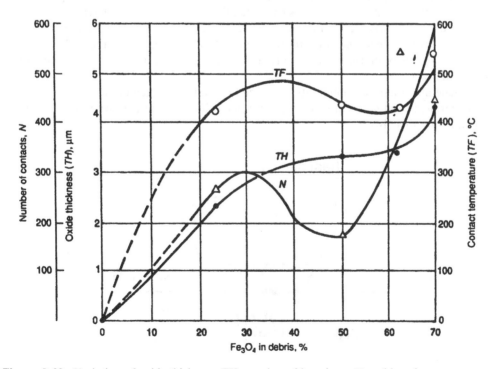

Figure 3.49 Variation of oxide thickness, TH, number of junctions, N, and junction temperature, TF, as a function of the percentage of Fe_3O_4 in the wear debris. Data are for unlubricated sliding between self-mated steel. (From Ref. 98, reprinted with permission from ASM International.)

The formation of oxides on a metal surface tends to reduce the wear. For example, in unlubricated sliding experiments with Cu, the author has observed an order of magnitude or more reduction in wear rate with the development of a Cu oxide on the surface. However, this is not always the case. A two order of magnitude increase in the wear rate of some steels has been observed in air over that obtained in vacuum (107). It should be recognized that the term oxidation is used to imply any chemical reaction altering the composition of the surface. It is not limited to effects from exposure to oxygen, though this is a very common one in many engineering applications. Alternate terms for oxidative wear are chemical wear and corrosive wear.

3.6. THERMAL WEAR PROCESSES

Thermal wear processes are those processes in which the primary cause of the wear is directly related to frictional and hysteretic heating as a result of relative motion. For most materials, thermal wear processes are generally limited to situations involving frictional heating as a result of relative sliding. However, with viscoelastic materials, thermal wear can occur as a result of hysteretic heating that is associated with any type of motion. Melting, thermal cracking, and thermal mounding or thermoelastic instability (TEI) are the most common forms of these processes but not the only ones. For example, evaporation and sublimation are other forms of thermal wear processes. All of these processes are related to the surface and near-surface temperature distributions that arise as a result of heating. These are usually characterized in terms of two temperatures. One temperature is the nominal temperature of the surface. The other is the maximum temperature at the asperity tips or junctions, which is called the flash temperature. With frictional heating the flash temperature is greater than the surface temperature. It can be several hundreds of degrees or more higher than the surface temperature and can reach the order of a few thousand degrees centigrade under some circumstances. Also, it is often the more important of the two.

Generally, these two temperatures are computed using different models (98,108,109). The linear heat conduction models used for a pin sliding on a disk shown in Figs. 3.50 and 3.51 illustrate this (109).

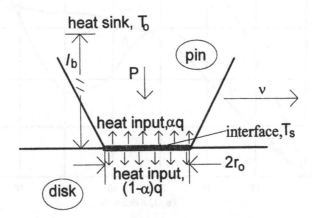

Figure 3.50 Model used for the bulk temperature increase of the surface. l_b is defined as the equivalent linear diffusion distance for bulk heating. It is the effective distance from the interface to a region that can be considered as a heat sink.

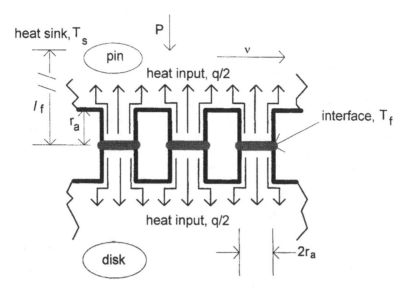

Figure 3.51 Model used for determining flash temperature. l_f is defined as the equivalent linear diffusion distance for flash heating. It is the effective distance from the junction interface to a region that can be considered as a heat sink.

The model for the bulk temperature is based on the apparent area of contact as shown in Fig. 3.50. The heat generated per unit area per unit time is given by

$$q = \frac{\mu P \nu}{A_a} \tag{3.68}$$

where μ is the coefficient of friction; P, the load; A_a, the apparent area of contact; ν, is the sliding velocity. The model assumes that this is shared between the two bodies, a fraction, α, going into the pin and $(1-\alpha)$ going into the disk. The heat flow into the pin and disk is different and as a result two different models are used to describe the temperature distribution in these bodies. The pin experiences a continual source of heat and the heat flow is described by the first law of heat flow. The disk is described by time-dependent equations for heat flow for the injection of heat. The quantity of heat that is injected is $2(1-\alpha) qr_o/\nu$. For self-mated materials, this model results in the following equation for the surface temperature, T_s (109).

$$T_s = T_0 + 2\alpha\beta\mu T^* \Phi_p \bar{P} \tag{3.69}$$

where

$$\alpha = \frac{2}{4 + \beta(\pi\Phi_p)^{1/2}} \tag{3.70}$$

$$T^* = \frac{ap}{k} \tag{3.71}$$

$$\bar{P} = \frac{P}{A_a p} \tag{3.72}$$

$$\Phi_p = \frac{r_0 \nu}{2a} \tag{3.73}$$

In these equations, p is the hardness, k is the thermal conductivity, and a is the thermal diffusivity. Φ_p is the Peclet Number. Essentially this is the ratio of the time it takes for the temperature to reach a maximum at a depth of half the width of the contact to the time it takes for the heat source to move half the contact width.[*] For a stationary heat source the Peclet Number is 0. For Peclet Numbers below 0.1, stable temperature distributions are established in both bodies during the time of contact. As a result the heat flow into both the pin and the disk can be considered as from a stationary source. In this case, the heat is uniformly divided between the two bodies, α is 0.5. For Peclet Numbers above 0.1, the thermal distribution in the disk is not stabilized during the contact time and as a result more heat tends to flow into the disk. For Peclet Numbers above 100, almost all the heat flows into the disk. For intermediate values, the portion of the heat going into the disk increases with increasing speed, that is increasing Peclet values.

β in these equations is a dimensionless linearization factor introduced to account for the fact that the heat flow is three-dimensional, not linearly as assumed by the model. It is essentially the ratio of the heat diffusion distance into the surface to r_0. The heat diffusion distance is nominally the depth below the surface where there is no increase in temperature. For steel β has been found to be approximately 6 (109). Assuming that the lateral diffusion of the heat is proportional to thermal diffusivity, its value for other materials can be approximated by

$$\beta = \frac{5.5 \times 10^{-5} \, \text{m}^2}{a} \, \frac{}{\text{s}} \tag{3.74}$$

The model for the flash temperature is based on the real area of contact, as illustrated in Fig. 3.51. In this case, both surfaces are described by time-dependent heat flow equations. For self-mated materials, the model results in the following equations for the flash temperature, T_f:

$$T_f = T_s + \mu T^* \beta \Phi_p \left(\frac{r_a}{r_0}\right) \tag{3.75}$$

$$T_f = T_s + \frac{\mu T^* \beta \Phi_p \bar{P}^{1/2}}{n^{1/2}} \tag{3.76}$$

Equation (3.76) results from the additional assumptions that the asperities are plastically deformed, that is, that the real area of contact is P/p. In this equation, n is the number of junctions and can be estimated by the following (109):

$$n = \left(\frac{r_0}{r_a}\right)^2 \bar{P}(1 - \bar{P}) + 1 \tag{3.77}$$

It has been found that changes in the real area of contact primarily result from changes in the number of junctions formed and not from changes in size of the junctions (37,38,110–114). Studies have shown that the typical radius of junctions is of the order of 10^{-5}–10^{-6} m

[*]The time it takes for the temperature to reach a maximum at a depth h is (h^2/a).

Table 3.7 Temperature Equations for Frictional Heating [a]

Surface temperature, T_s
Stationary heat source surface (1)

$$T_s = T_o + 2\alpha_s \beta \mu T_1^* \Phi_{p_1}^* \bar{P}$$

$$\Phi_{p_1}^* \equiv \frac{\nu r_o}{2a_1}$$

Moving heat source surface (2)

$$T_s = T_o + \frac{4(1-\alpha_s)\mu}{\pi^{1/2}} T_2^* \Phi_{p_2} \bar{P}$$

$$\alpha_s = \frac{2}{4+\pi^{1/2}\left(\frac{T_1^*}{T_2^*}\right)\left(\frac{\Phi_{p_1}^*}{\Phi_{p_2}^{1/2}}\right)}$$

Flash temperature, T_f
General

$$T_f = T_B + 2\alpha_f \mu T_1^* \beta_2 \left(\frac{r_a}{r_0}\right) \Phi_{p_1}$$

$$\alpha_f = \frac{a_1 k_1}{a_1 k_1 + a_2 k_2}$$

For $A_A = P/p$

$$T_f = T_B + \frac{2\alpha_f \mu T_1^* \beta_1 \bar{P}^{1/2} \Phi_{p_1}}{n^{1/2}}$$

$$n = \left(\frac{r_0}{r_a}\right)^2 \bar{P}(1-\bar{P}) + 1$$

$$T^* = \frac{ap}{K} \qquad \bar{P} = \frac{P}{A_A p}$$

$$\Phi_p = \frac{\nu r_0}{2a} \qquad \beta = \frac{5.5 \times 10^{-5} \text{m}^2/\text{s}}{a}$$

Symbols: P, load; p, hardness of softer surface; a, thermal diffusivity; k, thermal conductivity; A_A, apparent area of contact; ν, sliding velocity; μ, coefficient of friction; r_0, radius of apparent contact area; r_a, radius of junctions (approximately 10^{-5} m; n, number of junctions; Φp, Peclet Number.
[a]Based on the Lim/Ashby temperature relationships for self-mated materials (Ref. 109).

but can be much larger and smaller in some circumstances (110–114). For typical situations, a nominal value of 10^{-5} m is often used for thermal calculations (109). Generalized forms of the equations for surface and flash temperatures are given in Table 3.7. In this table, an equivalent Peclet Number for the stationary heat source surface, Φ^*_p, is defined for consistency. It can be seen that the distribution of heat or the heat partition between the surfaces is affected by differences in thermal properties between the two surfaces.

Oxide and other layers on surfaces can also have a significant effect on frictional heating and the apparent conductivity of a surface. This is shown by the following equation for the effective value of the thermal conductivity of a surface with a thin layer on it (109):

$$k_e = \frac{k_s k_1}{(1 - z/\beta r_a)k_1 + (z/\beta r_a)k_s} \tag{3.78}$$

k_e is the effective conductivity; k_s is the conductivity of the substrate; k_1 is the conductivity of the layer; and Z is the thickness of the layer.

Actual temperatures tend to be lower than those predicted by these equations, primarily because heat can be dissipated by other mechanisms, such as convection, radiation, and cooling by lubricants. Such effects, particularly cooling by lubricants, can result in significantly lower temperatures. Temperature increases under lubricated conditions are generally negligible, except for thermoelastic instability.

Instead of determining the heat partition at the interface, that is, α and $(1 - \alpha)$, the actual temperature can be determined by using the values obtained for each surface, assuming that all the heat goes into that surface. It has been shown that

$$\frac{1}{T} = \frac{1}{T_1^*} + \frac{1}{T_2^*} \tag{3.79}$$

T_1^* and T_2^* are the temperatures obtained for surfaces 1 and 2, assuming all the heat goes into that surface; T is the actual surface temperature (115).

Most thermal wear processes can be grouped into three general types. One group is comprised of those processes, which are simply related to the maximum temperature. Melting, softening, evaporation and sublimation would be examples of this type. The second group is comprised of those processes, which are directly related to thermal gradients. Thermal fatigue and thermal cracking are examples of this type. Those processes, which result from thermoelastic instability, comprise the last group. All these types of processes require significant temperature rise. How high a rise is significant depends on the materials and mechanism. For example, for the first type of mechanism, a rise of less than 100°C can be significant for some polymers, while a rise in excess of a 1500°C is required for melting of metals and intermediate temperatures for the other types of mechanisms.

With the first type of thermal mechanisms, wear scars typically exhibit features that are suggestive of melting, liquid flow, and thermal degradation. Examples of these features are shown in Fig. 3.52. The following Eq. (3.80), is one proposed for melt wear of a pin sliding against a disk (109). The model is illustrated in Fig. 3.53. It is based on a linearization model for heat flow from a stationary source, similar to the one used to develop Eq. (3.68). It assumes that a portion of the heat is conducted through the pin, maintaining the temperature differential, and a portion of the heat is absorbed as latent heat into the melted layer. The depth rate of wear $\dot{h}$ (units of length per unit time) is given by

$$\dot{h} = K \left(\frac{k}{\beta r_0 L} \right) [(2\alpha T^* \beta \mu \bar{P} \Phi_p') - (T_m - T_0)] \tag{3.80}$$

In this equation, L is the latent heat for melting and T_m is the melting temperature. Φ_p' has the same form as the Peclet Number and is the defined as $(2r_0 v / a_{pin})$ (see Table 3.7). K is the fraction of the molten layer that is lost from the contact per unit time. The corresponding equation for flash temperature melting is Eq. (3.81).

$$\dot{h} = K \left(\frac{k}{\beta r_a L} \right) \left[\left(2\alpha_f T^* \beta \mu \left(\frac{r_a}{r_0} \right) \Phi_p' \right) - (T_m - T_0) \right] \tag{3.81}$$

Noting that Φ_p' is equal to the Peclet Number for the junction contact, $v r_a / a$, times (r_0 / r_a), this equation can be rewritten as

$$\dot{h} = K \left(\frac{k}{\beta r_a L} \right) [(2\alpha_f T^* \beta \mu \ \Phi_p) - (T_m - T_0)] \tag{3.82}$$

where Φ_p is the Peclet Number for the junctions.

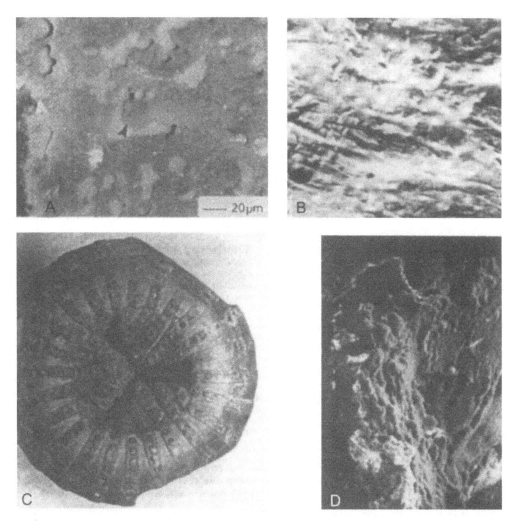

Figure 3.52 Examples of wear scars from situations in which melting has occurred. "A" shows the worn surface of an unfilled polymer, where melting has taken place as a result of sliding. Regions of melting are the large patches, such as the one indicated by the arrow. "B" also shows a polymer wear scar where the melting resulted from sliding. "C" shows a diamond drag bit on which the diamonds have been burned and flattened in an abrasive wear situation. "D" shows the worn surface of a polymer, where melting and charring has occurred as a result of repeated impacts. ("A" is from Ref. 183, "B" is from Ref. 184, "C" is from Ref. 185, and "D" is from Ref. 186, "B", and "C" reprinted with permission from ASME.)

The second class of thermal wear mechanism is mechanisms resulting from the thermal fluctuation, ΔT, caused by frictional heating. In some materials fracture can take place if ΔT or the thermal strain, ε_T, is large enough. More generally, repeated cycles of ΔT can result in the nucleation and propagation of cracks, that is thermal fatigue. As with most fatigue wear processes, these processes can be described by a power law relationship, such as,

$$W \propto \varepsilon_T^n \qquad\qquad\qquad (3.83)$$

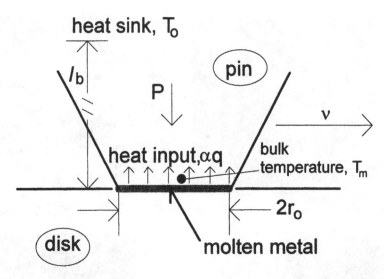

Figure 3.53 Model used for surface melting. l_b is defined as the equivalent linear diffusion distance for bulk heating. It is the effective distance from the interface to a region that can be considered as a heat sink.

or

$$\dot{W} \propto \sigma_T^n \qquad (3.84)$$

σ_T is the corresponding thermal stress and $\dot{W}$ is wear rate. The exponent is generally 1 or greater and can be large, for example, the order of 10. There is a wide range in the appearance of wear scars produced by this type of mechanism. Figures 3.54 and 3.55 show examples of wear scars resulting from thermal fracture and fatigue.

A model has been proposed for the type of thermal wear of ceramics illustrated in Fig. 3.55. (116) This model assumes that there are micro-cracks in the ceramic and that the severe wear shown in Fig. 3.55B results from the growth of these cracks. With this model, it is shown that magnitude of the wear rate in this region can be correlated with a thermal severity factor, which is the ratio of the temperature fluctuation, ΔT, to the thermal shock resistance of the material, ΔT_s. This is shown in Fig. 3.56. Analysis of these data results in the following approximate relationship between this factor, TS, and wear rate:

$$W \propto TS^8 \qquad (3.85)$$

The limited data in the mild region suggest a similar relationship with a much lower exponent.

In the model, the following equation for TS, where k_e is the effective conductivity of the contact, is developed:

$$TS = \frac{\mu P v}{\Delta T_s k_e r_0} \qquad (3.86)$$

In the model, it is assumed that for crack growth, the following condition, based on linear elastic fracture theory, must be satisfied:

$$1.12\sigma_T \sqrt[2]{\pi d} \geq K_{LC} \qquad (3.87)$$

d is the initial size of the crack and K_{LC} is the fracture toughness of the material. It also

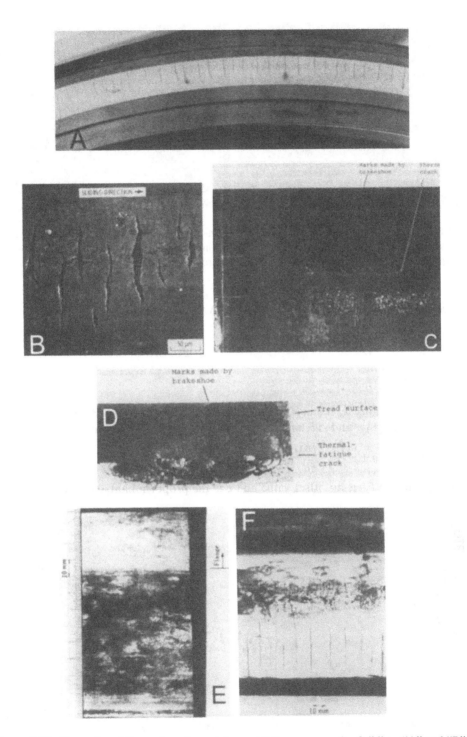

Figure 3.54 Examples of thermal cracks and thermal fatigue as a result of sliding. "A" and "B" are on the worn surfaces of metal seals. "C", "D", "E", and "F" are wear scars on metal train wheels. ("A" is from Ref. 117, "B" is from Ref. 187, "C"–"F" from Ref. 188, reprinted with permission from Elsevier Sequoia S.A.)

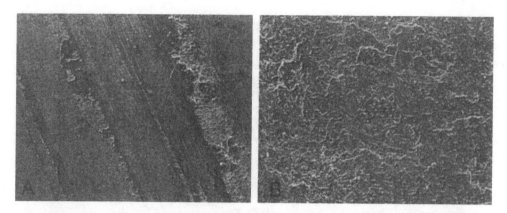

Figure 3.55 Examples of thermal wear scars on ceramics. The micrographs show the appearance of a wear scar on the zirconia specimen after sliding against an unlubricated alumina ball at two different speeds. "A" is 0.15 m/s and "B" is 0.40 m/s. (From Ref. 116, reprinted with permission from Elsevier Sequoia S.A.)

assumes the following relationships for σ_T and ΔT:

$$\sigma_T = \left(\frac{E\lambda}{1-\eta}\right)\Delta T \tag{3.88}$$

$$\Delta T = \frac{\mu P v}{r_0 k_e} \tag{3.89}$$

E is Young's Modulus; λ, the coefficient of thermal expansion; η, is Poisson's ratio. It is also assumed that K_{LC} and ΔT_s are related by the following equation:

$$\Delta T_s = \Delta T_{s0} + \frac{c(1-\eta)K_{LC}}{E\lambda\sqrt[2]{\pi d}} \tag{3.90}$$

In this equation, ΔT_{s0} is an offset value and c is the proportionality constant.

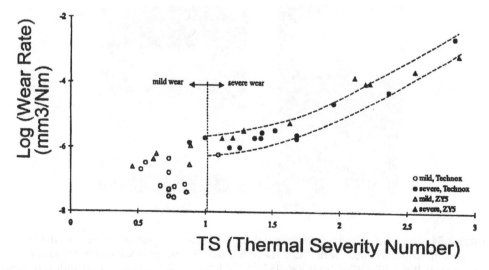

Figure 3.56 Wear rate as a function of the thermal severity number, TS. (From Ref. 116, reprinted with permission from Elsevier Sequoia S.A.)

The remaining class of thermal wear mechanisms is that associated with thermoelastic instability. The acronym TEI and the term, thermal mounding, are other names used for these processes. These processes essentially involve the collapse of the real area of contact to a few localized areas as a result of localized thermal expansion (117). These areas are referred to as hot spots or patches. Once formed, these sites can initiate other forms of wear, including other forms of thermal wear. The minimum number of hot spots is the minimum number required for mechanical stability, which in some cases can be as little as one. While not limited to these situations, TEI wear processes are often significant in the wear behavior of seals, electrical brushes, and brakes (28,108). In addition to wear, TEI processes can directly cause leakage in seals as a result of increased separation between surfaces. Wear scars associated with thermoelastic instability tend to exhibit localized heat-affected and thermally distressed areas, that is, hot spots or patches. Examples of such wear scars are shown in Fig. 3.57. The following scenario describes the evolution of these hot spots.

Assume that as a result of a nonuniform temperature distribution or nonhomogeneity in thermal properties, a region or regions in the apparent contact area begins to bulge above the mean level of the surface. As a result of this tendency, these areas will absorb more heat and experience increased wear. If conditions are such that the increase in heat is dissipated fast enough and the differential wear rate is large enough, the bulge will not form and conditions will tend to become stable and more uniform across the contact. However, if the increase in heat results in still higher local temperature and the increased wear rate is not high enough, the contact will become unstable. A bulge will form and continue to grow, until contact between the surfaces is limited to those regions. It has been found that the onset of this unstable behavior can be related to speed. There is a critical speed, $\nu*$, above which a contact becomes unstable and thermal bulges or patches will form and below which they do not form.

Unlike other types of thermal wear processes, which generally do not occur under lubricated conditions, TEI can occur under lubricated conditions. While the local collapse of a fluid film can lead to TEI behavior, less severe perturbations to the lubricant film can also cause the formation of thermal patches as a result of changes in viscous heating in the fluid (118,119).

Studies have indicated that stable arrangements or groups of hot patches can occur, each with their own critical speed. While stable, these groups are not necessarily stationary. For example, with seals, hot patches have been found to slowly precess around the seal (120). The critical speeds for the formation of these groups depend on the size and geometry of the contacting members. In addition to these factors, $\nu*$ is also a function of the relative conductivity of the surfaces, thermal and mechanical properties of the surface, wear, and lubricant properties but not directly of load. The following two equations have been obtained for $\nu*$. Equation (3.91) is for an unlubricated system and Eq. (3.92) for a lubricated system (117,118). Both are based on some limiting assumptions: no wear; a nonconductive, flat and rigid counterface; simple cup face seal configuration. However, they do provide some insight concerning the significance of some parameters affecting TEI behavior

$$\nu^* = \frac{4\pi k}{E\lambda\mu\chi} \tag{3.91}$$

$$\nu^* = \frac{2\pi\zeta}{\chi}\left(\frac{k}{\gamma\lambda}\right)^{1/2} \tag{3.92}$$

In both equations, χ is the spacing between the hot patches. ζ is the mean film thickness and γ is the viscosity of the lubricant, respectively. The lowest critical speed would occur for the

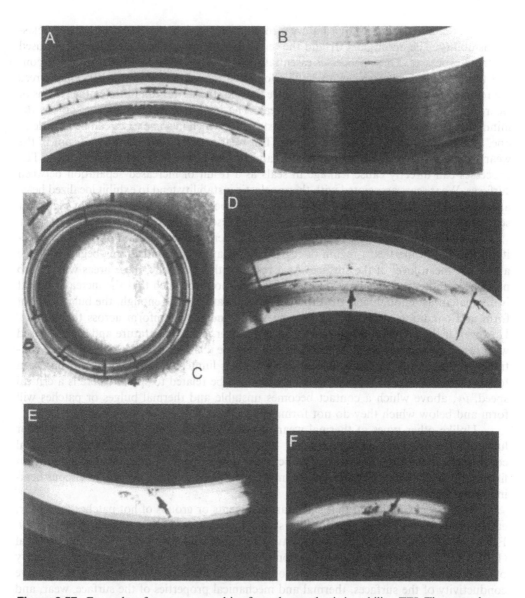

Figure 3.57 Examples of wear scars resulting from thermoelastic instability, TEI. The examples are from seals used in different applications. "A"–"D" are metal seals. "E" and "F" are carbon seals. The localized regions of damage and discoloration are the result of thermoelastic instability. (From Ref. 189, reprinted with permission from Elsevier Sequoia S.A.)

largest spacing between hot patches possible. For the cup configuration assumed by these models, this would be the circumference of the cup. The effects of wear and counterface conductivity on the value of $\nu*$ are significant. The modeling results shown in Figs. 3.58, 3.59, 3.60, and 3.61 illustrate their significance.

The effect of counterface conductivity has been modeled for a cup seal configuration (120). The results of this analysis are shown in Figs. 3.58 and 3.59. In Fig. 3.58, this effect is demonstrated as a function of the ratio of conductivities of the two surfaces. In Fig. 3.59,

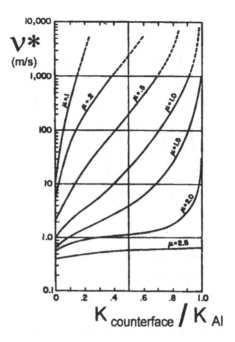

Figure 3.58 Variation of the critical disturbance velocity for a less conductive body sliding against a more conductive body. The properties of both bodies were assumed to be aluminum with the exception that the less conductive body was assumed to have a hypothetical reduced conductivity. (From Ref. 117, reprinted with permission from Elsevier Sequoia S.A.)

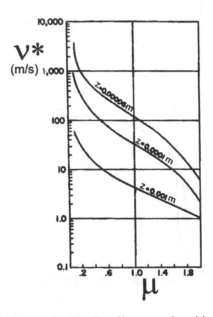

Figure 3.59 Effect of the thickness of a thin glass film, z, on the critical disturbance velocity in aluminum. (From Ref. 117, reprinted with permission from Elsevier Sequoia S.A.)

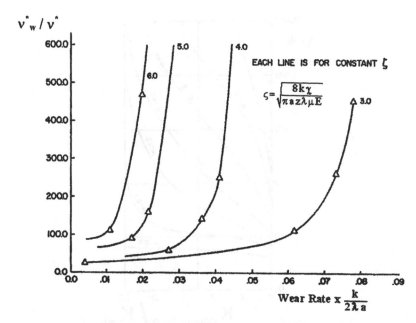

Figure 3.60 The effect of wear on the critical disturbance velocity of a scraper. Larger values of ς result in greater amounts of heat going to the counterface. (From Ref. 117.)

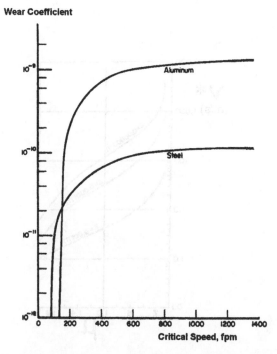

Figure 3.61 The effect of wear rate on the critical disturbance velocity for unlubricated self-mated steel and aluminum. (From Ref. 188, reprinted with permission from Elsevier Sequoia S.A.)

it is illustrated by the effect of thin insulating layer on $\nu*$. It was also found in these analyses that the motion of the hot spots is affected by the partition of heat. The effect of wear against a conductive counterface on $\nu*$ was modeled for a flat blade sliding against a rotating drum (119–122). The normalized results of that model are shown in Fig. 3.60. The results for a steel and aluminum blade are shown in Fig. 3.61. This graph shows that high wear rates can significantly increase $\nu*$.

An important aspect of TEI thermal wear processes is that they can occur under conditions where there is only a moderate rise in surface temperature, for example, TEI behavior has been observed in situations where the temperature rise of metal surfaces is 100°C or less (123). The other types of thermal mechanism typically require significantly higher temperatures. For these, the severity of thermal wear can be reduced by cooling and using materials whose properties are less sensitive to increases in temperature.

3.7. TRIBOFILM WEAR PROCESSES

Many investigators have identified tribofilms and their importance to wear and friction behavior (22,107,124–133). Tribofilms are layers of compacted wear debris that form on surfaces during sliding. Such films are also called transfer films, third-body films or simply third-bodies. The term transfer film is commonly used when the composition of the material in the layer is the same as the counterface. The term third-body is a generic term for any interface layer or zone which has different material properties than the surfaces and across which velocity differences are accommodated (134). When used to refer to a tribofilm, it generally implies a mixture of wear debris in the layer. Tribofilms act as a lubricant layer between the surfaces, providing separation and accommodating relative motion between the two surfaces. Relative motion with these layers is accomplished by shear within the layer or slip between the layer and the surface. Examples of tribofilms on wear surfaces are shown in Figs. 3.62 and 3.63.

Tribofilm wear processes are wear processes in which mass loss from the surfaces or tribosystem occurs through loss of material from tribofilms. As material is lost from these films fresh wear debris from the surface enters the layer to maintain the film. This process is illustrated in Fig. 3.64. Before being lost from the film, debris material is circulated within the layer and between the surfaces. When a stable film is formed, equilibrium requires that the amount of material entering into the layer is the same as lost from the layer. Therefore, once a stable film is formed, wear behavior can be described with the same models and relationships used for debris-producing mechanisms, such as adhesion or repeated-cycle deformation mechanisms, by considering the film as a lubricant. Conceptually, if WR_D is the wear rate of a debris-producing mechanism without a tribofilm present, the wear rate, WR, with the tribofilm present is

$$WR = K WR_D \tag{3.93}$$

where K is a proportionality constant, which can generally be incorporated into the empirical wear coefficients of the model.

Tribofilms have a significant effect on wear behavior (22,124,135). Generally when stable films are formed, a significant reduction in wear rate is seen. Such an effect can be seen in the data shown in Fig. 3.65. While such films are frequently cited as being key aspects in the wear of polymer–metal systems (124,125,129,130,135), such films can also occur in other sliding systems, for example metal–metal and metal–ceramic. (87,126,130–133,136). Examples of these are shown in Fig. 3.63. Generally, it is the softer

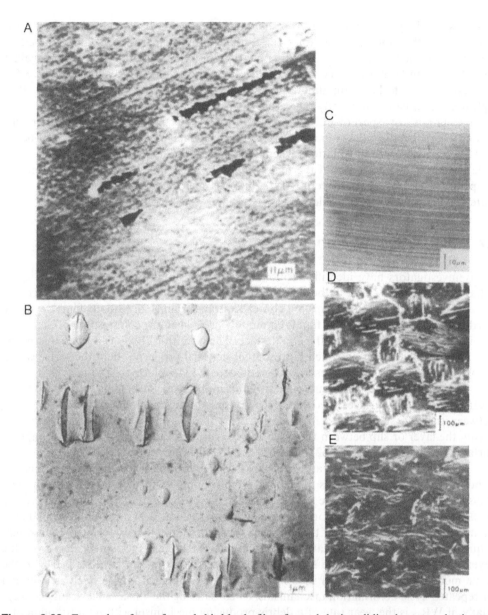

Figure 3.62 Examples of transfer and third body films formed during sliding between plastic and metal surfaces. "A" and "B" show disrupted polymer transfer films formed on a metal counterface. "C" shows a continuous polymer transfer film formed on a metal counterface sliding against a fabric reinforced plastic. "D" shows the initial stages and "E" the final stages of the third-body film formed on the surface of the plastic in that case. ("A" and "B" from Ref. 128, reprinted with permission from Butterworth Heinemann Ltd.; "C" and "D" from Ref. 140, reprinted with permission from ASME.)

material that will form the film. It should be recognized that while the material in these layers originates from the sliding members, the properties may be different since they can experience high shear and deformation, as well as elevated temperature in the formation process.

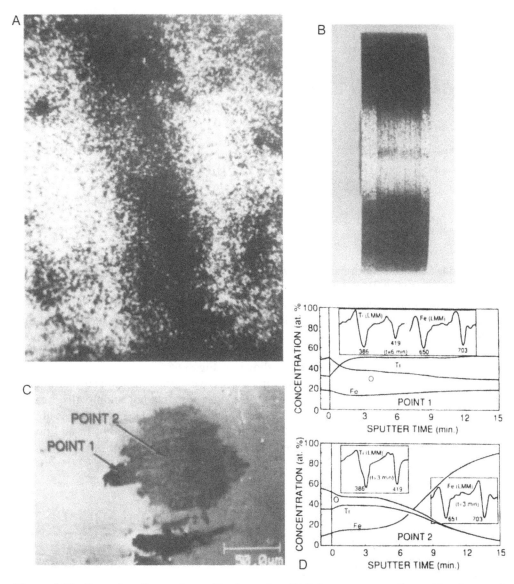

Figure 3.63 Examples of nonpolymer film formation. "A" shows an autoradiographic micrograph of the wear track on a Ni surface, sliding against a ferrite counterface. The dark region indicates the existence of a ferrite layer on the surface of the Ni. "B" shows the graphite film that is formed during rolling contact between graphitic Al counterfaces. "C" shows the transfer film formed on a steel surface in sliding contact with a TiN counterface. The Auger spectra shown in "D" confirm the presence of the film. ("A" from Ref. 133, reprinted with permission from Elsevier Sequoia S.A.; "B" from Ref. 131, reprinted with permission from ASME; "C" and "D" from Ref. 190, reprinted with permission from ASME.)

Initially, the films tend to form in patches but with continued sliding, the coverage becomes more uniform. During this phase, the thickness of the deposition might change as well. At some point, a stable film with a characteristic thickness is established. Studies have indicated that the more complete the coverage, the better the wear performance.

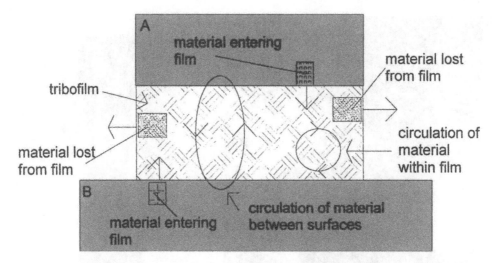

Figure 3.64 Schematic illustrating the flow of material associated with tribofilm wear processes.

Stable and beneficial tribofilms generally do not form under lubricated conditions. This is because lubricants tend to inhibit the adhesion of the wear debris to the surfaces and thus inhibit film formation (125,135). Because of this behavior, the wear rate of sliding system, which benefits from tribofilm formation, can increase with the introduction of a poor lubricant. An example of this behavior is shown in Fig. 3.66.

Because of the effect that material properties have on attachment, the formation and properties of tribofilms are characteristics of material pairs, not simply the wearing material. For example, in tribosystems, where tribofilms are involved, differences in wear have been observed with different counterface material (22,125,137–139). In addition to this, several other factors have also been identified as being significant in the formation

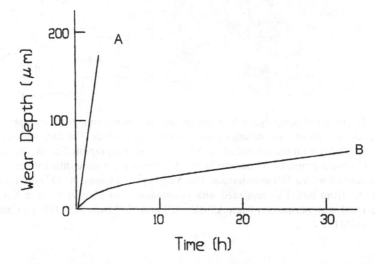

Figure 3.65 Reduction in plastic wear rate as a result of transfer film formation for unlubricated sliding against stainless steel. For "A", the stainless steel counterface is rough, 0.14 μm Ra, inhibiting transfer film formation. For "B", the counterface is smoother, less than 0.05 μm Ra, allowing transfer film formation. (From Ref. 125.)

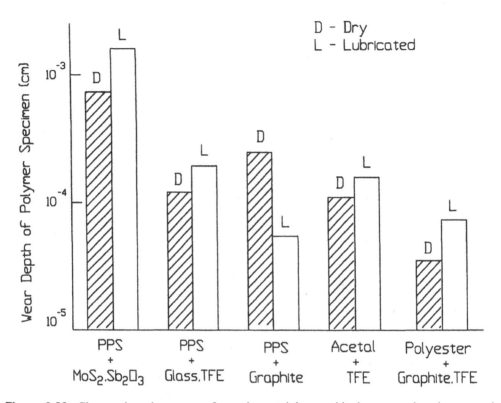

Figure 3.66 Changes in polymer wear for various stainless steel/polymer couples when water is used as a lubricant. (From Ref. 135.)

and development of such films. Roughness (22,125,127–130), load (124,140), speed (124,127,130,136,141), and type of motion (126) have all been found to influence these types of films. Several of these studies were done in the context of polymer–metal sliding systems, but there is no reason to indicate that these influences are limited to those systems. These studies suggest that there are optimum conditions associated with several of these parameters (128,129). An example of this for roughness is shown in Fig. 3.67. The proposed explanation for such behavior is that a certain degree of roughness promotes adhesion of the wear debris to the surface, in much the same way that it helps with the adhesion of coatings and platings. On the other hand, too coarse a roughness would result in larger wear debris, which would not adhere as well. In addition, a thicker film would be required to protect against the higher asperities and such films would tend to be unstable. These counter trends result in an optimum condition. Thus, some studies have concluded that a harder counterface is preferred to a softer one in that the optimum roughness condition will remain stable and not be altered by wear (125). It should be noted that the complete dependency of roughness is probably not explained by these rudimentary concepts. For example, it has been indicated that the flatness of the asperity tips may also be a factor (127).

The influence of speed on polymer film formation is shown in Fig. 3.68. As for roughness, there appears to be an optimum for speed as well. The reason proposed for this behavior is that a certain degree of softening of the polymer surface has to occur for significant transfer to occur. At low speed, the temperature is too low for softening; at higher speeds, however, the temperature increases and the flow characteristics of the softened sur-

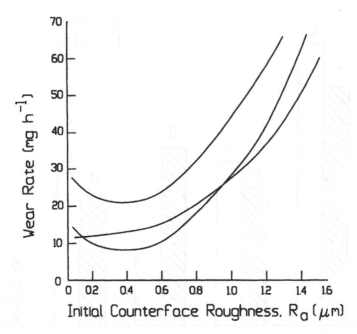

Figure 3.67 The effect of counterface surface roughness on transfer film formation in the case of filled PTFE sliding against steel. (From Ref. 128.)

face layer allow film formation to occur. At still higher speeds, the temperature is so high that the flow characteristics would degrade and film formation would not occur. As a consequence, it as been proposed that an important material property for transfer film formation is the rheological properties of the polymer (130). Similar concepts can be proposed for the effect of load. Increased load will promote adhesion and will also increase temperature. Excessive load will tend to result in larger wear debris, higher temperature, and more effectively remove material from the contact surfaces. Studies have shown that film

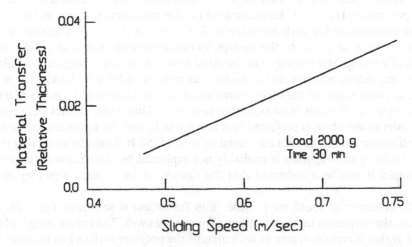

Figure 3.68 The effect of sliding speed on transfer film formation in the case of a polymer/polymer couple. (From Ref. 124).

formation can either decrease (124) or increase (140) with increasing load. Again, this would suggest that an optimum condition should exist for film formation.

Geometrical and shape elements, which can affect the trapping and displacement of debris in the contact region, can also have an effect of tribofilm formation (128).

3.8. ABRASIVE WEAR

Abrasive wear is wear caused by hard particles and protuberances. Abrasion and erosion are terms commonly used for abrasive wear situations. These two types of situations are illustrated in Fig. 3.69. When two surfaces are involved, the wear situation is generally referred to as abrasion. A distinction is usually made between two types of abrasion, two-body and three-body abrasion, because of significant differences in the wear behavior associated with these two situations. Two-body abrasion is when the wear is caused by protuberances on or hard particles fixed to a surface. Three-body abrasion is when the particles are not attached but between the surfaces (142). Filing, sanding, and grinding would be examples of two-body abrasion, as well as wear caused by magnetic media and paper; a rough, file-like metal surface sliding on a polymer surface would be another. Examples of three-body abrasion would be wear caused by sand or grit in a bearing and hard wear debris and abrasive slurries trapped between moving surfaces. The term erosion is generally applied to abrasive wear situations when only one surface is involved. Slurry erosion and solid particle erosion are common generic terms for such situations. Solid par-

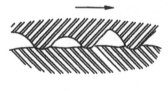

Two—Body Abrasion

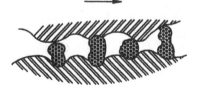

Three—Body Abrasion

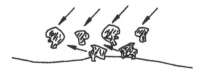

Erosion

Figure 3.69 Abrasive wear situations.

ticle erosion is when a stream of particles or fluid containing particles impacts a surface, causing wear. The wear caused by sand and grit in air streams are examples. An example of slurry erosion would be the wear of pipes through which slurries are pumped. Examples of abrasive wear scars are shown in Figs. 3.70 and 3.71.

In the following discussion of abrasive wear two-body abrasion by protuberances is considered to be equivalent to two-body abrasion by hard particles or abrasive grains attached to a surface.

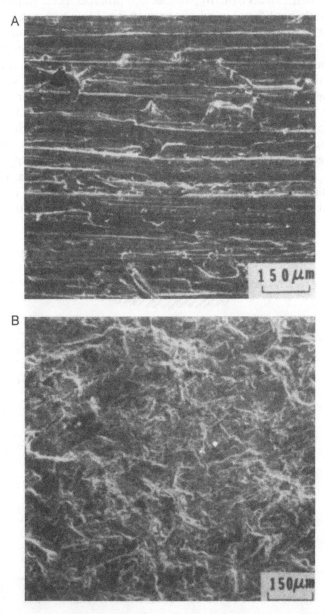

Figure 3.70 Examples of abrasive wear. "A", two-body abrasion. "B", particle impingement. "C" and "D", three-body abrasion. ("A", and "B" from Ref. 152 and "C" and "D" from Ref. 64; reprinted with permission from ASME.)

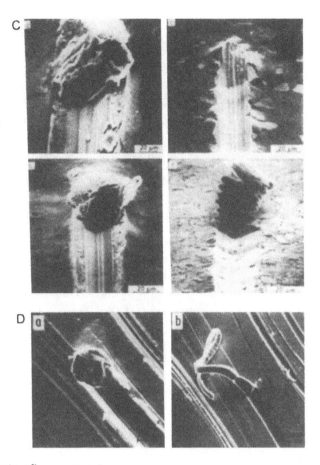

Figure 3.70 (*continued*)

When the abrasives are harder than the surface they are wearing, the dominant type of wear mechanism in abrasive wear is single-cycle deformation, though repeated-cycle deformation mechanisms, as well as chemical and thermal mechanisms may also be involved (143). When the abrasives are softer than the surface they are wearing, the dominant type of mechanism becomes repeated-cycle deformation (61,144). In abrasive wear situations, the significance of single-cycle deformation mechanisms does not decrease with sliding or duration, as they typically do in nonabrasive wear situations. Generally, these mechanisms remain the same unless there is a change with the characteristics of the particles involved, such as changes in size, sharpness, or amount. Such changes can take place as a result of particle wear and fracture and the accumulation of particles within the contact area with time. The atmosphere and fluid media in which abrasive wear takes place is often a factor in the abrasive wear. Wear rates tend to be higher when there is a chemical interaction with the wearing surface. This is generally attributed to chemical wear mechanisms, the modification of surface mechanical properties as a result of chemical interaction, that is, the Rebinder Effect (145), and synergistic effects between wear and corrosion (146). Synergism between wear and corrosion results from the fact that wear produces fresh surfaces, which are more readily oxidized. In turn, this increased oxidation results in higher wear rates.

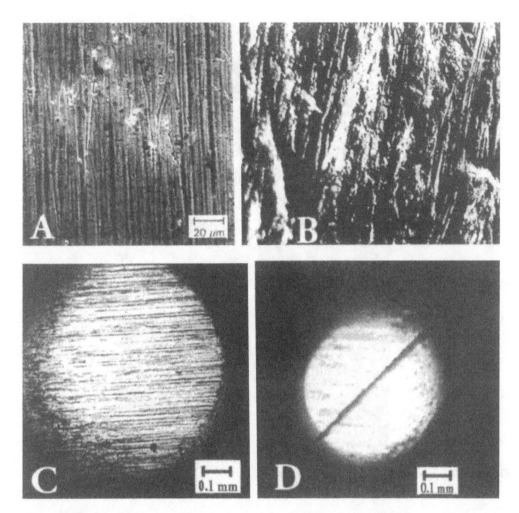

Figure 3.71 Expample of wear scars resulting from three-body abrasion. "A","B", and "C" illustrate various degrees of severe abrasive wear, while "D" is an example of mild abrasive wear. ("A" from Ref. 148, "B" from Ref. 191, "C" from Ref. 192, and "D" from Ref. 34. "A" and "B" reprinted with permission from ASME. "C" reprinted with permission from IBM. "D" reprinted with permission from Elsevier Sequoia S.A.)

It is generally found that one or more of the following equations can describe abrasion: (34,61,144,147–151)

$$V = KPS \tag{3.94}$$

$$V = \frac{KPS}{p} \tag{3.95}$$

$$V = \frac{KPS}{p^n} \tag{3.96}$$

In these equations, V is wear volume, P is load, S is sliding distance, and p is hardness. K is a wear coefficient, which is determined empirically. Equation (3.94) is the most broadly

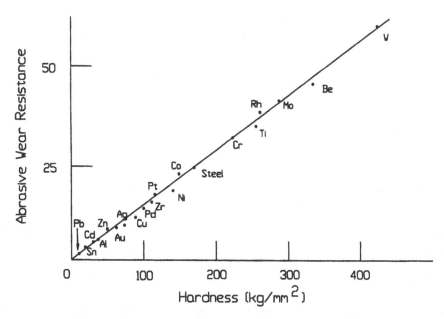

Figure 3.72 The effect of hardness on the abrasive wear rate of pure metals. Data are for two-body abrasion, when the abrasive is harder than the abraded surface. (From Ref. 193.)

applicable one. It is applies to most materials systems, independent of the relative hardness of the surface to the abrasives, and to two-body and three-body abrasion. In this equation, the wear coefficient is a function of the wearing material, the abrasives, the media or environment in which the abrasion takes place, and the freedom of the particles to move. Equation (3.95), which is the same as the equation used for single-cycle wear, Eq. (3.20), is generally applicable to all material systems and types of abrasion when the abrasive is harder than the wearing surface. As discussed in Sec. 3.3, K in this case tends to be dependent only on material type, not individual materials. This is illustrated by the data in Figs. 3.20 and 3.72. Otherwise the dependencies are the same as with Eq. (3.93). The situation is the same with Eq. (3.96). However, this equation applies to situations where the hardness of the surface and abrasives is similar or when the surface is harder. When similar, limited data indicate that n is around 10. When the surface is harder, n is around 5 (151). The change in the hardness dependency between these two equations is the result of the change in the type of dominant wear mechanism, that is, from single-cycle deformation to repeated-cycle deformation.

Nominal values for K in Eq. (3.95) are given in Table 3.8 for a variety of conditions. It can be seen that K ranges over several orders of magnitude and that some trends exist. One trend that is evident is that two-body abrasive wear situations generally have higher values of K than three-body conditions. The explanation for this is that in the three-body situation, the abrasive grain is free to move and therefore may not always produce wear. For example, it may roll and tumble along the surface instead of sliding and cutting out a groove. Or it may align itself so that the bluntest profile presents itself to the surface. This concept is illustrated in Fig. 3.73.

A second trend that is illustrated by the data given in Table 3.8 is that the larger the abrasive grain or particle, the larger the value of K. This same trend is also found in three-body abrasion (152). In addition to the intuitive one that larger grains can form larger

Table 3.8 *K* Values for Abrasive Wear

Condition	K	
	Dry	Lubricated
Two-body		
File	5×10^{-2}	10^{-1}
New abrasive paper	10^{-2}	2×10^{-2}
Used abrasive paper	10^{-3}	2×10^{-3}
Coarse polishing	10^{-4}	2×10^{-4}
$< 100\,\mu m$ particles	10^{-2}	
$> 100\,\mu m$ particles	10^{-1}	
Nominal range, dry and lubricated: <1 to $>10^{-4}$		
Three-body		
Coarse particles	10^{-3}	5×10^{-3}
Fine particles	10^{-4}	5×10^{-4}
Nominal range, dry and lubricated: $<10^{-2}$ to 10^{-6}		

groves, several reasons for this trend have been proposed. One of these is that surface roughness and debris clogging effects become less significant with larger grains. This is illustrated in Fig. 3.74. Another mechanism that has been proposed is that larger grains are more likely to fracture with multiple engagements, forming new particles with sharp edges, while smaller particles are likely to have their edges rounded by a wear process. Still another possibility is that with naturally occurring abrasive particles it is frequently difficult to separate size and sharpness. Therefore it has been suggested that in certain cases the larger particles may just be sharper. As with most aspects of wear all of these effects probably contribute to the overall trend with some being more significant than others in particular situations.

While the precise reasons for the dependency on size is not known, there appears to be a very definite relationship that applies to many situations (34,153). This trend is shown

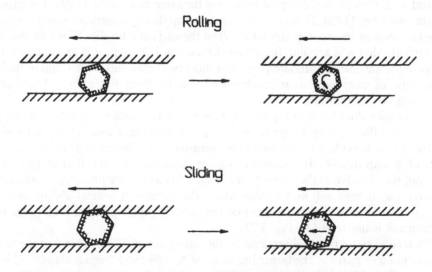

Figure 3.73 The effects of rolling and sliding actions in three-body abrasion.

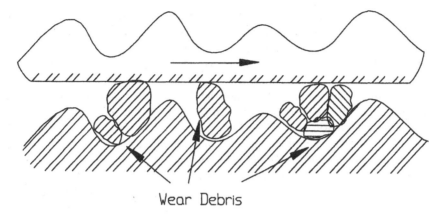

Wear Debris

Figure 3.74 The effect of the accumulation of wear debris in two-body abrasive wear.

in Fig. 3.75. It can be seen that in all cases there appears to be an almost linear relationship between size and wear, up to approximately 100 μm, but above that size wear tends to be independent of particle size. As with the general trend with size seen in Table 3.8, there is no established explanation for the transition above 100 μm. However, it has been proposed that in addition to the aspects mentioned earlier, particle loading could also play a role. As size increases, the number of particles involved at any instant may change, probably decreasing, which would tend to decrease the abrasive wear rate. As the number decreases, the load per particle would increase, which would tend to increase wear associated with each particle. These two effects would tend to offset each other. Under certain conditions, they could cancel and stable wear behavior as a function of size could result.

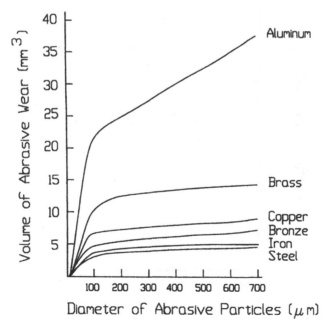

Figure 3.75 The influence of abrasive particle size on wear. The data are for SiC particles. (From Ref. 153.)

The third trend, which can be seen in the data given in Table 3.8, is the influence of lubrication. Lubrication tends to increase abrasive wear. This is consistent with the effect of lubrication on single-cycle deformation wear discussed in Sec. 3.3, where two mechanisms are proposed for this. One is a change from plowing to cutting. Basically, cutting results in more material removal than plowing or plastic deformation. By reducing, the coefficient of friction lubrication can increase the likelihood or amount of cutting taking place by lowering the critical attack angle for cutting. The variation in critical attack angle with the coefficient of friction is shown in Fig. 3.76. The second way lubrication can affect abrasive wear is the prevention of clogging by wear debris (154). While both are likely to be involved in the case of abrasion, the primary mechanism for this trend is generally accepted to be the accumulation of wear debris, which shares the load and protects the surface from the abrasive grains. The presence of a liquid lubricant at the interface helps to flush the wear debris from the interface and to reduce the shielding effect. The simplest illustration of this behavior is the build-up of debris that occurs on polishing and sanding papers and on files without lubrication. When these surfaces become sufficiently contaminated, the effective abrasive action decreases.

Experiments with dry, silicon carbide abrasive paper show that the wear rate decreases to 0 with time and that the effect occurred sooner with finer grain papers (155). While two effects probably contribute to the total behavior, that is, clogging of

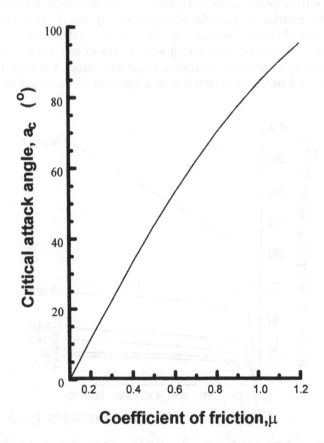

Figure 3.76 Critical angle of attack for cutting as a function of the coefficient of friction (based on Eq. (3.27)).

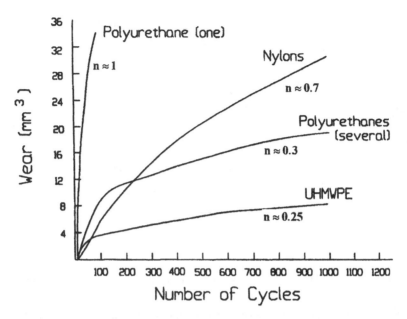

Figure 3.77 Effect of debris accumulation on abrasive wear. The equivalent value of the exponent in Eq. (3.97) is shown on the graphs. (From Ref. 156, reprinted with permission from Elsevier Science Publishers.)

the surface of the paper and wear of the grains, the latter point suggests clogging as being a significant factor. Finer grain paper would be more easily clogged than larger grain paper. A common practice in filing and polishing (sanding) is to use a lubricant to reduce clogging of the paper or file.

The significance of contamination by wear debris on the abrasive wear process is also illustrated in a study of abrasive wear of polymers (156). The following type of relationship was found for wear:

$$V = Kx^n \tag{3.97}$$

with $n < 1$. With $n < 1$ a decreasing wear rate occurs. Graphically this behavior is illustrated in Fig. 3.77 for several plastics and different values of n. This behavior was correlated to the build-up of a polymer layer on the abrasive surface which prevented some of the abrasive grains from contacting the wear surface. As the layer became thicker, more and more grains would be buried. This effect is illustrated in Fig. 3.78.

In addition to size, the wear coefficients in the equations for abrasive wear are also affected by other attributes of the particles. One is their sharpness or angularity. Wear coefficients generally are higher for angular particles than rounded particles. The number of particles involved is also a factor. In general, there tends to be a saturation level in terms of the number of particles, above which wear rate does not increase. This is generally attributed to the fact that below a certain number of particles part of the load is supported by asperity contact. This is shown in Fig. 3.79. Difference in particle friability and wear resistance can also affect values of these coefficients (34,157).

These same general trends also apply to the wear coefficient when the surface is harder than the abrasives, that is K in Eq. (3.95). However, the exact relationships can be different. For example, with a harder surface, particle size is not a factor above

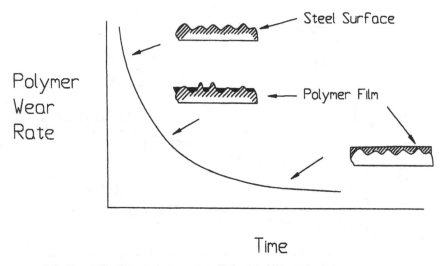

Figure 3.78 The effect of polymer film formation on abrasive wear rate.

10 µm, while for softer surfaces, this only occurs above 100 µm (34,158). Difference in particle friability and wear resistance would also tend to be more important when the surfaces are harder than when they are softer.

As shown previously in Fig. 3.20, a one-to-two order of magnitude reduction in wear rate is typically found when the hardness of the abraded surface exceeds the hardness of the abrasive. This is a very significant fact for practical handling of abrasive wear situations. Basically, to achieve low wear rates in abrasive wear situations, the goal is to select a material which is harder than the abrasives encountered.

In erosive situations, particles are not pressed against the surface as in abrasion; they impact the wearing surface. The load between the particle and the surface is an impulse load, which can be described in terms of the momentum and kinetic energy of the particle.

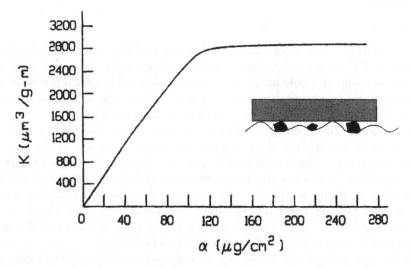

Figure 3.79 The effect of the amount of abrasives, α, on the abrasive wear coefficient, showing a saturation effect. (From Ref. 34.)

Because of this difference, it is necessary to modify the equation for single-cycle wear for applicability to erosion situations. A simple way of extending this equation to particle erosion is as follows (70).

The equation for single-cycle wear, Eq. (3.20), relates wear to the normal load. The first step in the derivation is to convert from normal load to frictional load, F. This is done by means of Amontons' Law, Eq. (1.1), namely,

$$F = \mu P \tag{3.98}$$

Equation (3.20) then becomes

$$V = \frac{K(Fx)}{\mu p} \tag{3.99}$$

where the product Fx represents the energy dissipated by sliding during the impact. The total kinetic energy of a particle stream of total mass, M, and particle velocity, v, is given by

$$E = \frac{1}{2} Mv^2 \tag{3.100}$$

As a result of the impact with the surface, a fraction, β, of the energy is dissipated in the form of wear. Equating this loss to Fx, the following expression is obtained:

$$V = \frac{K\beta Mv^2}{2\mu p} \tag{3.101}$$

In erosion, it has been established that the angle at which the stream impinges the surface influences the rate at which material is removed from the surface and that this dependency is also influenced by the nature of the wearing material (70,159). This is shown in Fig. 3.80. Such a dependency is to be anticipated. This can be seen by considering the

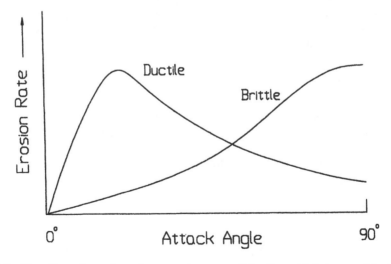

Figure 3.80 The effect of attack angle on erosion rates of ductile and brittle materials. (From Ref. 194.)

impact of a single particle with a surface. The angle determines the relative magnitude of the two velocity components of the impact, namely the component normal to the surface and the one parallel to the surface. The normal component will determine how long the impact will last, that is, the contact time, t_c, and the load. The product of t_c and the tangential velocity component determine the amount of sliding that takes place. The tangential velocity component also provides a shear loading to the surface, which is in addition to the normal load related to the normal component of the velocity. Therefore, as the angle changes, the amount of sliding that takes place also changes, as does the nature and magnitude of the stress system. Both of these aspects influence the way a material wears. These changes would also imply that different types of materials would exhibit different angular dependencies as well.

As can be seen in Fig. 3.80, the effect of angle on erosion rate is significantly different for ductile and brittle materials. With brittle material, the maximum erosion rate occurs at normal impact, while for ductile materials it occurs at some intermediate and generally much smaller angle. These differences can be understood in terms of the predominant modes of damage associated with these types of materials.

As discussed in Sec. 3.3, brittle fracture tends to increase the amount of wear over that caused by displacement, that is by cutting and plowing. As indicated in Fig. 3.20, this could be by as much as $10 \times$. As a general rule, brittle materials are more likely to fracture under normal impact conditions, that is, impacting velocity perpendicular to the surface, than ductile materials. Consequently, as the erosive condition moves from a more grazing situation to a more normal impact, brittle materials would experience a greater tendency to experience brittle fracture, which would tend to increasingly mask the ductile or cutting contributions. For brittle materials, the erosion rate would then be expected to monotonically increase with the angle.

For ductile materials, cutting and plowing are the predominant modes and fracture is negligible. The model for single-cycle deformation indicates that the wear due to cutting and plowing is proportional to the product of load and distance (see Eq. (3.20)). Since load increases with angle and sliding decreases with angle, an intermediate angle should exist where the product of the two is maximum.

This angular dependency is contained in β in Eq. (3.101). Assuming that β can be separated into an angular factor, Φ, and a factor independent of angle, β', and combining several of the material-sensitive parameters and numerical factors into one, K_e, the following expression can be obtained:

$$V = \frac{K_e \Phi M v^2}{p} \tag{3.102}$$

Examining this equation for erosive-wear volume, it can be seen that it does not provide an explicit dependency on duration or exposure. However, such a dependency is implicitly contained in M, the total mass of particles. If Q is particle mass per unit time, then M is Qt, where t is the time of exposure to the particle stream. Including this into Eq. (3.102), the following form is obtained for particle erosion:

$$V = \frac{K_e \Phi v^2 Q t}{p} \tag{3.103}$$

Another variation of Eq. (3.102) is frequently encountered in the literature. Comparison of erosive wear situations and resistance to erosion is often done in terms of the relative amount of material removed from the surface to the amount of abrasive particle causing the wear (160). With d as the density of the particles, the following equation can be obtained:

$$\frac{V}{V_a} = \frac{K_e d \Phi v^2}{p} \tag{3.104}$$

where V_a is the volume of abrasive used to produce the wear.

A compilation of values for the erosive wear coefficient, K_e, is given in Table 3.9. Comparing these values to the values for wear coefficients for abrasion, Table 3.8, it can been seen that they are very similar. This is consistent with the underlying hypothesis that the same wear mechanisms occur in both situations.

Equations for solid particle erosion, which are equivalent to Eqs. (3.94) and (3.96) for abrasion, can be developed in a similar manner. These are:

$$V = K_e \Phi v^2 Q t \tag{3.105}$$

$$V = \frac{K_e \Phi v^2 Q t}{p^n} \tag{3.106}$$

Equation (3.106) applies when the surface is harder than the particles. In general, the erosive wear coefficients in these equations for particle erosion have similar sensitivities to their counterparts for abrasion, that is, they can be affected by characteristics of the abrasives, type of material or simply material, atmosphere, and fluid media (161).

In controlling abrasive wear, the most significant feature is that once the wearing surface becomes harder than the abrasive, wear rates are dramatically reduced. The effect here is equivalent to the use of lubricants to control adhesive wear. Both give orders-of-magnitude improvement. Further discussion and examples of abrasion and particle erosion can be found in Chapter 9, and in Chapters 5 and 7 of Engineering Design for Wear: Second Edition, Revised and Expanded.

3.9. WEAR MAPS

Wear maps are graphical techniques used to characterize various aspect of wear behavior in terms of independent operational parameters of the tribosystem, such as speed and load. Various forms of wear maps are typically used to identify ranges of these parameters with wear mechanisms, wear rates, and acceptable operating conditions. Generally, they are two-dimensional graphs where the axes are the independent operational parameters. Curves are plotted on these graphs to separate regions of different wear behavior and to represent conditions of constant wear rate. In addition to the generic name of wear map such plots are also referred to as wear mechanism maps, wastage maps, material performance maps, wear transition maps, wear rate maps, and contour wear maps, depending on their nature and use. Examples of different types of wear maps are shown in Figs. 3.81–3.84 and 3.86–3.90. Figures 3.81–3.84 are examples of ones used for sliding wear; Figs. 3.86 and 3.87, for tool wear; and Figs. 3.88, 3.89, and 3.90, for solid particle erosion. While maps of these types can be developed on a purely theoretical or experimental basis, most are primarily empirical-based. However, theoretical considerations are often involved to facilitate the construction and to minimize the amount of data required (106,109,162–170).

Table 3.9 K_e Values for Erosion

Target material	K_e
Soft steel	$8 \times 10^{-3} - 4 \times 10^{-2}$
Steel	$1 \times 10^{-2} - 8 \times 10^{-2}$
Hard steel	$1 \times 10^{-2} - 1 \times 10^{-1}$
Aluminum	$5 \times 10^{-3} - 1.5 \times 10^{-2}$
Copper	$3 \times 10^{-3} - 1.3 \times 10^{-2}$

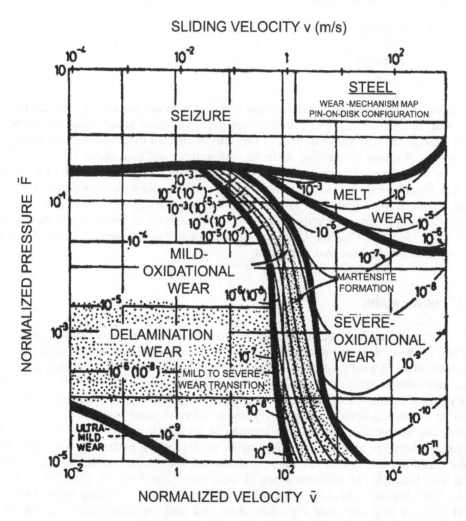

$\tilde{F}$ = load/(apparent area × hardness) $\tilde{V}$ = velocity × contact diameter/2 × thermal diffusivity

Figure 3.81 Wear map developed for unlubricated sliding between self-meted steel. (From Ref. 109, reprinted with permission from Elsevier Science Publishers.)

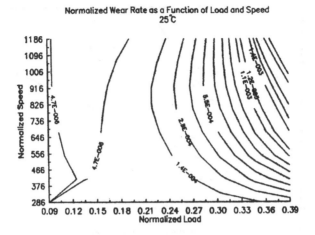

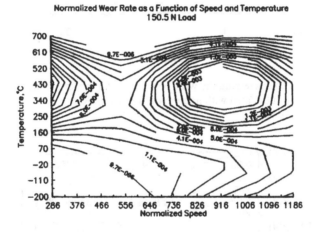

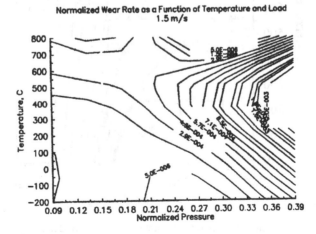

Figure 3.82 Wear mechanisms maps used to characterize the unlubricated sliding wear behavior of self-mated 440C stainless steel. (From Ref. 163, reprinted with permission from Elsevier Sequoia S.A.)

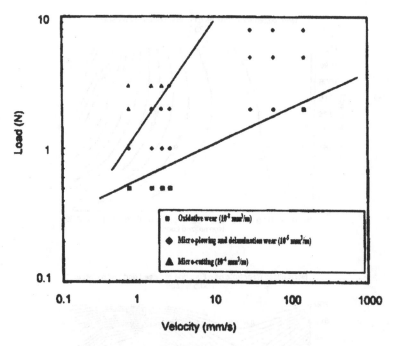

Figure 3.83 Wear map characterzing the wear behavior of Cr ion-implanted iron sliding against a hard steel counterface and lubricated with liquid paraffin. (From Ref. 168, reprinted with permission from Elsevier Sequoia S.A.)

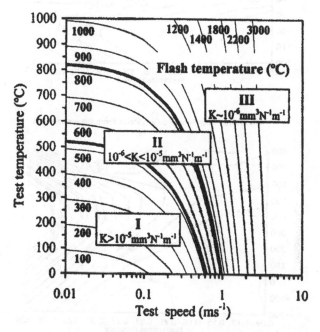

Figure 3.84 Example of wear maps used to characterize the unlubricated wear behaviour of silicon nitride ceramics sliding against steel. These maps are referred to as wear transition maps and show the transition in wear behavior of the ceramic as function of ambient temperature and speed. (From Ref. 136, reprinted with permission from Elsevier Sequoia S.A.)

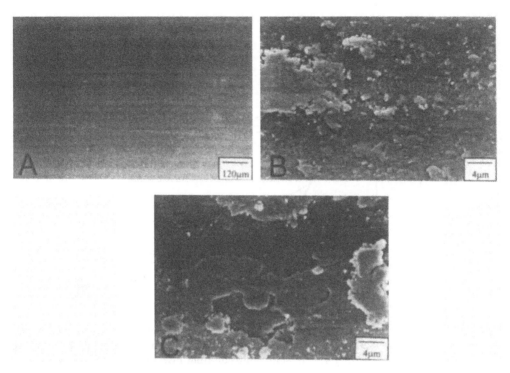

Figure 3.85 Examples of wear scars on silicon nitride ceramics in the different regions identified in wear transition maps, such as illustrated in Fig. 3.84. "A" is for 22°C and 0.5 m/s; "B" is for 200°C and 0.5 m/s; "C" is for 22°C and 3.5 m/s. (From Ref. 136, reprinted with permission from Elsevier Sequoia S.A.)

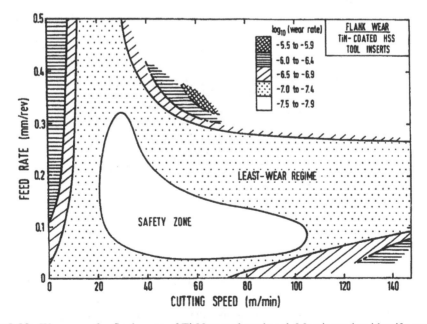

Figure 3.86 Wear map for flank wear of Ti-N coated steel tool. Map is used to identify acceptable regions of operation. (From Ref. 167, reprinted with permission from Elsevier Sequoia S.A.)

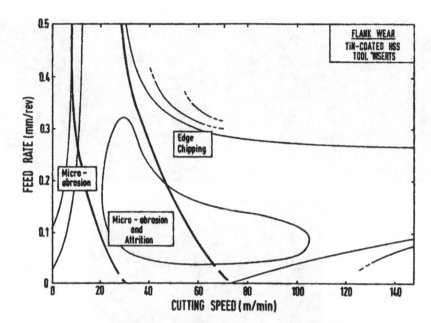

Figure 3.87 Wear map for flank wear of Ti-N coated steel tools, identifying regions of different wear behavior. These regions are superimposed on those used to identify acceptable performance (see Fig. 3.86). (From Ref. 167, reprinted with permission from Elsevier Sequoia S.A.)

Figure 3.81 shows the first wear map that was developed to illustrate the wear map concept (109). This map is one proposed for unlubricated or dry sliding between steels, using normalized pressure and normalized velocity as the axes. In it the boundaries between regions of different wear mechanisms are identified, as well as the locus of pressures and velocity conditions for a constant normalized wear rate within those regions.

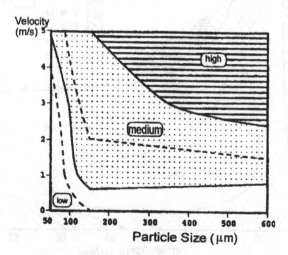

Figure 3.88 Example of wear maps used to characterize erosion. This type of map is referred to as velocity-particle size wastage maps. Axes are particle velocity and particle size. (From Ref. 164, reprinted with permission from Elsevier Sequoia S.A.)

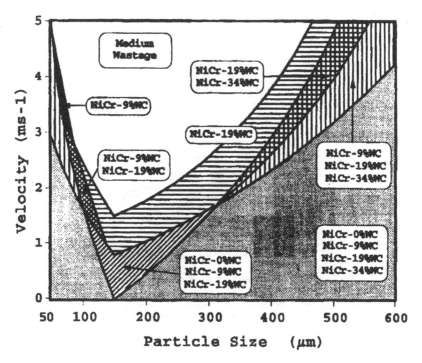

Figure 3.89 An example of the types of wear maps used to characterize erosion. This type is referred to as materials performance maps. In these, the different regions are used to identity wastage conditions for individual materials. In this particular map, the various shaded regions are regions of low wastage for the individual materials. The clear region is a region of medium wastage for all the materials. (From Ref. 164, reprinted with permission from Elsevier Sequoia S.A.)

Another example of a wear map is the contour wear maps developed for 440C stainless steel for use in high-pressure oxygen turbopumps, shown in Fig. 3.82 (163). In these maps contour lines of constant wear rate are plotted. For these applications, it was desirable to characterize wear behavior in terms of three operating parameters, load, speed, and ambient temperature. To accomplish this, wear maps were developed for different combinations of these parameters, as illustrated in the figure. In this case, the wear map can be thought of as a three-dimensional wear space with axes of load, speed, and temperature, where surfaces of constant wear rate can be identified. The two-dimensional graphs can then be thought of as planes in that space and the contour lines are the intersection of those planes with these surfaces.

A less complex wear map than these two examples is shown in Fig. 3.83. In this case load and speed are the operating variables of interest (168). This wear map is an example of a wear mechanism map. In this map, three regions of different wear mechanisms, each with a characteristic order-of-magnitude wear rate, are identified. These regions were identified by physical examination of the worn surfaces and wear rate determination.

Wear transition maps are illustrated in Fig. 3.84. In this case, ambient temperature and speed are the axes (165). In this map, three different regions of wear behavior are identified and correlated with different ranges of a wear coefficient for a ceramic slider. In this tribosystem, wear behavior can be correlated to flash temperature and isothermal contours for flash temperature are also plotted on the map. For this tribosystem, the three different

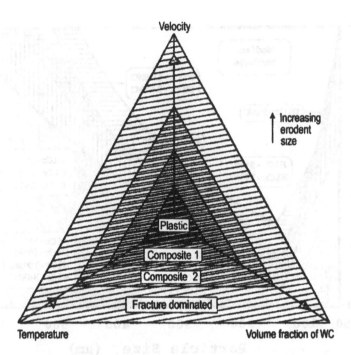

Figure 3.90 An example of four-variable wear map used to describe erosion behavior. This map is used to characterize wear behavior in terms of wear mechanisms and use erodent size, erodent velocity, temperature, and material composition as parameters controlling the wear. (From Ref. 164, reprinted with permission from Elsevier Sequoia S.A.)

wear regions are a result of differences in the formation of tribofilms on the surface of the ceramic, as shown in Fig. 3.85. In region I, there is no evidence of transfer and a tribofilm is not formed. In regions II and III, higher flash temperature promotes adhesion of tool steel wear particles to the ceramic surface. In region II, there is partial film formation. In region III, the film is more uniform and extensive.

Wear maps are also used to characterize tool wear and to determine optimum operating conditions for least tool wear (162,166,167,169). Example of a wear rate map used for this purpose is shown in Fig. 3.86 (167). In this case, the axes are feed rate and cutting speed. The boundaries of the regions are based on wear rate. Mechanism information can also be placed on the map to provide a wear mechanism map. This is shown in Fig. 3.87, where three regions of different dominant mechanism are identified. It is can be seen in this map that the least wear region includes two different mechanism regions. With tools, such maps are used to characterize both flank wear, as illustrated in the figures, and crater wear (169).

The use of wear maps in solid particle erosion is illustrated in Figs. 3.88, 3.89, and 3.90. Three different types of wear maps are used (164). One is referred to as wastage maps and the axes are generally particle size and velocity. Figure 3.88 is an example of this. In this case, the maps identify regions of high medium, and low wastage rates, which were based on the depth of wear, x, in a standard test. High wastage rate was equivalent to $x \geq 8\,\mu m$; medium, $4\,\mu m \leq x < 8\,\mu m$; low, $x \leq 4\,\mu m$. These levels were based on the approximate levels of wastage that can typically be tolerated in a fluid bed conveyer. Material performance maps are developed from this by overlaying the wastage maps obtained for different materials. In this case, it was for different metal matrix composites (MMC). An example of such a wear map is shown in Fig. 3.89. In this type of application, ambient temperature is also a factor and these types of curves were developed for a series of temperatures, which spanned the application range. A third form of a wear map that is used to characterize solid particle erosion is the ternary map shown in Fig. 3.90. This is a wear mechanism map in which four different modes of wear are identified. The fractured dominated and plastic mechanisms are mechanisms associated with the wear of the reinforcing particle, while the remaining two are wear modes associated with the composite (171). This map shows the combined effect of temperature, velocity, composition, and particle size on type of mechanism involved (164,172).

A common example of a wear map is the PV diagrams (pressure–velocity graphs) often used to describe the wear behavior of engineering plastics. In this case, the PV Limit curve separates pressure and velocity combinations into two regions of wear behavior, one that is generally considered to be acceptable for applications and the other that is not (173).

REFERENCES

1. M Peterson. Mechanisms of wear. In: F Ling, E Klaus, R Fein, eds. Boundary Lubrication. ASME, 1969, pp 19–38.
2. V Tipnis. Cutting tool wear. In: M Peterson, W Winer, eds. Wear Control Handbook. ASME, 1980, p 901.
3. K Budinski. Tool material. In: M Peterson, W Winer, eds. Wear Control Handbook. ASME, 1980, p 950.
4. L Kendall. Friction and wear of cutting tools and cutting tool materials. In: P Bau, ed. Friction, Lubrication, and Wear Technology, ASM Handbook. Vol. 18. Materials Park, OH: ASM International, 1992, p 613.
5. J Vleugels, O Van Der Biest. Chemical wear mechanisms of innovative ceramic cutting tools in the machining of steel. Wear 225–229:285–294, 1999.
6. V Venkatesh. Effect of magnetic field on diffusive wear of cutting tools. Proc Intl Conf Wear Materials ASME 242–247, 1977.
7. E Usui, T Shirakashi. Analytical prediction of cutting tool wear. Wear 100:129–152, 1984.
8. W Schintlmeister, W Walgram, J Kanz, K Gigl. Cutting tool materials coated by chemical vapour deposition. Wear 100:153–170, 1984.
9. O Vingsbo. Wear and wear mechanisms. Proc Intl Conf Wear Materials ASME 620–635, 1979.
10. R Bayer. Wear in electroerosion printing. Wear 92:197–212, 1983.
11. C Livermore. Technology. The Industrial Physicist, Dec 2001/Jan 2002. American Institute of Physics, pp 20–25.
12. J Burwell, C Strang. Metallic wear. C. Proc Roy Soc A 212:470, 1953.
13. K Ludema. Selecting materials for wear resistance. Proc Intl Conf Wear Materials ASME 1–6, 1981.
14. D Tabor. Wear - A critical synoptic view. Proc. Intl. Conf. Wear Materials ASME 1–11, 1977.
15. M Peterson. Mechanisms of wear. In: F Ling, E Klaus, R Fein, eds. Boundary Lubrication. ASME, 1969, p 19–38.
16. E Rabinowicz. Friction and Wear of Materials. New York: John Wiley and Sons, 1965.
17. D Rigney, W Glaeser, eds. Source Book on Wear Control Technology. ASME, 1978.
18. D Scott, ed. Wear. Treatise on Material Science and Technology. Vol. 13. New York: Academic Press, 1979.
19. R Bayer. Wear Analysis for Engineers. HNB Publishing, 2002.

20. R Bayer, R Schumacher. On the significance of surface fatigue in sliding wear. Wear 12:173–183, 1968.
21. V Jain, S Babadur. Tribological behavior of unfilled and filled poly(amide-imide) copolymer. Proc Intl Conf Wear Materials ASME 385–389, 1987.
22. A Hollander, J Lancaster. An application of topographical analysis to the wear of polymers. Wear 25(2):155–170, 1973.
23. N Saka. Effect of microstructure on friction and wear of metals. In: N Suh, N Saka, eds. Fundamentals of Tribology. Cambridge, MA: MIT Press, 1980, pp 135–172.
24. N Suh. The delamination theory of wear. Wear 25:111–124, 1973.
25. V Jain, S Bahadur. Experimental verification of fatigue wear equation. Proc Intl Conf Wear Materials ASME 700–705, 1981.
26. A Bower, K Johnson. The influence of strain hardening on cumulative plastic deformation in rolling and sliding line contact. J Mech Phys Solids 37(4):471–493, 1989.
27. K Johnson. Proceedings of the 20th Leeds–Lyon Symposium on Tribology. Elsevier, 1994, p 21.
28. R Burton, ed. Thermal Deformation in Frictionally Heated Systems. Elsevier, 1980.
29. J Burwell, C Strang. Metallic wear. Proc Roy Soc A 212:470, 1953.
30. T Kjer. A lamination were mechanism beased on plastic waves. Proc Intl Conf Wear Materials ASME 191–198, 1987.
31. W Campbell. Boundary lubrication. In: F Ling, E Klaus, R Fein, eds. Boundary Lubrication. ASME 87–118, 1969.
32. R Bayer. Tribological approaches for elastomer applications in computer peripherals. In: Denton R, K Keshavan, eds. Wear and Friction of Elastomers. STP 1145, ASTM International, 1992, pp 114–126.
33. R Bayer. A general model for sliding wear in electrical contacts. Wear 162:913–918, 1993.
34. R Bayer. A model for wear in an abrasive environment as applied to a magnetic sensor. Wear 70:93–117, 1981.
35. B Bricoe. The wear of polymers: An essay on fundamental aspects. Proc Intl Conf Wear Materials ASME 7–16, 1981.
36. R Bayer, T Ku. Handbook of Analytical Design for Wear. New York: Plenum Press, 1964.
37. J Greenwood, J Williamson. Contact of nominally flat surfaces. Proc Roy Soc A 295:300–319, 1966.
38. J Greenwood, J Tripp. The elastic contact of rough spheres. J Appl Mech Trans ASME March, 153–159, 1967.
39. F Bowden, D Tabor. The Friction and Lubrication of Solids. New York: Oxford U. Press, Part I, 1964, and Part II, 1964.
40. R Holm. Electric Contact Handbook. New York: Springer, 1958.
41. W Glaeser. Lecture Notes, Wear Fundamentals Course for Engineering, International Wear of Materials Conference, ASME, 1989.
42. E Rabinowicz. Surface interaction. Friction and Wear of Materials. New York: John Wiley and Sons, 1965, p 50.
43. D Kuhlmann-Wilsdorf. In: D Rigney, ed. Fundamentals of Friction and Wear of Materials. ASM, 1981.
44. B Bhushan, B Gupta. Friction, wear and lubrication. Handbook of Tribology. McGraw-Hill, 1991, p 2.4.
45. J McFarlane, D Tabor. Proc R Soc Lond A 202:244, 1950.
46. J Ferrante, J Smith, J Rice. Microscopic Aspects of Adhesion and Lubrication. Tribology Series. New York: Eslevier Science Publishing Co., 1982, p 7.
47. J Archard. Contact and rubbing of flat surfaces. J App Phys 24:981–988, 1953.
48. E Rabinowicz. Adhesive wear. Friction and Wear of Materials. New York: John Wiley and Sons, 1965, pp 139–140.
49. E Rabinowicz. Wear coefficients–metals. In: M Peterson, W Winer, eds. Wear Control Handbook. ASME, 1980, pp 475–506.

50. K Ludema. Introduction to wear. In: P Blau, ed. Friction Lubrication and Wear Technology, ASM Handbook. Vol. 18. Materials Park. OH: ASM International, 1992, p 175.
51. E Rabinowicz. Adhesive wear. Friction and Wear of Materials. New York: John Wiley and Sons, 1965, pp 139–164.
52. B Bhushan, B Gupta. Handbook of Tribology. Section 3.3. McGraw-Hill, 1991.
53. E Rabinowicz. Adhesive wear. Friction and Wear of Materials. New York: John Wiley and Sons, 1965, pp 151–163.
54. E Rabinowicz. Adhesive wear. Friction and Wear of Materials. New York: John Wiley and Sons, 1965, pp 159–162.
55. R Errichello. Friction, lubrication, and wear of gears. In: P Blau, ed. Friction, Lubrication, and Wear Technology, ASM HandBook. Vol. 18. Materials Park, OH: ASM International, 1992, pp 535–545.
56. P Engel. Impact wear. In: P Blau, ed. Friction, Lubrication and Wear Technology, ASM Handbook. Vol. 18. Materials Park, OH: ASM International, 1992, pp 263–270.
57. P Blau. Rolling contact wear. In: P Blau, ed. Friction, Lubrication and Wear Technology, ASM Handbook. Vol. 18. Materials Park, OH: ASM International, 1992, pp 257–262.
58. D Evans, J Lancaster. The wear of polymers. In: D Scott, ed. Treatise of Materials Science and Technology. Vol. 13. Academic Press, 1979, pp 86–140.
59. E Rabinowicz. Abrasive and other types of wear. Friction and Wear of Materials. New York: John Wiley and Sons, 1965, p 169.
60. M Moore, F King. Abrasive wear of brittle solids. Proc Intl Conf Wear Materials ASME 275–284, 1979.
61. M Moore, P Swanson. The effect of particle shape on abrasive wear. Proc Intl Conf Wear Materials ASME 1–11, 1983.
62. H Hokkirigawa, Z Li. The effect of hardness on the transition of abrasive wear mechanism of steel. Proc Intl Conf Wear Materials ASME 585–594, 1987.
63. N Suh, H-C Sin, N Saka. Fundamental aspects of abrasive wear. In: N Suh, N Saka, eds. Fundamentals of Tribology. Cambridge, MA: MIT Press, 1980, p 493–518.
64. K Zum Gahr. Formation of wear debris due to abrasion. Proc Intl Conf Wear Materials ASME 396–405, 1981.
65. K Zum Gahr, D Mewes. Microstructural influence on abrasive wear resistance of high strength, high toughness medium carbon steels. Proc Intl Conf Wear Materials ASME 130–139, 1983.
66. E Rabinowicz. Abrasive and other types of wear. Friction and Wear of Materials. New York: John Wiley and Sons, 1965, p 179.
67. R Bayer, W Clinton, C Nelson, R Schumacher. Engineering model for wear. Wear 5:378–391, 1962.
68. T Tallian. On competing failure modes in rolling contact. ASLE Trans 10(4):418–439, 1967.
69. P Engel. Impact Wear of Materials. Tribology Series. Chapter 3. New York: Elsevier Science Publishing Co., 1978.
70. E Rabinowicz. The wear equation for erosion of metals by abrasive particles. Proceedings of the Fifth International Conference on Erosion by Solid and Liquid Impact, 38–5, Cambridge, UK: Cavendish Laboratory, 1979.
71. K Zum, K Gahr, H Franze. Rolling-sliding wear on precipitation hardened structures of an austenitic steel. Proc Intl Conf Wear Materials ASME 23–32, 1987.
72. V Sastry, D Singh, A Sethuramiah. A study of wear mechanisms under partial elastohydro-dynamic conditions. Proc Intl Conf Wear Materials ASME 245–250, 1987.
73. Roller Bearings, Part I and Part II. Lubrication (Jul–Sept and Oct–Dec), Beacon, NY: Texaco, Inc., 1974.
74. C Lipson. Machine Design, 1/8/70. Cleveland, OH: Penton Publ. Co., pp 130–134.
75. P Engel. Impact Wear of Materials. Tribology Series. New York: Elsevier Science Publishing Co., 1978.

76. V Jain, S Babadur. Tribological behavior of unfilled and filled poly(amide-imide) copolymer. Proc Intl Conf Wear Materials ASME 389–396, 1987.

77. N Saka. Effect of microstructure on friction and wear of metals. In: N Suh, N Saka, eds. Fundamentals of Tribology. Cambridge, MA: MIT Press, 1980, pp 135–172.

78. V Jain, S Bahadur. Experimental verification of fatigue wear equation. Proc Intl Conf Wear Materials ASME 700–706, 1981.

79. M Shaw, F Macks. Analysis and Lubrication of Bearings. New York: McGraw-Hill, 1949.

80. R Morrison. Machine Design, 8/1/68. Cleveland, OH: Penton Publ. Co., pp 102–108.

81. E Zaretsky, ed. Life Factors for Rolling Bearings. STLE SP-34. STLE, 1999.

82. G Lundberg, A Palmgren. Dynamic capacity of roller bearings. Acta Polytechnica. Mech. Eng. Series. 1(3), 1947.

83. S Timoshenko, J Goodier. Theory of Elasiticity. New York: McGraw-Hill, 1951.

84. P Engel, T Lyons, J Sirico. Impact wear theory for steels. Wear 23:185–201, 1973.

85. A Rosenfield. Modelling of dry sliding wear. Proc Intl Conf Wear Materials ASME 390–393, 1983.

86. A Atkins, K Omar. The load-dependence of fatigue wear in polymers. Proc Intl Conf Wear Materials ASME 405–409, 1985.

87. W Glaeser. A case of wear particle formation through shearing-off at contact spots interlocked through micro-roughness in "adhesive wear". Proc Intl Conf Wear Materials ASME 155–162, 1987.

88. E Zaretsky, ed. Lubrication. Life Factors for Rolling Bearings. STLE SP-34. Section 5.6.2. STLE, 1999, pp 210–214.

89. M Reznikovskii. In: D James, ed. Abrasion of Rubber. Maclaren, 1967.

90. G Hamilton, L Goodman. The stress field created by a circular sliding contact. J App Mech 33(2):371–376, 1966.

91. J Hailing. A contribution to the theory of mechanical wear. Wear 34(3):239–250, 1975.

92. V Jain, S Bahadur. Development of a wear equation for polymer-metal sliding in terms of fatigue and topography of sliding surface. Proc Intl Conf Wear Materials ASME 556–562, 1979.

93. R Bayer. Prediction of wear in a sliding system. Wear 11:319–332, 1968.

94. J Lancaster. Plastics and Polymers 41(156):297, 1973.

95. A Hollander, J Lancaster. An application of topographical analysis to the wear of polymers. Wear 25(2):155–170, 1973.

96. P Clayton, K Sawley, P Bolton, G Pell. Wear behavior of bainitic steels. Proc Intl Conf Wear Materials ASME 133–144, 1987.

97. P Hurricks. Wear 19:207, 1972.

98. T Quinn. Oxidational wear. In: P Blau, ed. Friction, Lubrication, and Wear Technology, ASM Handbook. Vol. 18. Materials Park, OH: ASM International, 1992, pp 280–289.

99. H Uhlig. Mechanism of fretting corrosion. J App Mech 21:401, 1954.

100. G Yoshimoto, T Tsukizoe. On the mechanism of wear between metal sufaces. Wear 1:472, 1957–58.

101. T Tsukizoe. The effects of surface topography on wear. In: N Suh, N Saka, eds. Fundamentals of Tribology. Cambridge, MA: MIT Press, 1980, pp 53–66.

102. E Rabinowicz. Friction, Friction and Wear of Materials. Section 4.12. New York: John Wiley and Sons, 1965, pp 86–89.

103. T Quinn, J Sullivan. A review of oxidational wear. Proc Intl Conf Wear Materials ASME 110–115, 1977.

104. T Quinn, J Sullivan, D Rowson. New developments in the oxidational theory of the mild wear of steels. Proc Intl Conf Wear Materials ASME 1–11, 1979.

105. D Caplan, M Cohen. The effect of cold work on the oxidation of iron from 100 to 650°C. Corrosion Sci 6:321, 1966.

106. A Skopp, M Woydt, KH Habig. Tribological behavior of silicon nitride materials under unlubricated sliding between 22°C and 1000°C. Wear 181–183:571–580, 1995.

107. M Sawa, D Rigney. Sliding behavior of dual phase steels in vacuum and in air. Proc Intl Conf Wear Materials ASME 231–244, 1987.

108. W Winer, H Chang. Film thickness, contact stress and surface temperatures. In: M Peterson, W Winer, eds. Wear Control Handbook. ASME. 1980, pp 81–142.

109. S Lim, M Ashby. Wear-mechanism maps. Acta Metal 35(1):1–24, 1987.

110. E Rabinowicz, Surface interactions. Friction and Wear of Materials, Chapter 3. New York: John Wiley and Sons, 1965.

111. J Greenwood. The area of contact between rough surfaces and flats. Trans ASME J Lub Tech 89:81–91, 1967.

112. I Kraghelski, N Demkin. Wear 3:170, 1960.

113. C Allen, T Quinn, J Sullivan. Trans ASME J Tribol 107:172, 1985.

114. T Quinn, W Winer. Wear 102:67, 1985.

115. J Archard. The temperatures of rubbing surfaces. Wear 2:438, 1958–1959.

116. H Metselaar, A Winnubst, D Schipper. Thermally induced wear of ceramics. Wear 225–229:857–861, 1999.

117. R Burton. Thermal deformation in frictionally heated contact. In: R Burton, ed. Thermal Deformation in Frictionally Heated Systems. Elsevier, 1980, pp 1–20.

118. B Banerjee. The influence of thermoelastic deformations on the operation of face seals. In: R Burton, ed. Thermal Deformation in Frictionally Heated Systems. Elsevier, 1980, p 89–110.

119. T Dow, R Stockwell. Experimental verification of thermoelastic instabilities in sliding contact. Trans ASME Ser F July:359–364, 1977.

120. R Burton, V Nerlikar, S Kilaparti. Thermoelastic instability in a seal-like configuration. Wear 24:169–198, 1973.

121. J Barber. The influence of thermal expansion on the friction and wear process. Wear 10:155, 1967.

122. S Heckmann, R Burton. Trans ASME Ser F 99:247, 1977.

123. R Burton. Thermal deformation in frictionally heated contact. In: R Burton, ed. Thermal Deformation in Frictionally Heated Systems. Elsevier, 1980, p 6.

124. V Jain, S Bahadur. Material transfer in polymer-polymer sliding. Proc Intl Conf Wear Materials ASME 487–493, 1977.

125. B Mortimer, J Lancaster. Extending the life of aerospace dry-bearings by the use of hard, smooth counterfaces. Proc Intl Conf Wear Materials ASME 175–184, 1987.

126. P Blau. Effects of sliding motion and tarnish films on the break-in behavior of three copper alloys. Proc Intl Conf Wear Materials ASME 93–100, 1987.

127. N Eiss, M Bayraktaroglu. The effect of surface roughness on the wear of low-density polyethylene. ASLE Trans 23(3):269–278, 1980.

128. B Briscoe. Wear of polymers: An essay on fundamental aspects. Trib Intl 14(4):231–243, 1981.

129. B Briscoe, M Steward. Paper No. C27178, Proceedings of Trib. 1978 Conference on Material Performance and Conservation. I. Mech, Eng., 1978.

130. S Rhee, K Ludema. Mechanisms of formation of polymetric transfer films. Proc Intl Conf Wear Materials ASME 482–487, 1977.

131. P Rohatgi, B Pai. Seizure resistance of cast aluminum alloys containing dispersed graphite particles of different sizes. Proc Intl Conf Wear Materials ASME 127–133, 1977.

132. P Heilmann, J Don, T Sun, W Glaeser, D Rigney. Sliding wear and transfer. Proc Intl Conf Wear Materials ASME 414–425, 1983.

133. F Talke. An autoradiographic investigation of material transfer and wear during high speed/low load sliding. Wear 22:69–82, 1972.

134. M Godet. Third-bodies in tribology. Wear 136(1):29–46, 1990.

135. R Bayer, J Sirico. Influence of jet printing inks on wear. IBM J R D 22(1):90–93, 1978.

136. J Gomes, A Miranda, J Vieira, R Silva. Sliding speed-termperature wear transition maps for Si_3N_4/iron alloy couples. Wear 250:293–298, 2001.

137. J Theberge. A guide to the design of plastic gears and bearings. Machine Design, 2/5/70. Cleveland, OH: Penton Publ. Co., 114–120.

138. M Wolverton, J Theberge. How plastic composites wear against metals. Machine Design, 2/6/86, 67–71, 1986.
139. LNP Design Guide for Internally Lubricated Thermoplastics, LNP Corp, 1978.
140. J Lancaster, D Play, M Godet, A Verrall, R Waghorne. Paper No. 79–Lub-7, Joint ASME-ASLE Lubrication Conference Dayton, Oh, 10/79.
141. T Tsukizoe, N Obmae. Wear mechanism of unidirectionally oriented fiber-reinforced plastics. Proc Intl Conf Wear Materials ASME 518–525, 1977.
142. J Tylczak. Abrasive wear. In: P Bau, ed. Friction, Lubrication and Wear Technology, ASM Handbook. Vol. 18. Materials Park, OH: ASM International, 1992, p 184.
143. J Tylczak. Abrasive wear. In: P Bau, ed. Friction, Lubrication and Wear Technology, ASM Handbook. Vol. 18. Materials Park, OH: ASM International, 1992, pp 184–190.
144. J Larsen-Basse, B Premaratne. Effect of relative hardness on transitions in abrasive wear mechanisms. Proc Intl Conf Wear Materials ASME 161–166, 1983.
145. B Bhushan, B Gupta. Physics of tribological materials. Handbook of Tribology. McGraw-Hill, 1991, p 3.18.
146. B Madsen. Corrosive wear. In: P Blau, ed. Friction, Lubrication and Wear Technology, ASM Handbook. Vol. 18. Materials Park, OH: ASM International, 1992, pp 271–279.
147. E Rabinowicz. Abrasive and other types of wear. Friction and Wear of Materials. New York: John Wiley and Sons, 1965, pp 168–173.
148. P Swanson, R Klann. Abrasive wear studies using the wet sand and dry sand rubber wheel tests. Proc Intl Conf Wear Materials ASME 379–389, 1981.
149. P Anstice, B McEnaney, P Thornton. Wear of paper slitting blades: the effect of slitter machine settings. Trib Intl 14(5):257–262, 1981.
150. P Engel, R Bayer. Abrasive impact wear of type. J Lub Tech 98:330–334, 1976.
151. R Bayer. The influence of hardness on the resistance to wear by paper. Wear 84:345–351, 1983.
152. A Misra, I Finnie. A classification of three-body abrasive wear and design of a new tester. Proc Intl Conf Wear Materials. ASME 313–318, 1979.
153. G Nathan, W Jones. Wear 9:300, 1966.
154. E Rabinowicz. Abrasive and other types of wear. Friction and Wear of Materials. New York: John Wiley and Sons, 1965, pp 172–173.
155. T Mulhearn, L Samuels. The abrasion of metals: A model of the process. Wear 5:478–498, 1962.
156. J Thorp. Abrasive wear of some commerical polymers. Trib Intl 15(2):59–68, 1982.
157. P Blau, ed. Friction, Lubrication, and Wear Technology, ASM Handbook. Vol. 18. Materials Park, OH: ASM International, 1992; p 188.
158. E Rabinowicz. Abrasive and other types of wear. Friction and Wear of Materials. New York: John Wiley and Sons, 1965, pp 177–179.
159. I Finnie. Erosion of surfaces by solid particles. Wear 3:87–103, 1960.
160. Standard Test Method For conducing Erosion Tests by Solid Particle Impingement Using Gas Jets. ASTM G76.
161. T Kosel. Solid particle erosion. In: P Bau, ed. Friction, Lubrication, and Wear Technology, ASM Handbook. Vol. 18. Materials Park, OH: ASM International, 1992, pp 198–213.
162. S Lim, Y Liu, S Lee, K Seah. Mapping the wear of some cutting-tool materials. Wear 162–164:971–974, 1993.
163. A Slifka, T Morgan, R Compos, D Chaudhuri. Wear mechanism maps of 440C martensitic stainless steel. Wear 162–164:614–618, 1993.
164. M Stack, D Pena. Mapping erosion of Ni-Cr/WC-based composites at elevated temperatures: Some recent advances. Wear 251:1433–1441, 2001.
165. J Gomes, A Miranda, J Vieira, R Silva. Sliding speed-temperature wear transition maps for Si_3N_4/iron alloy couples. Wear 250:293–298, 2001.
166. C Lim, P Lau, S Lim. The effects of work material on tool wear. Wear 250:344–348, 2001.

167. S Lim, C Lim, K Lee. The effects of machining conditions on the flank wear of tin-coated high speed steel tool inserts. Wear 181–183:901–912, 1995.
168. D Yang, J Zhou, Q Xue. Wear behavior of Cr implanted pure iron under oil lubricated conditions. Wear 203–204:692–696, 1997.
169. C Lim, S Lim, K Lee. Wear of TiC-coated carbide tools in dry turning. Wear 225–229: 354–367, 1999.
170. S Wilson, T Alpas. Thermal effects on mild wear transitions in dry sliding of an aluminum alloy. Wear 225–229:440–449, 1999.
171. M Stack, J Chacon-Nava, M Jordan. Mater Sci Technol 12:171–177, 1996.
172. M Stack, D Pena. Solid particle erosion of Ni-Cr/WC metal matrix composites at elevated temperatures: construction of erosion mechanism and process control maps. Wear 203–204:489–497, 1997.
173. R Lewis. Paper No. 69AM5C-2. 24th ASLE Annual Meeting, Philadelphia, 1969.
174. A Fischet. Sliding abrasion tests. Proc Intl Conf Wear Materials. ASME 729–734, 1989.
175. L Chen, D Rigney. Transfer during unlubricated sliding wear of selected metal systems. Proc Intl Conf Wear Materials. ASME 437–446, 1985.
176. A Sedricks, T Mulhearn. Wear 7:451, 1964.
177. F. Aleinikov. The influence of abrasive powder microhardness on the values of the coefficients of volume removal. Soviet Phys Tech Phys 2:505–511, 1957. F Aleinikov. The effect of certain physical and mechanical properties on the grinding of brittle materials. Soviet Phys Tech Phys 2:2529–2538, 1957.
178. R Bayer. Wear of a C ring seal. Wear 74:339–351, 1981–1982.
179. R Bayer, E Hsue, J Turner. A motion-induced sub-surface deformation wear mechanism. Wear 154:193–204, 1992.
180. R Bayer. Impact wear of elastomers. Wear 112:105–120, 1986.
181. G Laird, W Collins, R Blickensderfer. Crack propagation and spalling of white case iron balls subjected to repeated impacts. Proc Intl Conf Wear Materials. ASME 797–806, 1987.
182. M Hamdy, R Waterhouse. The fretting-fatigue behavior of a nickel-based alloy (inconel 718) at elevated temperatures. Proc Intl Conf Wear Materials. ASME 351–355, 1979.
183. H Voss, K Friedrich, R Pipes. Friction and wear of PEEK–composites at elevated temperatures. Proc Intl Conf Wear Materials. ASME 397–406, 1987.
184. M Kar, S Bahadur. Micromechanism of wear at polymer-metal sliding interface. Proc Intl Conf Wear Materials. ASME 501–509, 1977.
185. W Jamison. Tools for rock drilling. In: M Peterson, W Winer, eds. Wear Control Handbook. ASME, 1980, pp 859–890.
186. R Bayer, P Engel, E Sacher. Contributory phenomena to the impact wear of polymer. In: R Deanin, A Crugnola, eds. Toughness and Brittleness of Plastics. ACS, 1976, pp 138–145.
187. R Bill, L Ludwig. Wear of seal materials used in aircraft propulsion systems. In: R Burton, ed. Thermal Deformation in Frictionally Heated Systems. Elsevier, 1980, pp 165–189.
188. T Dow. Thermoelastic effects in brakes. In: R Burton, ed. Thermal Deformation in Frictionally Heated Systems. Elsevier, 1980, pp 213–222.
189. J Netzel. Observations of thermoelastic instability in mechanical face seals. In: R Burton, ed. Thermal Deformation in Frictionally Heated Systems. Elsevier, 1980, pp 135–148.
190. T Singer, S Fayculle, P Ehni. Friction and wear behavior of tin in air: the chemistry of transfer films and debris formation. Proc Intl Conf Wear Materials. ASME, 1991, pp 229–242.
191. Y-J Liu, N-P Chen, Z-R Zhang, C-Q Yang. Wear behavior of two parts subjected to 'gouging' abrasion. Proc Intl Conf Wear Materials ASME 410–415, 1985.
192. R Bayer. Mechanism of wear by ribbon and paper. IBM J R D 26, 1978, 668–674.
193. N Kruschov, M Babichev. Investigation into the Wear of Materials. Moscow: USSR Acad. Sci., 1960.
194. L Ives, A Ruff. Election microscopy study of erosion damage in copper. In: W Adler, ed. Erosion: Prevention and Useful Applications. STP 664, ASTM, 1979, pp 5–35.

4
Wear Behavior and Phenomena

4.1. GENERAL BEHAVIOR

Some trends in wear behavior are shown in Figs. 4.1–4.7. Individually, these trends do not necessarily represent general behavior, since they are based on published data from specific wear tests and generally involve a limited number of materials and conditions (1–16). While limited, the data summarized in these figures illustrate the broad range of behavior that can be encountered in different wear situations. Several of these figures contain wear curves, that is, the graphical relationship between wear and usage (distance of sliding, time, number of cycles, etc.). Others illustrate the dependency of wear on various factors, such as load, roughness, speed, hardness, etc., which are significant to the design engineer. As can be seen by comparing the various graphs in these figures, a variety of plotting techniques is used, including linear, log–log, and semi-log, to summarize the behavior. The need to use such a variety of formats illustrates the variety of relationships and sensitivities that are associated with wear. The general character of the curves is often some form of nonlinear behavior. In several of these figures, a power relationship (x^n) between wear and the parameter is indicated; in others, transitions or max/min behavior is seen. At the same time, linear relationships or regions of linear behavior can often be found in these figures as well.

A frequently encountered behavior is the development of a period of stable wear behavior after some initial wear has taken place (1,6,8–14,17–20). A period of stable wear behavior is one in which there is a stabilization of wear mechanisms. Typically, in situations where the apparent contact area does not change with wear, this is also a period of lower and constant wear rate, after a initial period of higher and changing wear rate. This type of behavior is shown in Figs. 4.8 and 4.9. In such situations, the initial period is usually referred to as break-in. Break-in behavior results from surface and near-surface changes as a result of relative motion and wear and the emergence of different mechanisms. This break-in effect is in addition to and different from the run-in effect associated with conforming contacts. With nominally conforming contacts, there can be an apparent break-in period when a linear wear measure is used, such as scar depth or width. In this case, the depth wear rate decreases as true conformity is established by wear, even if there is no other change occurring. However, other break-in type changes are also common with run-in. An additional discussion of break-in behavior can be found in Sec. 4.4.

The morphology of the wear scar is generally different in the break-in and stable wear periods. Stable wear behavior is generally characterized by stable morphology. Volume wear rates are often constant in stable wear periods, as indicated previously. However, short-term cyclic variations and slowly decreasing wear rates are also possible in these periods (4,21).

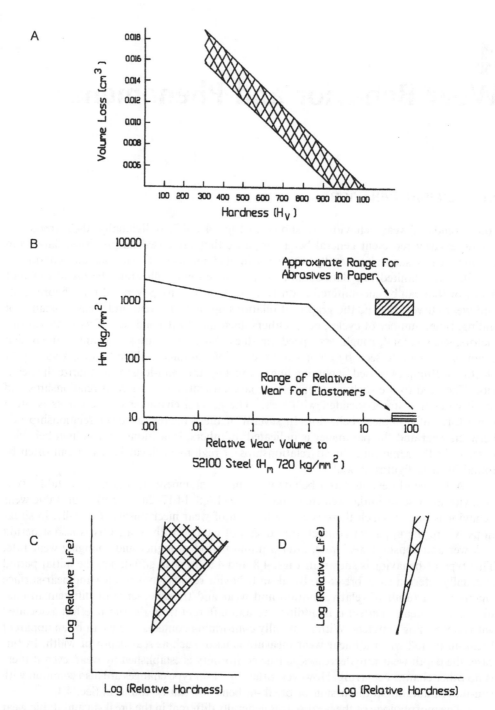

Figure 4.1 The effect of hardness on wear in several situations. "A", two-body abrasion of cast irons and steels; "B", sliding against paper; "C", sliding contact; "D", rolling contact. ("A" from Ref. 9; "B" from Ref. 75.)

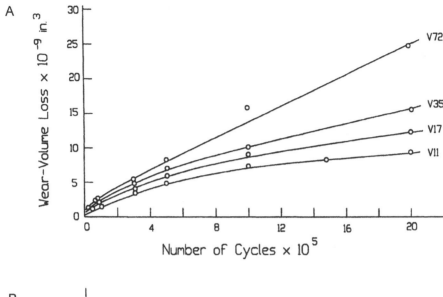

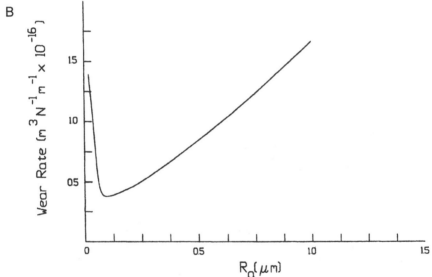

Figure 4.2 The effect of roughness in sliding. "A", steel against steel; "B", plastic against stainless steel. ("A" from Ref. 10; "B" from Ref. 16.)

This distinction between initial and long-term wear behavior is significant for several reasons. One is that run-in and break-in are precursors to stable wear behavior. For the designer or engineer, this means that suitable break-in may be required to obtain the stable period of low wear rate needed for long life. If this break-in does not occur, higher wear rates and unstable behavior might persist, resulting in reduced life. It is also important in engineering because it is sometimes necessary to take into account the magnitude of the wear associated with this initial period. It is also significant in terms of its relationship to wear studies. To the investigator, this stable period provides a convenient region for wear study. It is typically the type of region

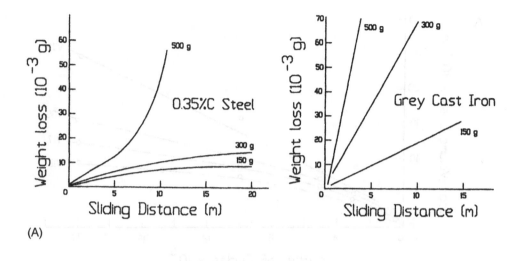

(A)

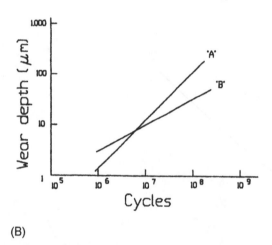

(B)

Figure 4.3 The influence of materials on wear in different situations. "A", abrasion of borided materials; "B", lubricated sliding against 52100 steel; "C", Au–Au sliding; "D", solid particle erosion. ("A" from Ref. 9; "B" from Ref. 4; "C" from Ref. 101; and "D" from Ref. 9.).

where wear rates, wear coefficients, and mechanisms are likely to be studied. However, tests have to be of sufficient duration that all run-in and break-in behavior have ceased.

In some wear situations, it is also possible that initial wear rates might be lower than longer-term wear rates. This is generally the result of initial surface films or layers, which act as lubricants and are gradually worn away. Such an effect is more common with unlubricated tribosystems than with lubricated tribosystems.

The effect of break-in and run-in illustrates another general aspect of wear behavior that needs to be recognized. This is that the immediate or current wear behavior can be influenced by earlier wear. In addition to the effect of break-in and run-in on longer-term wear, the influence of wear debris is another illustration. Since wear debris can be trapped in the wear region and cause further wear, its characteristics can influence current wear

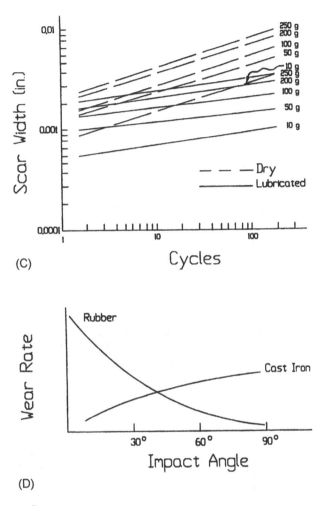

Figure 4.3 (*continued*)

behavior, while the prior wear behavior will determine the characteristics of the debris. For example, the occurrence of initial coarse debris as the result of a momentary overload condition or lubrication failure might inhibit the development of a mild, stable wear region that is more typical of the wear system. The momentary introduction of a small amount of abrasive into a system might trigger such a sequence as well.

Wear behavior is frequently characterized as being mild or severe (14,16,22–25). Mild wear is generally used to describe wear situations in which the wear rate is relatively small and the features of the wear scar are fine. Severe wear, on the other hand, is associated with higher wear rates and scars with coarser features. For reference, Fig. 4.10 shows wear scars that are representative of mild and severe wear. Most materials can exhibit both mild and severe behavior, depending on the specifics of the wear system in which they are used. Frequently, the transition from mild to severe are abrupt. Figure 4.11 illustrates such a transition for polymers. Mild wear behavior is generally required for engineering applications. In cases where severe wear behavior must be accepted, maintenance is high and lives are short.

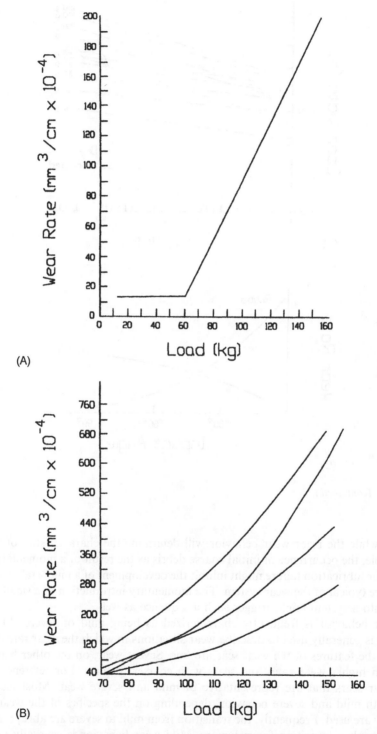

Figure 4.4 The effect of load. "A", "B", and "C", unlubricated sliding against steel; "D", general sliding and rolling. ("A" and "B" from Ref. 5; "C" from Ref. 9.)

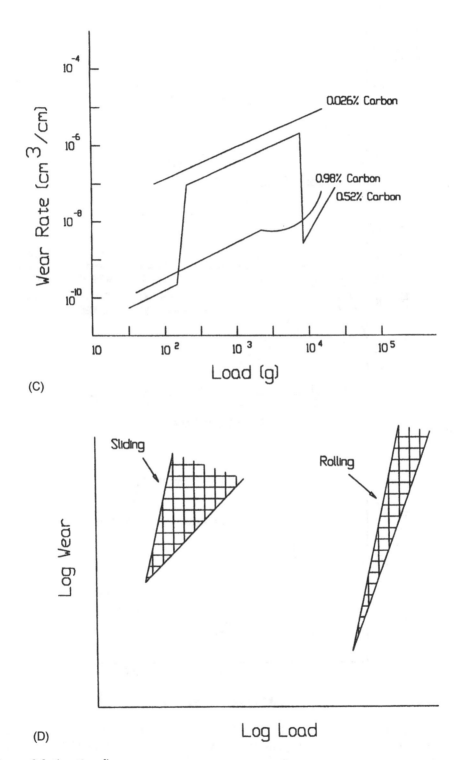

(C)

(D)

Figure 4.4 (*continued*)

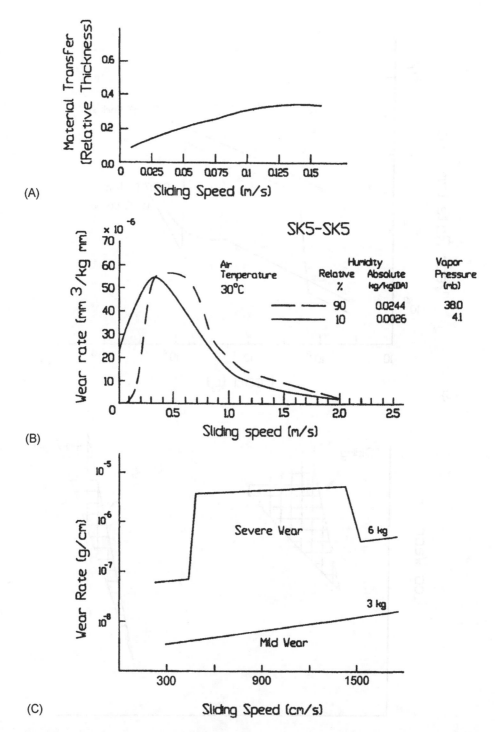

Figure 4.5 The effect of speed. "A", PTFE sliding against polyethylene; "B", unlubricated steel against steel; "C", unlubricated iron against steel. ("A" from Ref. 6; "B" from Ref. 13; "C" from Ref. 9. "A" and "B" reprinted with permission from ASME. "C" reprinted with permission from ASM International.)

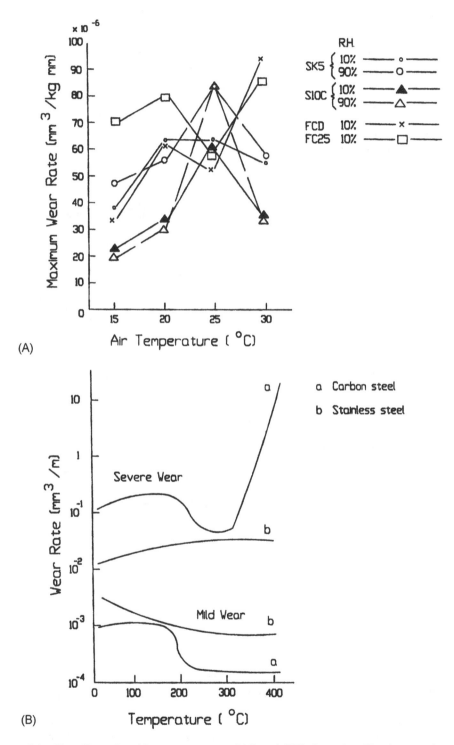

Figure 4.6 The effect of ambient temperature. "A" and "B" show the effect in several cases of unlubricated sliding between metal interfaces. ("A" from Ref. 13; "B" from Ref. 14.)

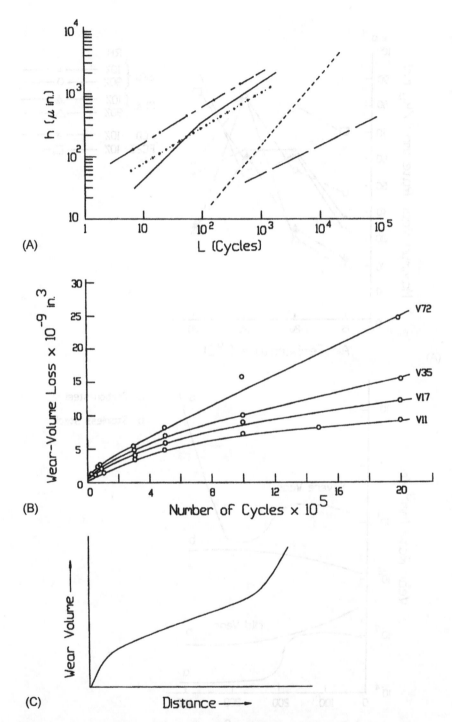

Figure 4.7 Wear behavior as a function of duration. "A", ceramic/steel sliding; "B", steel/steel sliding; "C", block-on-ring tests; "D", polymer composite/cermet sliding; "E", impact; "F", erosion; "G", slurry abrasion. ("A" from Ref. 77; "B" from Ref. 10; "C" from Ref. 15; "D" from Ref. 7; "E" from Ref. 78; "F" from Ref. 50; and "G" from Ref. 79.)

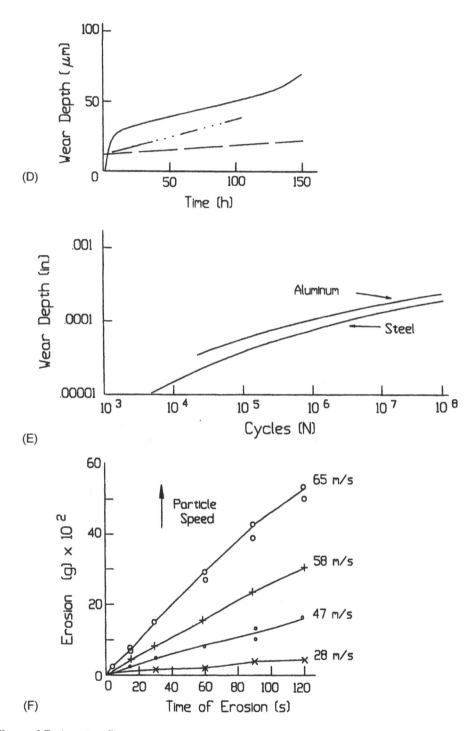

Figure 4.7 (*continued*)

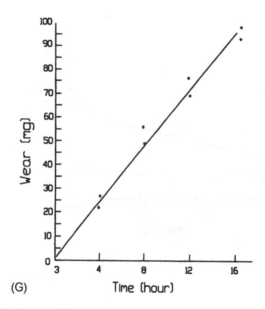

(G)

Figure 4.7 (*continued*)

There are several factors, which contribute to the general complex and varied nature of wear behavior. One factor is the number of basic wear mechanisms. Depending on the mechanism and the parameter considered, there are a mixture of linear and nonlinear relationships possible, as well as transitions in mechanisms. Consequently, a wide variety of behaviors is to be expected for different wear situations. A second factor is that wear mechanisms are not mutually exclusive and can interact in different ways. A final contributor to complex wear behavior is the modifications that take place on the wearing surfaces. As is obvious from the examination of worn surfaces, wear modifies the surface in addition to removing the material. Modifications to the topography are generally immediately apparent (e.g., scratches, pits, smearing, etc.). While less obvious, the composition of the surfaces can also be modified, as well as the mechanical properties of the surfaces. These changes to tribosurfaces are significant factors in the break-in behavior referred to previously. As a generalization, these

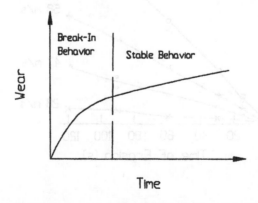

Figure 4.8 Wear curve showing the effect of break-in behavior.

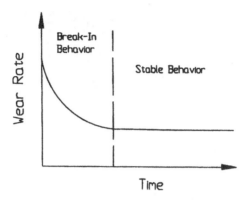

Figure 4.9 Wear rate behavior as a result of break-in behavior.

surface modifications can be influenced by a wide variety of parameters associated with the wearing system (e.g., relative humidity, nature of the relative motion, active components of a lubricant, etc.). Since the wear mechanisms are functions of the surface parameters, the dependencies of surface modifications on this larger set of parameters can result in more complex relationships for wear. In addition, since wear can influence surface modifications, a compounding of effects can take place. Interactions and trends with wear mechanisms, wear transitions, and modifications of tribosurfaces are discussed in further detail in the following sections.

The complex nature and range of wear behavior possible can generally be simplified and reduced to a practical level for engineering because of the limited range of tribosystem parameters that need to be considered. However, the range of behavior shown in Figs. 4.1–4.7, along with these observations regarding the many factors associated with wear behavior, suggests the following. As an overview, it is appropriate to consider wear behavior generally as nonlinear, with linear behavior possible under certain conditions and narrow ranges of parameters.

4.2. MECHANISM TRENDS

One factor that contributes to the complex nature of wear behavior is the possibility of different wear mechanisms. Depending on the mechanism and the parameter considered, there are a mixture of linear and nonlinear relationships possible, as well as transitions. For example, the simple model for adhesive wear gives a linear-dependency on sliding, while a model for fatigue wear gives a nonlinear dependency. In an abrasive wear situation, theory supports a transition in wear behavior when the abraded material becomes harder than the abrasive. In addition, not all mechanisms depend on the same parameters in the same way. For example, the model for corrosive wear indicates an explicit dependency on sliding speed; the models for the other modes do not contain an explicit dependency on speed. Consequently, a wide variety of behaviors is to be expected for different wear situations.

A contributing element to this complexity is that wear mechanisms are not mutually exclusive. Frequently wear scar morphology indicates the simultaneous or parallel occurrence of more than one mechanism (16,26–29). An illustration of this is shown in Fig. 4.12. In this sliding wear scar damage features suggestive of both single-cycle deformation and repeated-cycle deformation wear are present. The overall wear behavior of such a system

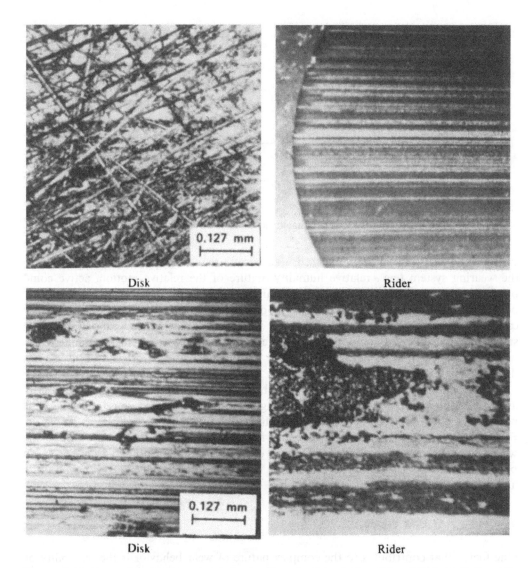

Figure 4.10 Examples of mild and severe wear scar morphologies for sliding. "Left side" shows an example of mild wear scars obtained in a pin-on-disk test; "Right side" shows severe wear scars. (From Ref. 80, reprinted with permission from ASME.)

could then be represented as the sum of individual wear processes, for example,

$$W_{\text{total}} = W_{\text{s-c-d}} + W_{\text{r-c-d}} \tag{4.1}$$

where W_{total} is the total wear, $W_{\text{s-c-d}}$ is the wear due to single-cycle deformation, and $W_{\text{r-c-d}}$ is the wear due to repeated-cycle deformation. Utilizing the expressions developed for single-cycle deformation and fatigue wear, Eqs. 3.25 and 3.44, and assuming that α is the fraction of the real area of contact that is wearing by single-cycle deformation, the following equation can be proposed for this system:

$$W_{\text{total}} = 2k \frac{\tan \theta}{\pi} \frac{\alpha P}{p} S + \Omega M G \Gamma^t (1 - \alpha) P^{1+t/3} S \tag{4.2}$$

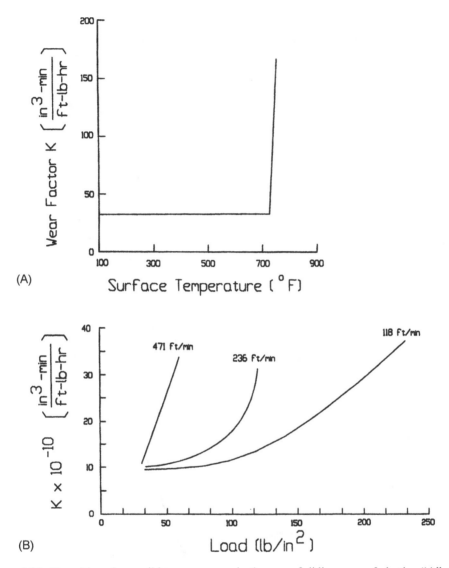

Figure 4.11 Transitions from mild to severe wear in the case of sliding wear of plastics. "A", poly-imide/steel couple in a thrush washer test; "B", TFE composite/steel couple in a journal bearing test. ("A" from Ref. 23; "B" from Ref. 81.)

Note that in this equation, P is the normal load and S is the distance of sliding. The other symbols are as previously defined for Eqs. (3.25) and (3.44). Since each of the individual mechanisms do not depend on the same parameters in the same manner, the dependency of W_{total} on these parameters would depend on the relative contribution of the individual mechanisms to the total wear. For example, if α is very small, that is to say single-cycle deformation is minor, the load dependency would be nonlinear and the wear behavior would not be sensitive to the sharpness and size of the asperities. If α was near unity (i.e., fatigue is minor), the wear would be sensitive to the asperities' sharpness and the load dependency would approach a linear one. For intermediate values of α, there would be both a nonlinear dependency on load and a dependency on asperity size and shape, which

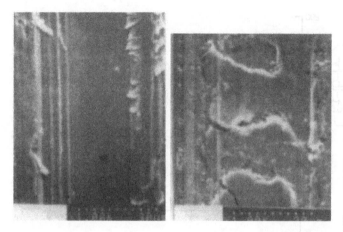

Figure 4.12 Sliding wear scars on ceramics, showing evidence of single-cycle deformation and repeated-cycle deformation wear. (From Ref. 82; reprinted with permission from ASME.)

would be different from the ones associated with either mechanism. This trend is illustrated in Fig. 4.13, where it is assumed that α can be related to the load range. For light loads, it was assumed that fatigue is negligible but at high loads, it predominates. Consequently, the possibility of the simultaneous occurrence of several mechanisms can lead to a wider range of behavior than those based on the individual mechanisms.

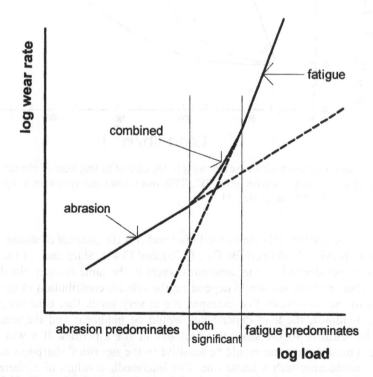

Figure 4.13 Example of combined wear behavior when there are two mechanisms present.

The individual mechanisms can also interact in a sequential fashion, giving rise to another possible factor for complex wear behavior. For example, fatigue wear can weaken the surface by the formation of cracks and allowing an adhesive event to remove the wear particle. Mathematically, the wear may be described by the equation for adhesive wear, where the wear coefficient K is now dependent on the fatigue parameters of the system in addition to the normal parameters associated with adhesion. In Eq. 3.7 for adhesive wear, K is the probability that a given junction will result in wear. Assuming that this probability is proportional to the fatigue wear rate (see Eq. (3.44)), the following equation may be proposed for such a system:

$$W = \frac{\Omega M G \Gamma^t P^{2+t/3} K'}{3p} S \tag{4.3}$$

where K' is the constant of proportionality. Comparison of this equation with Eq. 3.7 illustrates the complexity that the concept of one wear mechanism initiating another introduces. Additional dependencies are introduced (e.g., the fatigue parameter t), and other dependencies change (e.g., the dependency on load is no longer linear). Another example of this type of sequential interaction would be either fatigue or adhesive wear mechanisms forming debris, which then acts as an abrasive.

In considering these ways in which the basic wear mechanisms may interact in a given situation, two points should be noted. In the case of parallel interaction, modification of the parameters effecting one of the wear modes may have little or no effect on the overall wear behavior. However, in the sequential interaction, it should always have an effect on the overall behavior. With the first example, changing parameters to reduce fatigue wear would have negligible effect on the wear in the low load range, where abrasion predominates. In the second example, the overall wear would be reduced since it would tend to reduce the effective probability of an adhesive failure. The second point to note is that it is also possible to have both types of interactions (i.e., parallel and sequential) occur in a given wear system. This confounded type of interaction can also contribute to the complex nature of wear behavior.

While such interactions may make it necessary to consider more than one type of mechanism as significant, it is frequently not necessary to do so. It is generally possible to consider one mechanism or type of mechanism as being the dominant and controlling mechanism within limited ranges of tribosystem parameters, as illustrated by the wear maps discussed in Sec. 3.9. In addition to different wear mechanisms being dominant mechanisms in different ranges of operating parameters, wear mechanisms also differ in their severity. This is illustrated in Fig. 4.14, where nominal ranges of a normalized and dimensionless wear rate, Ω, for different mechanisms for sliding are plotted.[*] The dimensionless wear rates are based on wear rates from wear situations in which the mechanism is considered to be the dominant type. There are several equivalent ways of defining this rate, which is shown by the following equations:

$$\Omega = \left(\frac{p}{P}\right) W' = \left(\frac{p}{\sigma}\right) h' \tag{4.4}$$

---

[*]This wear coefficient, Ω, is equivalent to the wear coefficient, K, of a linear wear relationship, $W = K \, PS/p$, where P is load, S is sliding distance, p is hardness, and W is wear volume.

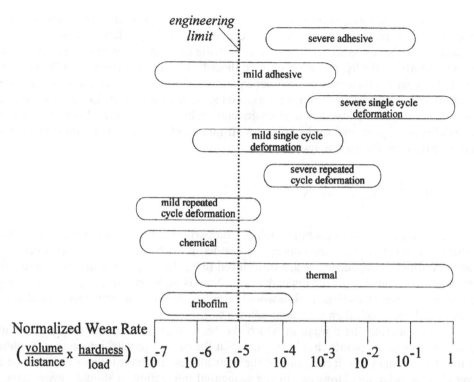

Figure 4.14 Nominal empirical ranges for normalized wear rates [(wear volume × hardness)/(distance × load)] for different types of generic wear mechanisms.

$$\Omega = \left(\frac{p}{P}\right)\frac{\dot{W}}{v} = \left(\frac{p}{\sigma}\right)\frac{\dot{h}}{v} \tag{4.5}$$

In these equations, W is wear volume; h, wear depth; p, hardness; P, load; v, velocity; σ, contact pressure. The dot indicates wear rate with respect to time and the apostrophe indicates wear rate with respect to sliding distance. In this figure, the limiting value for Ω, which can be tolerated in engineering, is also indicated. Applications requiring long life and low wear values often require values of Ω one to two orders of magnitude lower than this. It can be seen from this figure that adhesive, severe single-cycle deformation, severe repeated-cycle deformation, and thermal mechanisms tend to be undesirable forms of wear. Wear rates associated with these mechanisms tend to be higher than those of the other types and generally unacceptable for two reasons. One is their intrinsic severity, as indicated by the ranges of Ω. The other is that these forms of wear tend to occur at higher velocities, pressures and loads than the milder forms of wear, essentially compounding their undesirability in applications.

There is an overall trend in wear behavior with contact stress. In general, the severity of wear increases with increasing stress. Not only does the severity of individual wear mechanism tend to increase with increasing stress, increasing stress tends to lead to the occurrence of more severe wear mechanisms. Both trends are indicated in the wear maps shown in Figs. 3.81–3.83 and in 4.15. Empirical models for sliding, rolling, and impact wear also illustrate such a trend, as well as the general nature of the wear mechanisms. For many wear situations, it is possible to correlate wear severity with the ratio of a

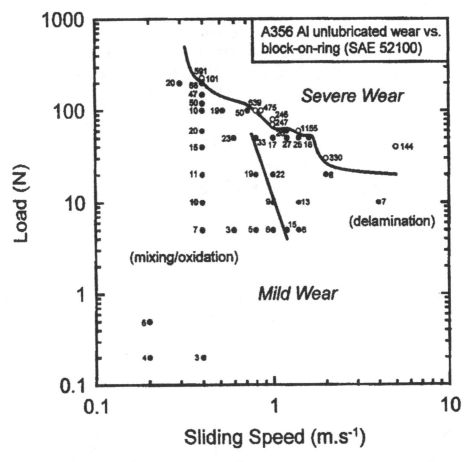

Figure 4.15 Empirical wear rate map for dry sliding wear of an aluminum block against a hard steel ring. Contact stress increases with load. (From Ref. 80, reprinted with permission from Elsevier Sequoia S.A.)

contact stress to a strength parameter, such as contact pressure to hardness (30); see Secs. 3 and Chapter 2 in *Engineering Design for Wear: Second Edition, Revised and Expanded* (EDW 2E).

Important corollaries to these two trends are that materials and other design parameters should be selected to avoid severe wear mechanisms from occurring and that stress levels should be reduced as much as possible to minimize wear.

4.3. TRIBOSURFACES

Since wear is primarily a surface phenomenon, surface properties are major factors in determining wear behavior. Changes to surfaces are a frequent factor in transitions in wear and friction behavior. Before discussing these, it is important to consider the general nature of a tribosurface. A tribosurface consists of the basic or nominal material of the surface plus any layers and films that are present. This is illustrated by the schematic cross-section for a typical metal surface shown in Fig. 4.16. There are numerous surface properties that are associated with wear and friction that can affect overall behavior

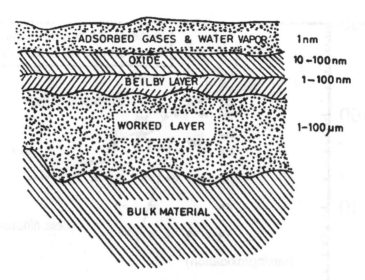

Figure 4.16 Illustration of an unlubricated metal surface. The worked layer is a region of the bulk material that is worked-hardened as result of maching. The Beilby Layer is an amorphous or micro-crytalline layer, resulting from melting and surface flow of molecular layers during machining and hardened by subsequent quenching. While all these layers are typical of most engineering surfaces, an oxide layer, Beilby Layer, and worked layer may not be present with all metals and maching processes. In general, the thickness and properties of these layers depend on the material, environmental exposure, and maching processes. (From Ref. 84, reprinted with permission from The McGraw-Hill Companies.)

and cause transitions. The models for the primary wear mechanisms indicate that geometrical, mechanical, physical, and chemical parameters are involved. Geometrical parameters include the overall shape of the contacting surfaces, as well as the distribution and shapes of asperities. Mechanical parameters would include elastic moduli, hardness, and fatigue parameters. Physical parameters could be work hardening characteristics, diffusion constants, and lattice parameters. Composition and polarity of the surface are examples of chemical factors. Thicknesses and other properties of the various layers and films are additional factors.

While surface parameters influence wear, surface parameters can be influenced by wear. In effect, this means that wear and surface parameters are mutually dependent and a stable wear situation would be one in which the surface properties do not change with wear. If the wear process and the set of surface parameters are not mutually consistent, the wear behavior will be unstable until a mutually consistent condition can be established. This interdependency means that in addition to identifying the relationships between wear and the initial parameters of tribosurfaces, as was done in the treatment of the primary mechanisms, it is necessary to consider how the tribosurface may be modified as a result of wear. The principle types of modifications are treated in this section. The manner in which these modifications can result in transitions in wear behavior is discussed in the subsequent section.

Wear can cause geometrical changes both on a macro- and micro-scale. On a macro-scale, the nature of the contact between two bodies changes, effecting the distributions of stress and load across the contact region. An example of this would be a contact situation, which initially is a point contact (e.g., a sphere against a plane). As wear occurs, one of the

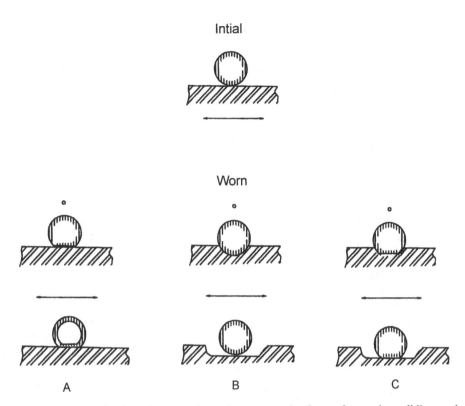

Figure 4.17 Changes in the contact configuration as a result of wear for a sphere sliding against a plane. "A", only the sphere wears; "B", only the flat wears; and "C", both the sphere and the flat wears.

three contact situations indicated in Fig. 4.17 evolves. In two of these cases, the final contact situation is a conforming contact. In the third, "b", it changes from a point contact to a line contact. This type of change is further illustrated in Fig. 4.18, which shows profilometer traces of a flat surface and a sphere after some wear was produced on the flat surface. Another example of a macro-geometrical change is that which takes place between two flat surfaces, which are initially misaligned. Initially, the contact is confined to the region near the edges; with wear, however, contact over the entire surface can be established. Such changes are extremely significant in situations in which the wear depends on stress, such as in mild sliding and impact wear situations (3,31). Such a change is related to run-in behavior.

One way for micro-geometrical changes to occur is the result of asperity deformation as a result of contact. An example of this is shown in Fig. 4.19, where a profilometer trace through a wear track on a flat surface is shown. A micrograph of the surface is also shown. In this case, the tips of the asperities in the worn region appear to be more rounded; asperity heights appear to be more uniform in the wear track as well. In general, initial wear in sliding and rolling systems tends to increase the radius of curvature of the asperities and to provide a more uniform distribution of asperity heights. These changes tend to increase the number of asperities involved in the contact as well as to reduce the stress associated with each junction. Initially, the asperity deformation tends to be in the plastic range, while subsequent engagements would likely

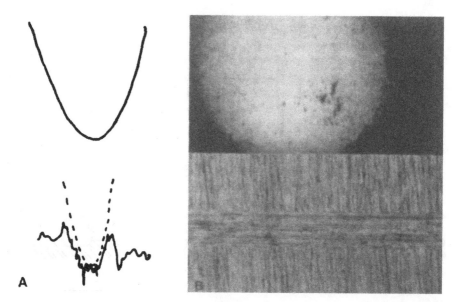

Figure 4.18 ("A") Profilometer traces of sphere and a wear scar, produced by the sphere, are superimposed, illustrating the conforming nature of the worn contact. ("B") Micrographs of the sphere and flat after wear are also shown. (Unlubricated sliding between a 52100 steel sphere and a 1050 steel flat.) (From Ref. 52, reprinted with permission from Elsevier Sequoia S.A.)

result in elastic deformation as a result of these changes. As material is worn from the surface, the general result is a different micro-geometry or topography, characteristic of the wear processes involved. An example of this would be the striations that result from abrasive wear; another might be adhesive wear fragments attached to the surface or the roughening on the surface caused by erosion. Micrographs illustrating the morphological features of worn surfaces are shown in Fig. 4.20 for a variety of conditions. Significant changes in surface topography are evident in these micrographs.

In addition to these geometrical changes associated with wear, other changes which influence the physical and mechanical properties of tribosurfaces can occur, as can changes in material composition and structure. An example of these types of changes is the

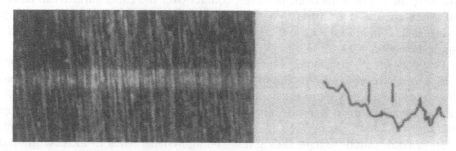

Figure 4.19 An example of asperity modifications in the early stages of wear. The micrograph and the profilometer trace are for a Monel C Platen, worn by a 52100 steel sphere in lubricated sliding. The wear track is located between the vertical lines on the trace. (From Ref. 52, reprinted with permission from Elsevier Sequoia S.A.)

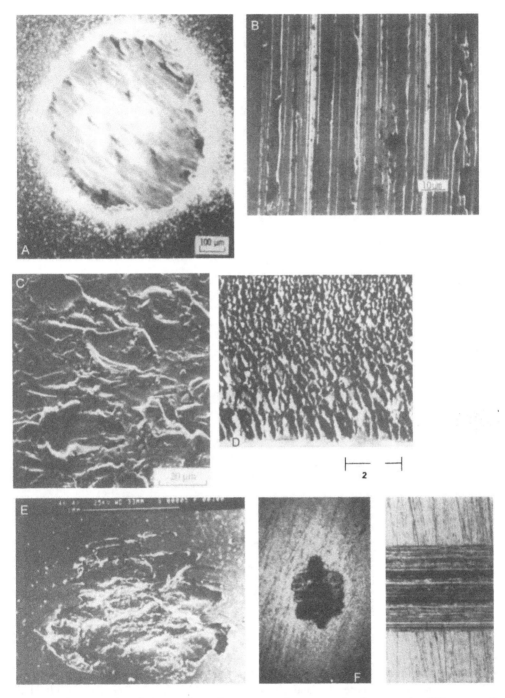

Figure 4.20 Examples of changes in surface topography as a result of wear under different conditions. "A", fretting; "B", sliding; "C", erosion; "D", erosion; "E", rolling; "F", sliding; "G" and "H", compound impact; "I", erosion; "J", slurry erosion. ("A" from Ref. 85; "B" from Ref. 83; "C" from Ref. 87; "D" from Ref. 88; "E" from Ref. 89; "F" from Ref. 52; "G" and "H" from Ref. 3; "I" from Ref. 90; and "J" from Ref. 91. "A", "B", "E", "I", and "J" reprinted with permission from ASME; "G" and "H" reprinted with permission from Elsevier Sequoia S.A.; and "C" and "D" reprinted with permission from ASM International.)

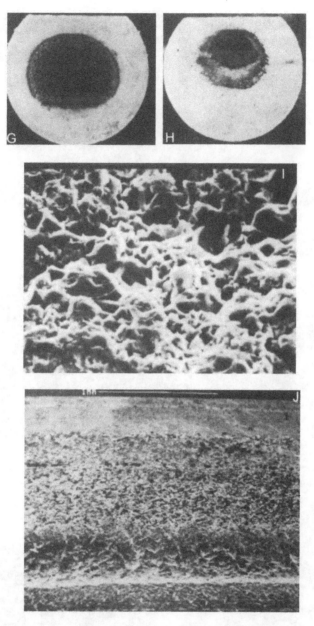

Figure 4.20 (*continued*)

oxidative or chemical wear process discussed in the section on wear mechanisms. In this case, an oxide or other type of reacted layer is formed on the surface, as a result of the wear. In general, the properties of the reacted layer will be different from those of the parent material or any initial oxide. Another way in which the chemical make-up of the tribosurface can be modified is conceptually illustrated in Fig. 4.21. Wear fragments from the counterface, or wear fragments from the reacted layer, are worked into the surface, forming a composite structure. An example of this type of phenomenon in the case of impact wear is shown in Fig. 4.22. Similar observations have been made for sliding wear (32–34).

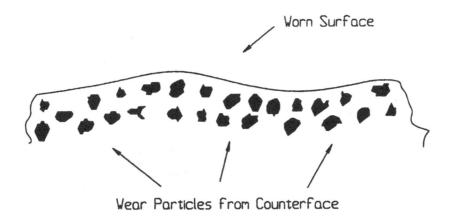

Figure 4.21 Mechanically mixed surface layer.

Structural changes can also take place as the result of the wearing action (e.g., plastic deformation and flow). In the case of metals, changes in dislocation density and grain size at or near the surface are frequently observed in wearing situations. Figure 4.23 shows some examples of this behavior. Frequently, these changes result in a harder, more brittle surface. When hardness changes as a result of wear, it is generally found that wear behavior is related to the modified hardness. This is also true, when hardness is affected by frictional heating, as described further in this section.

Another example is the formation of a layer composed of extremely fine grains and flow-like striations found in some sliding wear situations (Fig. 4.24). As is apparent in this figure, the morphology of this layer is very suggestive of fluid flow, and shows both laminar and turbulent characteristics in different regions. Careful examination of the

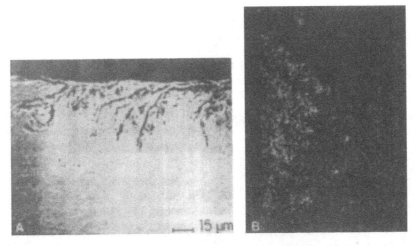

Figure 4.22 Example of mechanical mixing under impact conditions between a steel sphere coated with a thin layer of NI (5000 A) and a Cu flat. A cross-section through the wear scar in the Cu is shown in "A". In "B", an EDX dot map of NI is shown for this region, confirming the mixing. (From Ref. 92, reprinted with permission from Elsevier Sequoia S.A.)

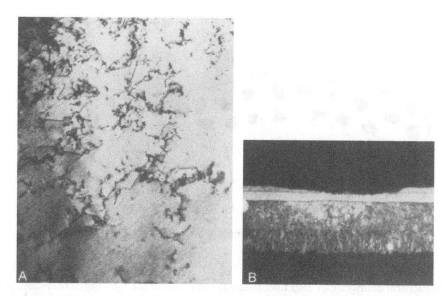

Figure 4.23 Examples of the dislocation networks formed and grain size changes produced during wear. "A" is the dislocation networks formed beneath the wear surface of a Cu specimen during erosion. "B" shows the grain growth occurring in Cu in a small amplitude sliding wear situation. ("A" from Ref. 93, reprinted with permission form ASTM; "B" from Ref. 94, reprinted with permission from ASME.)

micrograph shows the formation of what appears to be an extrusion lip. Because of this lip formation and the general characteristics of the layer, fine grains, and flow characteristics (which are very similar to behavior seen in metal working operations), this mode of wear has been referred to as extrusion wear (35).

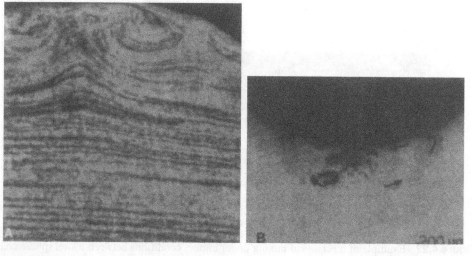

Figure 4.24 Example of subsurface flow during wear: "A" for sliding, "B" for impact. ("A" from Ref. 95, reprinted with permission from ASM International; "B" from Ref. 92, reprinted with permission from Elsevier Sequoia S.A.)

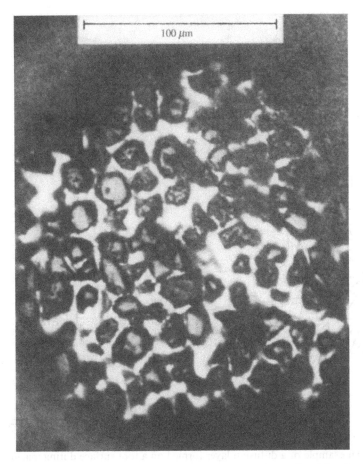

Figure 4.25 Example of preferential wear of a soft matrix. This micrograph is of a coating, which contained diamond particles in an Rh matrix, after it was worn by sliding against a paper surface coated with magnetic ink. (From Ref. 36.)

Preferential wearing of one phase of a multiphase material can also result in a change in the composition of the surface and influence wear behavior. An example of this would be the preferential removal of a soft matrix around a hard filler, grain, or particle (36). This situation is illustrated in Fig. 4.25. This type of action can also produce topological or roughness changes. A further way by which surface composition can change is shown in Fig. 4.26. This is by preferential diffusion of certain elements (37), either to the surface of a material or into the surface of the mating material. Solubility and temperature are strong factors in this type of mechanism. This is frequently a factor in wear situations involving high temperatures, such as in machine tool wear.

Up to this point, metals have been used to illustrate composition and structural changes in tribosurfaces; however, such changes are not confined to this class of materials. Similar changes can take place with other classes but they may be of different types, which depend on the basic nature of the material. With polymers, for example, changes in both the degree of crystallinity, chain length, and degree of cross-linking have been observed. The formation of different polymer structures has also been observed to occur as a result of wear (38–41).

Mild Steel

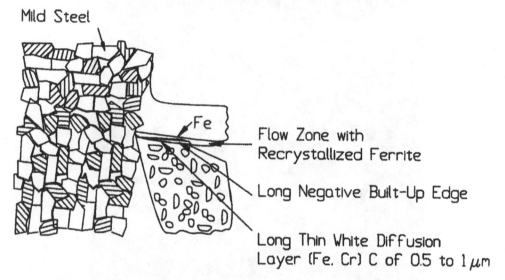

Figure 4.26 An example of a diffusion layer formed on a tool surface during machining. The diagram shows the location of the diffusion layer on the tool surface. (From Ref. 96.)

A factor that has to be included with the consideration of tribosurfaces is surface temperature. In addition to leading to thermal wear mechanisms, discussed in Sec. 3.6, surface temperatures can affect wear behavior in other ways. As indicated in that section, there are several factors which influence surface temperature, such as the heat energy generated at the surface, the thermal conductivities of the materials, heat conduction paths away from the interfaces, and ambient temperature. This is illustrated in Fig. 4.27. Because the heat or thermal energy is generated in the surface (e.g., frictional heating in sliding), surface temperatures are generally higher than elsewhere in the materials, which can influence the nature of the surface in two ways. One is simply related to the fact that most material properties are temperature dependent. As a result, the surface will exhibit material behavior appropriate to that elevated temperature. For example, the hardness of metals can decrease with surface temperature. This type of effect is particularly important in the case of polymer wear. With polymers degradation in wear performance is generally observed with the surface temperatures approaching and exceeding the glass transition temperature (23,42,43). A second way in which surface temperature can influence tribosurfaces is through the temperature dependencies that the surface modification processes have. Elevated temperature can increase reaction rates, influence phase changes, increase diffusion, and enhance flow characteristics of materials.

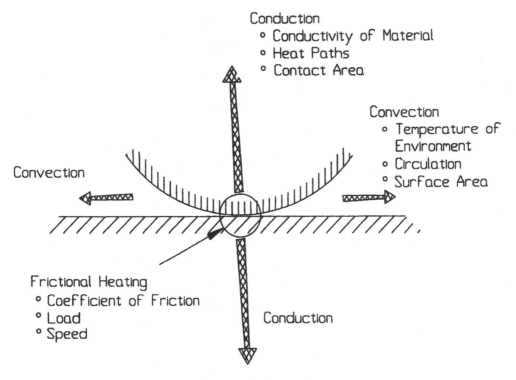

Figure 4.27 Factors affecting surface temperatures.

In addition to these possibilities, removal of existing layers and the formation of new layers or films, such as, transfer and third-body films, can modify tribosurfaces, and in turn affect wear behavior (7,16,24). Typically, a change in the coefficient friction can be found with changes in oxide structure, removal of other films, and the formation of tribofilms (44). Removal of existing films and oxides is often a contributing factor of break-in behavior (45). Many investigators have identified the formation of transfer and third-body films and their importance in wear and friction behavior (6,7,18,26,33,38,39,42,46–49). The formation of these tribofilms and related wear processes are described in Sec. 3.7.

While it is convenient for discussion to consider separately the various ways tribosurfaces can undergo changes as a result of wear, it must be recognized that under actual wearing conditions these changes can be going on simultaneously and in an interactive fashion. For example, a flow layer might also contain a mixture of material from both surfaces, as well as oxides of these materials. An illustration of what typical conditions might exist on worn surfaces is shown in Fig. 4.28.

4.4. WEAR TRANSITIONS

These changes to tribosurfaces, coupled with the possibility of different wear mechanisms, can result in dramatic changes or transitions in wear behavior (44). The break-in phenomenon discussed previously is an example of such a transition. Break-in transitions tend to be gradual and are usually attributed to the modification of the surfaces by such mechanisms as oxide formation, transfer film formation, and asperity profile modification that

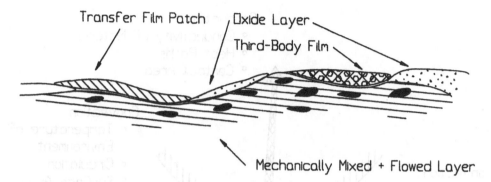

Figure 4.28 Possible state of a surface modified by wear.

are associated with the wearing action. Transitions, which are associated with changes in parameters, such as load, speed, relative humidity, or temperature, tend to be much more abrupt and are frequently associated with a transition between mild and severe wear. In either case, there can be changes in relative contributions and interactions of the several possible wear mechanisms, along with changes in the characteristics of the wearing surfaces. To illustrate the general nature of such transitions and the possible factors involved, several examples will be considered.

During a break-in period, the wear rate is higher than after the break-in period. In this sense, break-in behavior can be thought of as a transition from a severer to a milder mode of wear. A wear curve, showing typical break-in behavior, is presented in Fig. 4.29. In this case, the wear is the wear of the plastic member of a cermet-plastic sliding wear system. The micrographs in the figure show the appearance of the surfaces after the break-in period. The break-in period in this case is associated with the formation of transfer and third-body films on the surfaces. Break-in behavior is not limited to polymer–metal sliding systems but occurs for other systems and in other situations. However, the mechanism involved may be different. For example, the fretting wear behavior of a metallic system is plotted in Fig. 4.30. A break-in period is evident and oxide film formations, along with topological changes, are associated with this period. The appearance of the wear scar in the break-in period and the stable period are different. These are shown in Fig. 4.31. Another example for a metallic system is shown in Fig. 4.32. In this case, it is for a more normal or gross sliding situation. However, the explanation is the same, oxide formation with surface temperature being a driving factor. The insets show how the wear surface appears in both regions.

In addition to the occurrence of break-in other transitions can occur as a function of duration of the wearing action. An example of this is the behavior found in some four-ball wear tests with lubricated metal pairs. This is shown in Fig. 4.33. In this particular case, there appears to be several identifiable regions of wear behavior. The initial break-in period, a region of steady state wear, followed by a period of zero wear rate, ultimately leads to a region of rapid or accelerated wear. The appearances of the worn surfaces in these regions are different. The overall behavior is likely the result of a complex relationship between film formation, topological modification, changing wear mechanisms, and lubricant effects. A possible scenario is that oxide and other films form in the two initial periods, along with micro-smoothing of the topography, which leads to a period of very low wear rate. Ultimately, however, fatigue wear roughens the surface and disrupts the beneficial film, leading to accelerated wear.

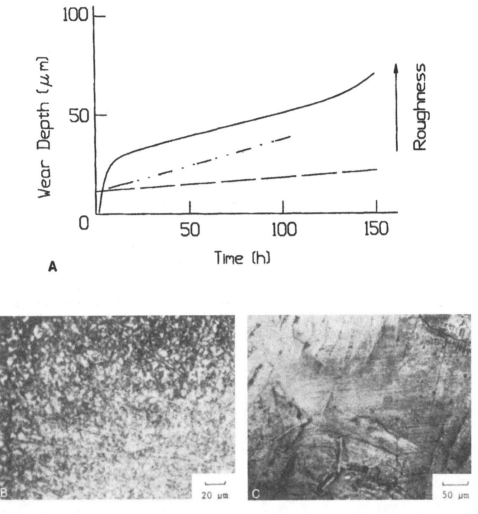

Figure 4.29 ("A") Wear curves for the plastic in a plastic-cermet sliding situation. The curves are for increasing roughness of the cermet. The mircrographs show the discontinuous transfer film on the counterface ("B") and the third-body film formed on the plastic ("C") during break-in that results in the reduction of wear rate. (From Ref. 7, reprinted with permission from ASME.)

Another example of a transition in wear behavior is shown in Fig. 4.34. This is for sliding wear of a metal–polymer system. In this case, an increase in wear rate is seen well beyond the break-in period, and is associated with the disruption of the third-body film. The data shown in the figure are for UHMWPE (ultra high molecular weight polyethylene) and were obtained in a thrust washer test. The same material, when tested in a pin-on-disk configuration, showed similar behavior (i.e., the occurrence of a transition well after break-in) (11,28). The reason for the increased wear was identified with the start of fatigue wear, as evidence by micro-cracks in the surface. These delayed contributions from fatigue wear are the result of the incubation period associated with this mechanism.

Another example of transitional behavior is the PV Limit generally associated with the wear of polymers in metal–polymer and polymer–polymer sliding systems. In this case,

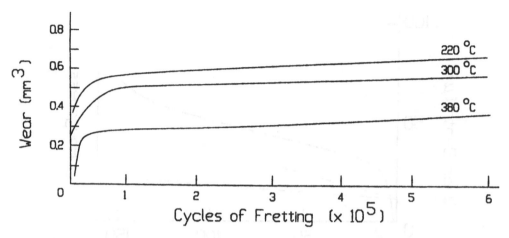

Figure 4.30 Break-in behavior in the case of fretting between two metals. (From Ref. 24.)

the transition is between mild and severe wear behavior. Sliding conditions above the PV Limit result in too high a temperature for the polymer, softening it to the extent that accelerated wear occurs. This behavior is illustrated in Fig. 4.35, which shows the PV Limit curve for a polyimide and the corresponding relationship between wear rate and surface

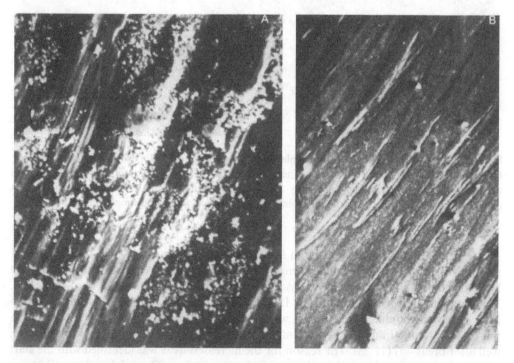

Figure 4.31 Examples of wear scar morphologies for the system referenced in Fig. 4.30. "A" illustrates wear scar morphology prior to the formation of a stable and continuous oxide film. "B" illustrates the appearance after the formation of such a film. (From Ref. 24, reprinted with permission from Elsevier Sequoia S.A.)

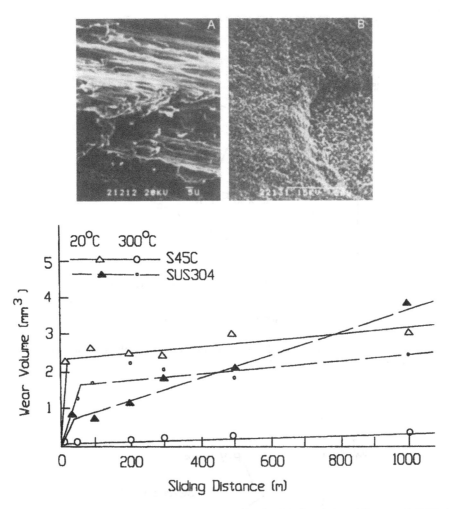

Figure 4.32 Example of break-in behavior as a result of oxide formation. Micrograph "A" shows the appearance of the wear scar prior to the formation of the oxide layer, and "B" shows the appearance after. (From Ref. 14, reprinted with permission from ASME.)

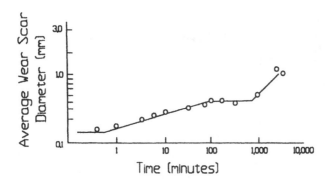

Figure 4.33 Time-dependent wear behavior of a lubricated metal couple in a four-ball wear test. (From Ref. 19.)

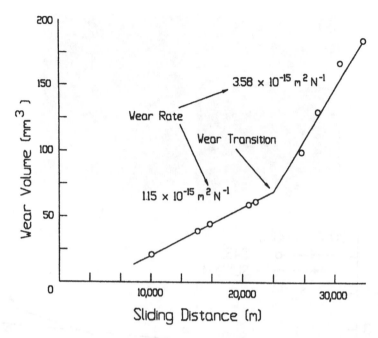

Figure 4.34 Transition in plastic wear behavior as a result of disruption of third-body film (unlubricated sliding between UHMWPE and steel. (From Ref. 16.)

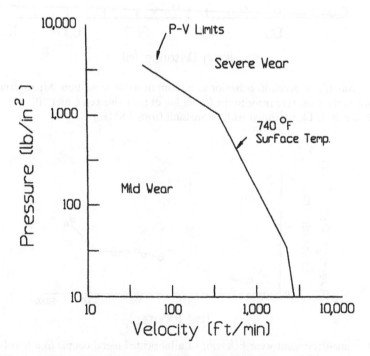

Figure 4.35 Wear behavior of polyimide sliding against a steel counterface. (From Ref. 23.)

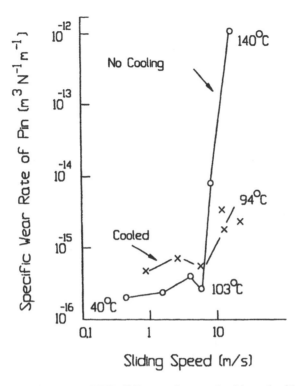

Figure 4.36 Wear behavior UHMWPE sliding against steel with and without cooling. (From Ref. 16.)

temperature. A similar transition for another polymer is indicated as a function of sliding speed in Fig. 4.36. The appearances of the polymer wear scar above and below this type of transition are different, as shown in Fig. 4.37. Above the transition, the morphology indicates the occurrence of softening and flow. The relationship between temperature, load,

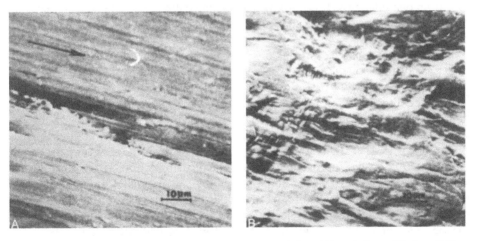

Figure 4.37 Examples of polymer wear scar morphology below (A) and above (B) the mild/severe wear transition in the case of sliding. (From Ref. 97, reprinted with permission from ASME.)

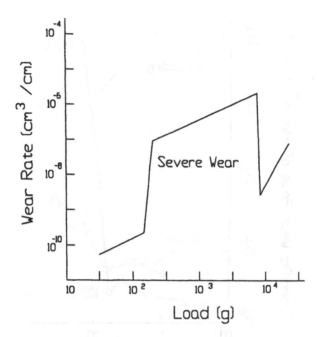

Figure 4.38 Transitions in the sliding wear rate for an unlubricated steel/steel couple, associated with changes in oxide structure. (From Ref. 9.)

and speed in this type of transition can be understood by recognizing that the product of the pressure and velocity is proportional to the energy dissipated at the sliding interface, which determines the surface temperature.

Metal and ceramic sliding systems also exhibit similar transitions related to temperature, load, and speed, as well as relative humidity. In general, these are usually associated with the formation of various oxides and films on the surface (5,13,14,24,33,50). Examples are shown in Figs. 38–42. Surface analysis techniques, such as microprobe, Auger and ESCA analysis have confirmed the existence of different oxides and films in these regions. The wear on one side of the transition is usually mild, while, on the other side, severe. Frequently, wear rates are found to change by as much as two orders of magnitude in these transitions.

Transitions can also be associated with stress levels. For example, in the case of the impact wear of thin polymer sheets, a stress limit has been identified with the transition from mild to a severe wear region (3,41,51). Above this limit, an over-stressed condition exists and the film experiences catastrophic wear. An example of the wear in this region is shown in Fig. 4.43. A much milder form of wear occurs below that limit. In this region of mild wear, further transitions can also occur. An example of this is the impact wear behavior of elastomers (51). During the initial period of wear, the elastomer progressively deforms, asymptotically approaching some limit of deformation. After a certain point, crack formation becomes evident and there is loss of material from the surface. The appearance of the wear scar before and after this transition is shown in Fig. 4.44, along with a graph of typical wear behavior below the critical stress. For materials, which have an elastic region and a plastic region, it is also common to find different wear behavior for loading conditions in the elastic range as compared to those in the plastic range. Frequently, severe wear behavior is associated with wear in the plastic situation. An example

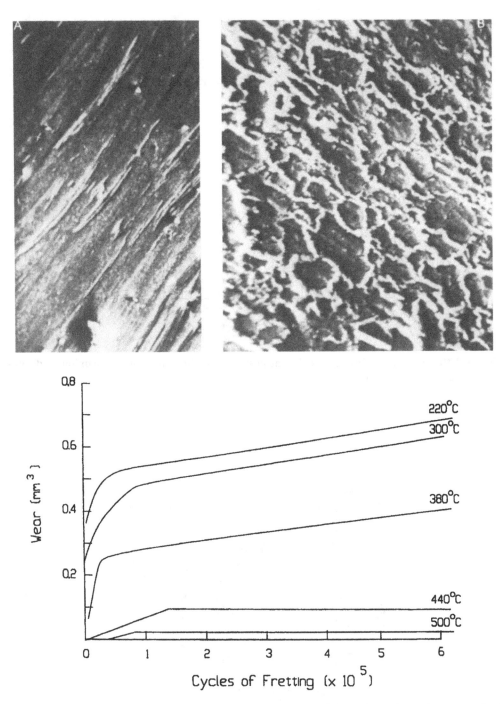

Figure 4.39 Differences in fretting water behavior of an unlubricated metal couple as a result of changes in oxide formation. The graph shows the influence of temperature on wear rate. The two micrographs illustrate the differences in wear scar morphology that can occur as a function of temperature. "A" is the condition after 120,000 cycles at 500°C; "B", 168,000 at 600°C. Wear was much lower at the higher temperature. (From Ref. 24, reprinted with permission from Elsevier Sequoia S.A.)

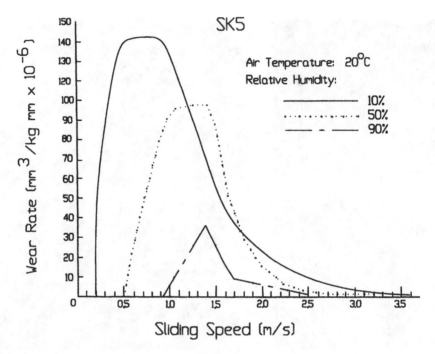

Figure 4.40 Wear rates of cast iron sliding against steel for different humidity conditions and without lubrication. (From Ref. 13.)

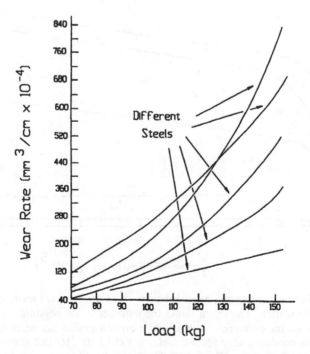

Figure 4.41 Wear rate as a function of load for different steels sliding against a steel counterface without lubrication. (From Ref. 5.)

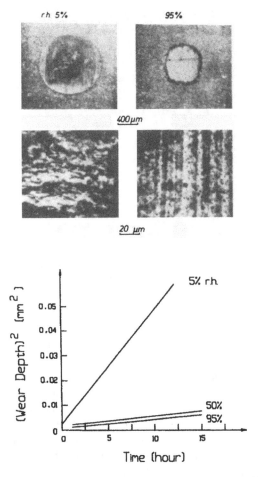

Figure 4.42 Wear of SiC fretting against unlubricated 52100 steel for different relative humidity conditions. The micrograph graphs show the wear scar morphologies occurring under very dry and very moist conditions. (From Ref. 50, reprinted with permission from ASME.)

of this is shown in Fig. 4.45. A wear scar, representative of the milder wear behavior generally found in elastic situations, is also shown in the figure for comparison. These suggest that the relative contributions from the various wear mechanisms are different above and below the yield point. Typically, single-cycle deformation and adhesion would tend to predominate at stress levels above the yield point (52).

A further example of a transition in wear behavior is shown in Fig. 4.46. In this case, the initial wear rate is low but with an increased number of operations, a dramatic increase in wear rate is seen. Examination of the wear scars in the two regions suggests that adhesion becomes more pronounced in the region of accelerated wear. Prior to this, several changes in the surface features were observed. One was that the surface became smoother, which contributes to increased adhesion. Discoloration of the surface was also observed to increase, suggesting increased surface temperature, which again would enhance adhesion. Wear tests with the same material system under similar stress conditions but at a much lower sliding speeds resulted in wear behavior which could be correlated with the initial

Figure 4.43 An example of "overstressed" impact wear behavior of polymer films. In this case, the film was a urethane coated fabric impacted by a steel hammer. (From Ref. 3, reprinted with permission from Elsevier Science Publishers.)

wear behavior in the application. It was concluded that the high sliding speed in the application was a significant factor in the surface modification that initiated the accelerated wear behavior. Several material systems were tried in this application. All the metal/metal metal and metal/ceramic systems tried exhibited this type of transition; metal/polymer systems did not but had wear rates well above the initial wear rates of these other systems.

Changes in the degree or amount of lubrication provided to a sliding or rolling system can also lead to transitions in wear behavior. The wear behavior as a function of amount of lubrication is shown in Fig. 4.47. In this case, the appearance of the wear scars looked similar in both regimes. However, there is often a significant difference in wear scar appearance of a well-lubricated surface and a poorly lubricated one. Typically, this is the result of the wear changing from mild to severe between these two conditions.

A further example of a transition in wear behavior is shown in Fig. 4.48 for a lubricated metallic system under combined rolling and sliding. There is a transition from mild to severe wear as a function of load and speed, analogous to PV behavior, and the transition is associated with oxide formation, softening of the metal, and lubricant rheology. The inset in the figure illustrates wear scar morphology on the two sides of the transition. Frequently, the wear above the transition in such a situation is referred to as scuffing.

Transitions in wear behavior can generally be associated with changes in the appearance of the wear scar and frequently in the nature of the debris that is produced (44,53). In addition, as will be further discussed in the section on friction, changes in frictional behavior are often associated with transitions in wear behavior. These two attributes, morphology and friction, are often key in identifying and understanding these transitions and observations regarding these are useful in engineering evaluations.

4.5. GALLING

Galling is a severe form of adhesive wear that can occur with sliding between metals and metals and ceramics. Localized macroscopic roughening and creation of highly deformed

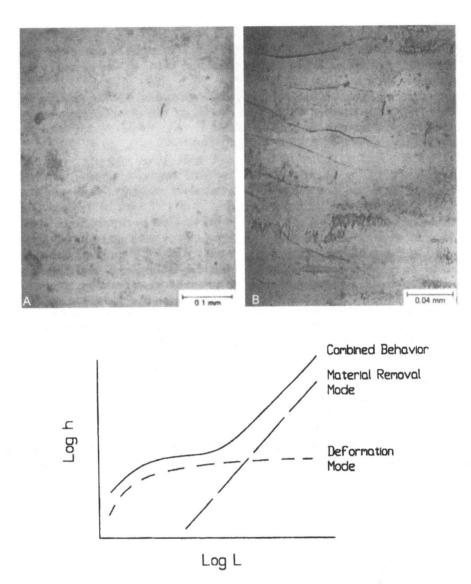

Figure 4.44 Impact wear behavior of elastomers. The micrographs show the initial ("A") and long-term ("B") appearances of the elastomer surface. (From Ref. 51, reprinted with permission from Elsevier Sequoia S.A.)

protrusions, resulting from adhesion and plastic deformation, are distinguishing characteristics of this wear mode. Examples of galling and galls are shown in Fig. 4.49. This type of wear is generally limited to unlubricated tribosystem, unless there is a breakdown with the lubrication. Galling is stress dependent, increasing with increasing stress. A minimum contact pressure is required for galling to take place (13,14). This minimum pressure depends on the material pair, as well as on parameters that affect adhesive wear behavior. The severity of the galling also depends on ductility. Galling tends to be more severe with ductile materials than with brittle materials.

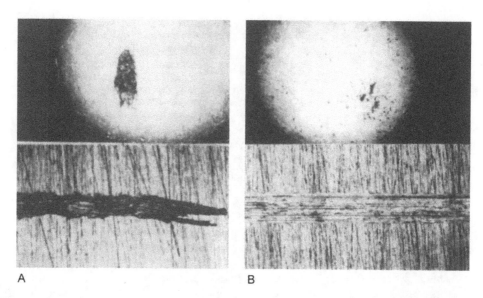

Figure 4.45 An example of severe wear behavior in a metal–metal sliding situation, when the elastic limit is exceeded, is shown in "A". "B" shows an example of milder wear when this limit is not exceeded. (From Ref. 52, reprinted with permission from Elsevier Sequoia S.A.)

The minimum contact pressure required for galling to take place is called the galling threshold stress. This threshold stress characterizes resistance to or susceptibility for galling between a pair of materials. Threshold stresses are generally determined in tests involving a single unidirectional rubbing action and examining the surfaces for evidence of galling. There is an ASTM (American Society for Testing and Materials) International standard test method, ASTM G98, for determining galling threshold stresses, which uses

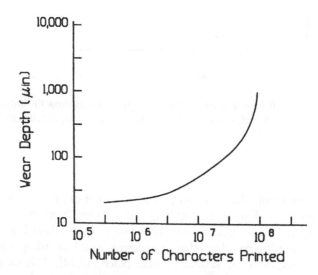

Figure 4.46 Wear behavior on an unlubricated stainless steel print band and a Cr-plated steel platen. Wear depth is measured behind the location of a character and results from sliding at this interface during printing.

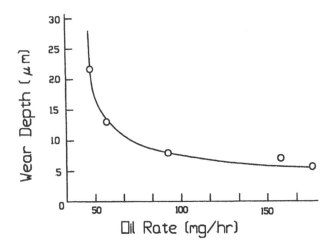

Figure 4.47 Wear of a steel/steel interface, subjected to combined impact and fretting, as a function of oil supply. (From Ref. 12.)

a single 360° rotation of a flat button pressed against a flat surface. A description of this test can be found in Sec. 9.2.9. Threshold stresses can also be determined using other methods (54–57). A tabulation of galling threshold stresses determined by such tests can be found in the appendix.

Surface changes caused by wear can affect susceptibility to galling. For example, changes, such as removal of protective layers and growth of the real area of contact with increased rubbing, are likely factors for lowering threshold values. There is evidence that such effects can be significant and lower threshold values when there is repeated contact. This is illustrated by the data shown in Table 4.1, where threshold values based on a single rotation and three rotations are listed.

The susceptibility to galling is also affected by surface roughness (54,56). Generally, increased R_a or center-line-average (CLA) roughness decreases the tendency for galling. This is shown in Fig. 4.50, where threshold values for a stainless steel–steel couple is plotted as a function of roughness. Extremely smooth surfaces, that is, $R_a < 0.25\,\mu m$, should be avoided. The tendency for galling can be affected by other aspects of surface roughness as well. Long-wavelength waviness tends to promote galling by localizing the contact. Galling tends to be most severe when there is a lay to the surface that is perpendicular to the sliding motion and least when the lay is parallel. A surface without lay, such as produced by grit blasting, is somewhere in-between. This influence of lay is caused by the effect that the lay has on the growth and size of junctions.

Material pairs, hardness, and ductility are also factors in galling behavior (56). Rankings of various metal combinations in terms of their susceptibility for galling are shown in Table 4.2. Hardness and ductility, through their effect on junction size and growth, are factors in galling. Higher hardness and fracture reduce the size of junctions. As a consequence, the severity of galling tends to be reduced by increases in hardness, reduced ductility, and brittle behavior. Galling thresholds tend to be high and galling mild for material couples that have these characteristics. Many surface hardening treatments provide all three attributes. Figure 4.51 illustrates the change in galling behavior that resulted from nitriding two stainless steel surfaces.

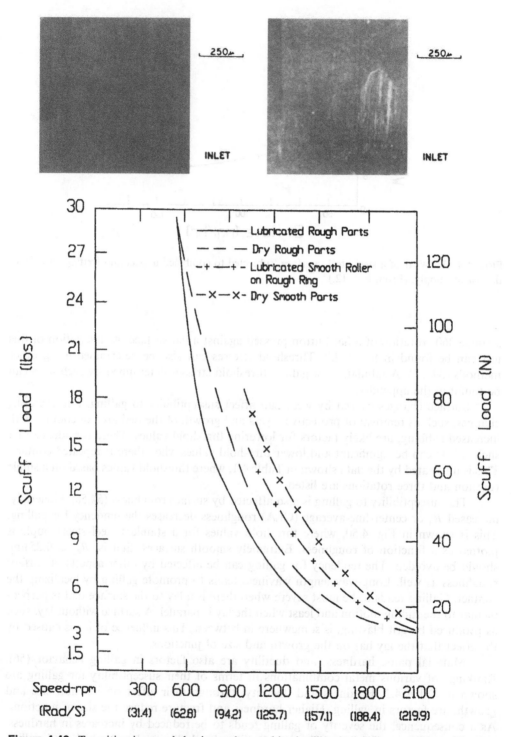

Figure 4.48 Transition in wear behavior under combined rolling and sliding conditions. The wear rate is much higher for loads and speeds above the curve. "A" shows wear scar morphology for conditions below the curve; "B" shows conditions above the curve. (From Ref. 98, reprinted with permission from ASME.)

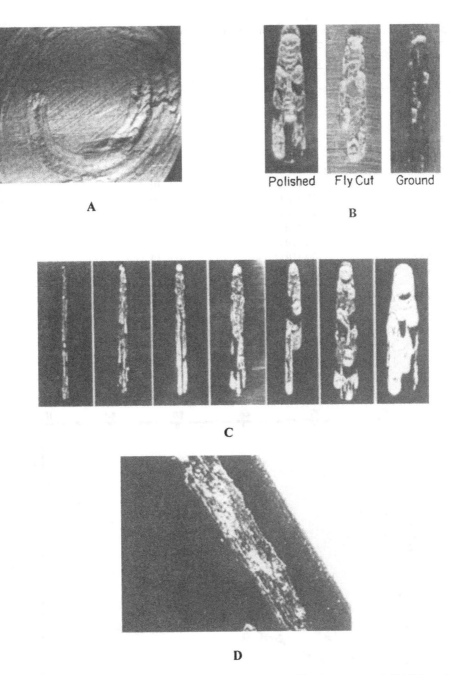

Figure 4.49 Examples of wear scars on metal surfaces where galling has occurred. ("A" from Ref. 54; "B" and "C" from Ref. 55, reprinted with permission from ASME.)

While not limited to these situations, galling tends to be a problem in two types of situations. One type of situation is when the contact is heavily loaded and the device is operating infrequently. A valve that tends to be closed most of the time and only occasionally opened would be an example of such a situation. The other type of situation is when

Table 4.1 Threshold Stresses for Galling as a Function of Cycles for Self-mated Stainless Steels

Material		Threshold stress (MPA)	
Stainless steel	Hardness	Single rotation[a]	Triple rotation
S20161	95 HRB	> 104	> 104
	28 HRC	> 104	> 104
S28200	96 HRB	166	7
S21800	92 HRB	> 104	48
T440C	55 HRC	124	14
T304	86 HRB	55	< 7
T430	98 HRB	10	< 7
S42010	50 HRC	> 104	21
T420	49 HRC	55	14
S24100	23 HRC	97	14
S45500	48 HRC	97	7
S66286	30 HRC	14	< 7
Type 303	85 HRB	138	< 7

[a]ASTM G98 Test Method.
Source: Ref. 73.

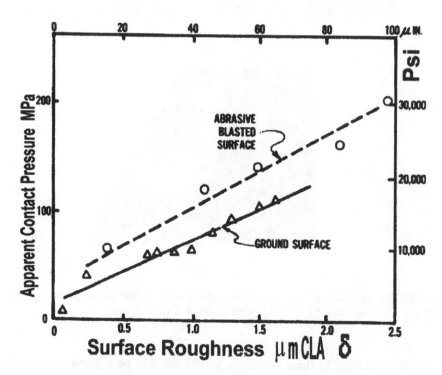

Figure 4.50 Effect of surface roughness on the galling threshold stress. The ordinate is the galling threshold stress. The data are for a 440C stainless steel (59HRC) against 1020 steel (90HRB). (From Ref. 54, reprinted with permission from ASME.)

Table 4.2 Galling Resistance of Material Pairs

	400 Series SS (soft)	400 Series SS (hard)	300 Series SS	Soft Steel	Hard Steel	Cast Iron	Ductile Iron	Hastalloy A and B	Hast-alloy C	Hast-alloy D	Stelite	Nitride	Cr Plate	Bronze (leaded)	Bronze A	Bronze B	Bronze D
400 Series SS (soft)	N	F	F	N	F	S	S	N	F	S	S	F	F	S	F	S	S
400 Series SS (hard)	F	S	F	S	S	S	S	F	S	S	S	S	S	S	S	S	S
300 Series SS	F	F	N	N	F	S	S	N	F	S	S	S	S	S	F	S	S
Soft steel	N	S	N	F	S	S	S	N	F	S	S	S	S	S	S	S	S
Hard steel	F	S	F	S	S	S	S	F	S	S	S	S	S	S	S	S	S
Cast iron	S	S	S	S	S	S	S	S	S	S	S	S	S	S	S	S	S
Ductile iron	S	S	S	S	S	S	S	S	S	S	S	S	S	S	S	S	S
Hastalloy A and B	N	F	N	N	F	S	S	N	F	S	S	S	S	S	F	S	S
Hastalloy C	F	S	F	F	S	S	S	F	F	S	S	S	S	S	S	S	S
Hastelloy D	S	S	S	S	S	S	S	S	S	S	S	S	S	S	S	S	S
Stellite	S	S	S	S	S	S	S	S	S	S	S	S	S	S	S	S	S
Nitride	F	S	S	S	S	S	S	S	S	S	S	S	S	S	F	S	S
Cr plate	S	S	S	S	S	S	S	S	S	S	S	S	?	S	S	S	S
Bronze (leaded)	S	S	S	S	S	S	S	S	S	S	S	S	S	S	S	S	S
Bronze A	F	S	F	S	S	S	S	F	S	S	S	F	S	S	F	S	S
Bronze B	S	S	S	S	S	S	S	S	S	S	S	S	S	S	F	S	S
Bronze D	S	S	S	S	S	S	S	S	S	S	S	S	S	S	S	S	S

Level of resistence: S, satisfactory; F, fair, N, little or none.
Source: Ref. 74.

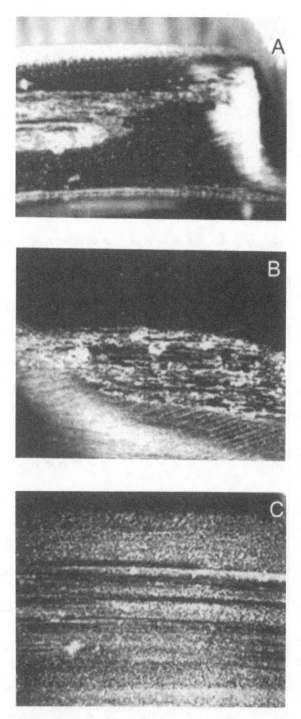

Figure 4.51 Wear scars showing the effect of increased hardness on galling, "A" and "B" are examples of the galling that took place on lug prior to hardening the surface. "C" shows the wear on the same surface after the surfaces were nitrided.

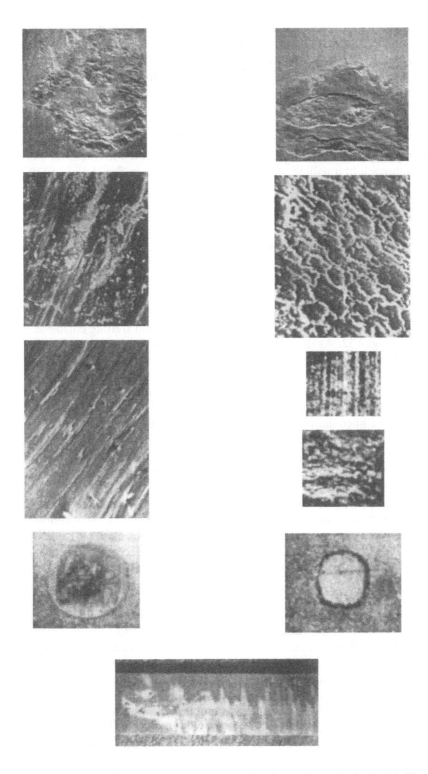

Figure 4.52 Examples of fretting wear scars on metal surfaces. (From Refs. 24, 50, 99, and 100, reprinted with permission from ASME, Elsevier Sequoia S.A., and ASTM.)

there are tight tolerances between almost or nearly conforming surfaces such as a piston in a cylinder. While nominal contact pressures in this type of situation are often small or 0, galling tends to occur as a result of high contact stresses caused by misalignment and errors-in-form. When galling occurs in these and other applications, the concern is generally not with material loss but binding that results from the size of the protuberances and debris that galling causes. In addition to the identification of galls in these situations, common manifestations of a galling problem are seizure, increased difficulty in operating a device, and scouring. The minimal criterion for avoiding galling in unlubricated situations is to design so that the maximum contact pressure is below the threshold galling stress. A design approach to a galling wear problem is discussed in Sec. 5.16 in EDW 2E.

4.6 FRETTING

When the amplitude of a reciprocating sliding motion is less than a few millimeters, the motion is generally referred to as fretting or fretting motion. The wear that results from this type of motion is called fretting wear, or simply fretting. If oxide formation is involved in the wear process, it is often called fretting corrosion. Fretting wear is the more general form and may occur with any material; fretting corrosion is generally limited to non-noble metals. Examples of fretting wear and fretting corrosion wear scars are shown in Fig. 4.52. As illustrated by the micrographs in the figure, fretting wear scar features tend to be fine. In addition, the morphology of fretting scars can be similar to that resulting from unidirectional and larger amplitude motion. However, as illustrated in these examples, striations, directionality, and evidence of adhesion tend to be less pronounced with fretting scars than with gross sliding. Also wear debris in fretting situations tends to be finer than in many other sliding situations. This modified appearance, the presence of fine and often oxidized debris, and the possibility of small amplitude vibrations are general indicators of fretting wear (58).

Fretting wear can involve the transition from non-abrasive to abrasive wear behavior (58). In the initial stages of fretting wear, it is a non-abrasive, typically resulting from adhesive, repeated-cycle deformation, or chemical wear mechanisms, or some combination of these. With the entrapment of work-hardened and oxidized wear debris from these mechanisms in the contact region abrasive wear behavior can become dominant. When this occurs, single-cycle deformation mechanisms tend to become significant and wear is generally accelerated by the abrasion. When this does not occur, longer-term behavior is characteristic of non-abrasive sliding wear. In such cases, repeated-cycle deformation or chemical wear mechanism likely dominates wear behavior. However, adhesive, thermal and tribofilm mechanisms are also possible as dominating and contributing mechanisms to the overall wear [(41,59–64) See Secs. 5.6 and 7.3 in EDW 2E].

While the mechanisms are the same for general sliding and fretting, there are some specific trends in fretting situations. For fretting amplitudes less than 1 mm, the dimensionless wear rate, Ω, defined in Eqs. (4.4) and (4.5) tends to dependen on the amplitude and load (58,60,61,63). Trends with amplitude and load are shown in Figs. 4.53 and 4.54, respectively. Analysis of the data suggests the following relationship between Ω and slip amplitude, ε, in the region between 10 and 100 μm. Above and below that range, it tends to be independent of amplitude. Wear has been observed with slip amplitudes as low as 0.06 μm (65,66).

$$\Omega \propto \varepsilon^{\approx 2} \tag{4.6}$$

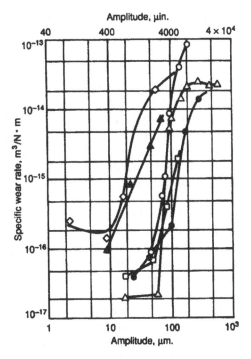

Figure 4.53 Effect of slip amplitude on wear rate. Each curve is from different investigations. (From Ref. 58, reprinted with permission from ASM International.)

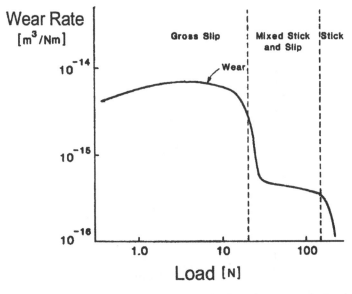

Figure 4.54 Example of the effect of load on slip amplitude and wear rate in fretting. (From Ref. 61, reprinted with permission from ASME.)

The data also suggest the following relationship with load, P:

$$\Omega \propto P^{0.3-1.5} \tag{4.7}$$

While the frequency of the motion is generally not a significant factor, it can be in some situations, particularly unlubricated situations (58,61). With unlubricated metals, wear rates tend to be inversely dependent on frequency below 10–20 Hz, because of increased time for oxide growth at lower frequencies. This behavior is shown in Fig. 4.55. Wear rates can also be affected by frequency as a result of the effect that frequency has on the power (energy per unit time) that is dissipated in the contact and consequently temperature. Generally, this is only significant in situations where there is the potential to develop high enough temperatures to affect wear behavior (41,62,63). In situations where this does not occur, there is no apparent effect over several orders of magnitude change in frequency. For example, under some conditions, wear rates of unlubricated steel couples have been found to be independent of frequency in the range between 100 and 2×10^5 Hz (67). An additional illustration of the typical trend with frequency can be found in Fig. 4.58.

While a small temperature increase of 20–30°C is likely in many fretting situations, larger increases, such as several hundred degrees, are also possible in others (41,62,63). In fretting situations, temperatures can be obtained by treating the interface as a stationary heat source. Equation (4.8) is an approximate relationship for the temperature rise, ΔT, based on a linearized form of the first law of heat flow. Ψ is the power dissipated and k_m's are the thermal conductivities. l_b's are the distance from the surface at which there is no temperature rise in each body. A is the apparent area for the bulk temperature rise and is the real area for the flash temperature.

$$\Delta T = \frac{\psi}{A} \left(\frac{l_{b1} l_{b2}}{k_{m1} l_{b2} + k_{m2} l_{b1}} \right) \tag{4.8}$$

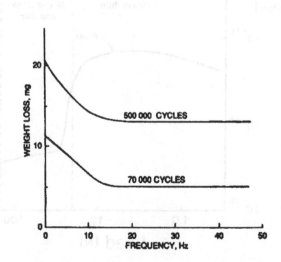

Figure 4.55 Example of the effect of frequency on fretting wear. (From Ref. 58, reprinted with permission from ASM International.)

Because of friction and the compliance of surfaces, small amplitude oscillatory displacements of bodies do not necessarily result in slip at the interface (68,69). There is a threshold displacement, which is required for slip to occur. Three regions of slip are generally identified in fretting situations. These are indicated on the graph in Fig. 4.54. Below the threshold is the stick region. In the stick region, there is no slip. However, some damage may occur as a result of plastic flow and repeated deflection. Above the threshold, there are two regions, the stick-slip and gross slip regions. The stick-slip region occurs at lower displacements than the gross slip region. In the gross slip region, there is slip across the entire contact and it is equal to the displacement of the bodies. In the middle region, there are limited and varying amounts of slip across the interface. Wear takes place in both these regions. Normalized wear rates tend to be lower in the mixed region as a result of reduced slip and tend to be more sensitive to frequency, as indicated in Fig. 4.58 (67). Fretting maps are used to identify combinations of parameters, most often displacement and load or load and shear force, which represent boundaries between the three regions of slip behavior. Examples of fretting maps are shown in Fig. 4.56. These maps indicate the general progression from one region to the other as a function of load, fretting displacement amplitude, and shear force. The progression of slip is illustrated in Fig. 4.57 for a point contact. In the simulation of a fretting contact between a sphere and a plane, stress analysis shows that there is a threshold for slip to take place and when it does, it does not occur uniformly over the entire contact surface (70,71). It starts in an annulus at the edge of the contact region. The width of this annulus grows with increasing displacement amplitude until slip occurs over the entire contact region. In the analysis, an applied shear force simulates displacement. The slip region is defined as the region over which the applied shear stress exceeds the product of the contact pressure and the static coefficient of friction.

Fretting fatigue is another phenomenon that results from fretting motion. It involves progressive damage to a solid surface and leads to the formation of fatigue cracks (72). It is a combination of normal structural fatigue, which results from cyclic strain and stress, and fretting wear. An example of a situation where fretting fatigue occurs is the contact between a clamp and a flexing beam. The flexing motion of the beam could result in slip between the clamp surface and the beam surface. In fretting fatigue, the wear that is caused by the fretting can accelerate the formation of fatigue cracks, which then propagate through the material, leading to fracture of the component. An example of typical fretting fatigue behavior as a function of frequency is shown in Fig. 4.58, along with the corresponding behavior of fretting wear in both the gross slip and partial slip regions.

The general methods to control and reduce sliding wear can be used to reduce and control fretting wear. An additional element to consider with fretting is the possible transition to abrasive wear. Materials and designs should be chosen to eliminate the accumulation of debris harder than the surfaces involved. This may be accomplished by providing paths for the debris to escape or lubrication methods designed to flush out debris. There is an additional way for resolving fretting problems. Fretting motions are generally not intended or required. They are either superimposed on an intended motion, such as fretting motions generated in an impact situation and at the reversal of directions or occur in nominally stationary situations, such as clamped joints or mated electrical contact, as a result of vibrations, actuation cycles, or thermal cycling. Consequently, the elimination or reduction of these motions offers another possibility for solving fretting wear problems. One approach would be to eliminate or isolate the cause or source of these motions. For example, in the case where these are caused by

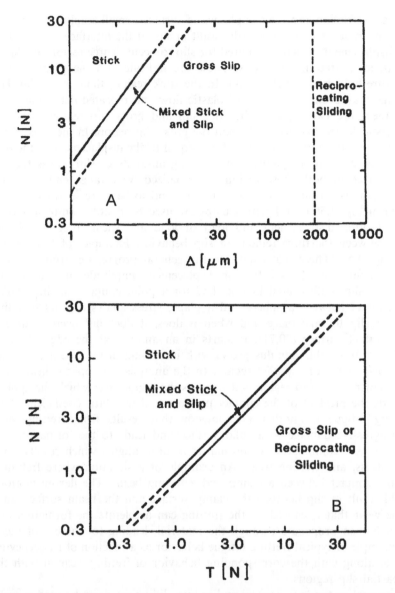

Figure 4.56 Example of wear maps used to characterize slip behavior in fretting situations. "A" is an example of a fretting map relating load and displacement amplitude to slip. "B" is an example of fretting map relating slip to the normal load and tangential load. (From Ref. 61, reprinted with permission from ASME.)

machine or structural vibrations, this might be done by the use of isolation dampeners, eliminating impact situations, tightening tolerances, or increasing stiffness. The other possibility is to decrease or eliminate the slip by changing the threshold for slip to occur. One possibility here is to increase the compliance of the interface. As a result of their ability to accommodate large strains, elastomers can often be used for this purpose by inserting them between the two surfaces. It is also possible to do this by increasing

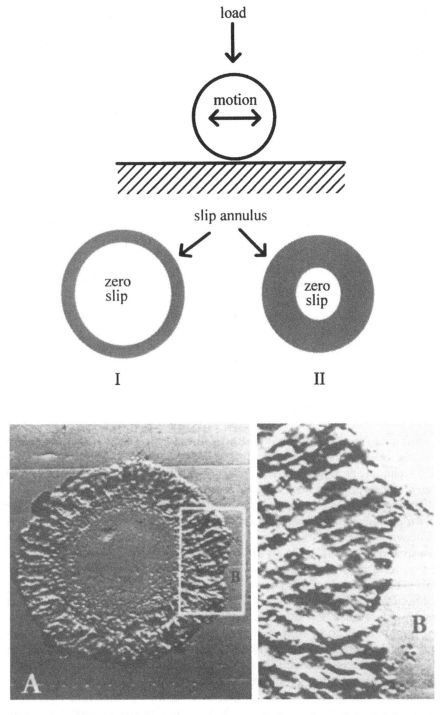

Figure 4.57 Slip annulus in fretting between a sphere and a flat surface. Higher friction and smaller amplitude result in a narrower annulus (I). Lower friction and larger amplitude result in a wider annulus (II). "A" is an example of a fretting wear scar where there was partial slip, showing the slip annulus. "B" is an enlargement of the wear in the slip annulus. ("A" and "B" from Ref. 65, reprinted with permission from ASTM.)

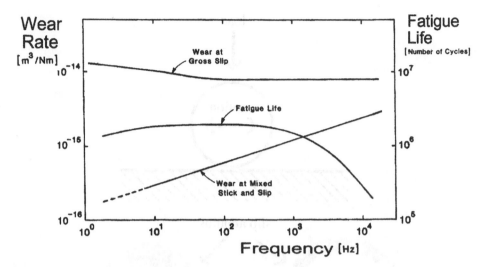

Figure 4.58 Example of the effect of frequency on wear rate and fretting fatigue life. The effect on wear rate is shown for complete slip and partial slip. (From Ref. 61, reprinted with permission from ASME.)

the load and the coefficient of friction. However, if these increases are insufficient to eliminate slip entirely, such changes could result in increased wear.

4.7. MACRO, MICRO, AND NANO TRIBOLOGY

The macro, micro, and nano classifications of tribology resulted from the needs of new technologies, such as magnetic disks storage and micro electro-mechanical device (MEMS) technologies. These newer applications tend to be more sensitive to wear and to different types of wear, than devices associated with older technologies. The basic distinction between these tribological categories is the order of magnitude of the dimension of the wear phenomenon involved. This is shown in Fig. 4.59. In addition to this, there are other differences as well. With macro-tribology, the concern is general with the accumulation of wear and wear behavior tends to be characterized in terms of wear rates. Wear behavior ranges from mild and fine to severe and coarse and encompasses a wide range of materials in macro-tribology. With nano and micro-tribology, the interest tends to be more with the occurrence of wear than with the accumulation of wear, as well as with the milder and finer forms of wear. For example, with repeated cycle deformation mechanisms, the concern is more likely to be with the first manifestation of wear, such as the appearance of surface cracks or fracture, then with the progressive loss of material. Milder forms of the various wear mechanisms are of more importance than coarser forms. Also atomic forms of wear become more significant in nano and micro-tribology than in macro-tribology. With nano and micro-tribology, the materials of interest tend to be more limited and different from those considered in macro-tribology. In summary, because of the uniqueness of the applications and the sensitivity to wear in these newer applications, the focus with nano and micro tribology tends to be on very specific forms of wear and wear mechanisms that occur with specific materials for a narrow range of conditions. Typically, the study of nano and micro tribological phenomena requires the use of more sophistication and state-

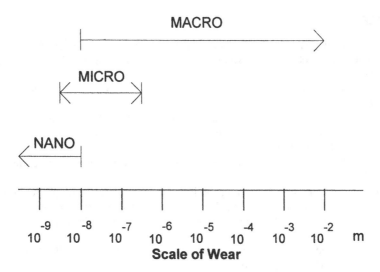

Figure 4.59 Schematic illustrating the magnitude of significant wear phenomena in macro, micro, and nano-tribology.

of-the-art instrumentation and testing apparatus than is generally required or used for studies in the macro-tribology range, particularly at the engineering level.

While there are these differences, the more fundamental and general understanding of tribological behavior and concepts, which are generally based on studies in the macro-tribology regime, tend to apply or can be extended to the nano and micro regimes. However, many more specific and less basic concepts do not. In general, the extrapolation and application of information, either data or concepts, obtained in one regime to another should be done with caution. In many situations, the data gathered in the nano and micro-tribology regimes or the macro-tribology regime are not useful or applicable in the other. For example, nano and micro-tribological evaluations of thin coatings and surface treatments for applications in newer technologies are often not applicable or useful in macro-tribology application. They tend to be too thin to significantly affect wear behavior in the macro-tribology applications. Some tribosystem trends between nano and macro-tribology are given in Table 4.3. Consideration of the effects of these differences on wear behavior generally helps in the extrapolation of information and data between macro, micro, and nano-tribological.

Table 4.3 Tribosystem Trends From Macro to Micro-Tribology

Bulk properties become less important; surface properties become more significant

Significant stresses move closer to the surface

Apparent area becomes less significant; real area becomes the primary consideration. In the extreme the apparent area equals the real area and there is one junction

Increasing focus on damage of or by individual asperities or at junctions

Single events become more important: single spall or crack; localized disruption of oxide or protective layers; adhesive failure at a junction; single scratch

Significance of materials and material properties change; the importance of materials and the significant of material properties tend to be different

---

REFERENCES

1. D Rigney, W Glaeser, eds. Source Book on Wear Control Technology. ASME, 1978.
2. R Bayer, R Schumacher. On the significance of surface fatigue in sliding wear. Wear 12:173–183, 1968.
3. P Engel. Impact Wear of Materials. Tribology Series. New York: Elsevier Science Publishing Co., 1978.
4. R Bayer. Prediction of wear in a sliding system. Wear 11:319–332, 1968.
5. P Clayton, K Sawley, P Bolton, G Pell. Wear behavior of bainitic steels. Proc Intl Conf Wear Materials ASME 133–144, 1987.
6. V Jain, S Bahadur. Material transfer in polymer-polymer sliding. Proc Intl Conf Wear Materials ASME 487–493, 1977.
7. B Mortimer, J Lancaster. Extending the life of aerospace dry-bearings by the use of hard, smooth counterfaces. Proc Intl Conf Wear Materials ASME 175–184, 1987.
8. R Bayer. The influence of hardness on the resistance to wear by paper. Wear 84:345–351, 1983.
9. T Eyre. An introduction to wear. In: D Rigney, W Glaeser, Source Book on Wear Control Technology. ASM, 1978, pp 1–10.
10. R Bayer, J Sirico. The influence of surface roughness on wear. Wear 35:251–260, 1975.
11. D Dowson, J Challen, K Holmes, J Atkinson. The influence of counterface roughness on the wear rate of polyethylene. Proceedings of the Third Leeds–Lyon Symposium on Trib. Guildford, UK: Butterworth Scientific, Ltd., 1976.
12. R Bayer. The influence of lubrication rate on wear behavior. Wear 35:35–40, 1975.
13. E Tsuji, Y Ando. Effect of air temperature and relative humidity on wear of carbon steels and cast irons. Proc Intl Conf Wear Materials ASME 94–99, 1977.
14. A Iwabuchi, K Hori, H Kudo. The effects of temperature, pre-oxidation and pre-sliding on the transition from severe wear to mild wear for s45c carbon steel and sus304 stainless steel. Proc Intl Conf Wear Materials ASME 211–218, 1987.
15. P Blau. Competition between wear processes during the dry sliding of two copper alloys on 52100 steel. Proc Intl Conf Wear Materials ASME 526–533, 1983.
16. J Anderson. High density and ultra-high molecular weight polythenes: Their wear properties and bearing applications. Trib Intl 15(1):43–47, 1982.
17. Tsukizoe T. The effects of surface topography on wear. In: N Suh, N Saka, eds. Fundamentals of Tribology. Cambridge, MA: MIT Press, 1980, pp 53–66.
18. P Blau. Effects of sliding motion and tarnish films on the break-in behavior of three copper alloys. Proc Intl Conf Wear Materials ASME 93–100, 1987.
19. P Schatzberg. Influence of Water and Oxygen in Lubricants on Sliding Wear, Paper presented at 25th ASLE Annual Meeting, 4–8 May 1970.
20. S Kang, K Ludema. The "breaking-in" of lubricated surfaces. Proc Intl Conf Wear Materials ASME 280–286, 1985.
21. Standard Test Method for Ranking Resistance of Plastic Materials to Sliding Wear Using a Block-On-Ring Configuration, ASTM G137.
22. N Saka. Effect of microstructure on friction and wear of metals. In: N Suh, N Saka, eds. Fundamentals of Tribology. Cambridge, MA: MIT Press, 1980, pp 135–172.
23. R Lewis. Paper No. 69AM5C-2, 24th ASLE Annual Meeting, Philadelphia, 1969.
24. P Hurricks. The fretting wear of mild steel from 200 to 500°C. Wear 30:189–212, 1974.
25. S Rhee, K Ludema. Transfer films and severe wear of polymers. Proceedings of the Second Leeds–Lyon Symposium on Trib. Guildford, UK: Butterworth Scientific, Ltd., 1974, pp 11–17.
26. N Eiss, M Bayraktaroglu. The effect of surface roughness on the wear of low-density polyethylene. ASLE Trans 23(3):269–278, 1980.
27. S Hogmark. Wear 31:39, 1975.
28. K Furber, J Atkinson, D Dowson. Proceedings of the Third Leeds–Lyon Symposium on Trib. Guildford, UK: Butterworth Scientific, Ltd., 1976, p 25.

29. P Blau. Competition between wear processes during the dry sliding of two copper alloys on 52100 steel. Proc Intl Conf Wear Materials ASME 526–533, 1983.
30. R Bayer. Wear Analysis for Engineers. Sections 2.2, 2.6.1 and 6.1. HNB Publishing, 2002.
31. R Bayer, T Ku. Handbook of Analytical Design for Wear. New York: Plenum Press, 1964.
32. W Glaeser. A case of wear particle formation through shearing-off at contact spots interlocked through micro-roughness in "adhesive wear". Proc Intl Conf Wear Materials ASME 155–162, 1987.
33. M Sawa, D Rigney. Sliding behavior of dual phase steels in vacuum and in air. Proc Intl Conf Wear Materials ASME 231–244, 1987.
34. J Lancaster, D Play, M Godet, A Verrall, R Waghorne. Paper No. 79-Lub-7, Joint ASME–ASLE Lubrication Conference, Dayton, OH, 10/79.
35. A Atkins, K Omar. The load-dependence of fatigue wear in polymers. Proc Intl Conf Wear Materials ASME 405–409, 1985.
36. D Roshon. Electroplated doamond-composite coating for abrasive wear resistance. IBM J R D 22(6):681–686, 1978.
37. V Venkatesh. Effect of magnetic field on diffusive wear of cutting tools. Proc Intl Conf Wear Materials ASME 242–247, 1977.
38. B Briscoe. Wear of polymers: an essay on fundemental aspects. Trib Intl 14(4):231–243, 1981.
39. B Briscoe, M Stewart. The effect of carbon and glass fillers on the transfers film behavior of ptfe composites. Tribology 1978. Institute of Mechanical Engineers, 1978, pp 19–23.
40. T Tsukizoe, N Obmae. Wear mechanism of unidirectionally oriented fiber-reinforced plastics. Proc Intl Conf Wear Materials ASME 518–525, 1977.
41. R Bayer, P Engel, E Sacher. Impact wear phenomena in thin polymer films. Wear 32:181–194, 1975.
42. S Rhee, K Ludema. Mechanisms of formation of polymeric transfer films. Proc Intl Conf Wear Materials ASME 482–487, 1977.
43. J Theberge. A guide to the design of plastic gears and bearings . Machine Design, 2/5/70, Cleveland, OH: Penton Publ. Co., pp 114–120.
44. P Blau. Friction and Wear Transitions of Materials. Park Ridge, NJ: Noyes Publications, 1989.
45. P Blau. Break-in, run-in, and wear-in. Friction and Wear Transitions of Materials. Chapter 5. Park Ridge, NJ: Noyes Publications, 1989, pp 268–351.
46. A Hollander, J Lancaster. An application of topographical analysis to the wear of polymers. Wear 25(2):155–170, 1973.
47. P Rohatgi, B Pai. Seizure resistance of cast aluminum alloys containing dispersed graphite particles of different sizes. Proc Intl Conf Wear Materials ASME 127–133, 1977.
48. P Heilmann, J Don, T Sun, W Glaeser, D Rigney. Sliding wear and transfer. Proc Intl Conf Wear Materials ASME 414–425, 1983.
49. F Talke. An autoradiographic investigation of material transfer and wear during high speed/low load sliding. Wear 22:69–82, 1972.
50. D Klaffke, K Habig. Fretting wear tests of silicon carbide. Proc Intl Conf Wear Materials ASME 361–370, 1987.
51. R Bayer. Impact wear of elastomers. Wear 112:105–120, 1986.
52. R Bayer, W Clinton, C Nelson, R Schumacher. Engineering model for wear. Wear 5:378–391, 1962.
53. E Rabinowicz. Friction and Wear of Materials. New York: John Wiley and Sons, 1965.
54. K Budinski. Incipient galling of metals. Proc Intl Conf Wear Materials ASME 171–178, 1981.
55. K Bhansali, A Miller. Role of stacking fault energy on the galling and wear behavior of a cobalt base alloy. Proc Intl Conf Wear Materials ASME 179–185, 1981.
56. M Peterson, K Bhansali, E Whitenton, L Ives. Galling wear of metals. Proc Intl Conf Wear Materials ASME 293–301, 1985.
57. P Swanson, L Ives, E Whitenton, M Peterson. Proc Intl Conf Wear Materials ASME 49–58, 1987.

58.   R Waterhouse. Fretting wear. In: P Blau, ed. Friction, Lubrication, and Wear Technology, ASM Handbook. Vol. 18. Materials Park, OH: ASM International, 1992, pp 242–256.
59.   T Kayaba, A Iwabuchi. The fretting wear of 0.45 percent carbon steel and austenitic stainless steel from 20 C up to 650 C in air. Proc Intl Conf Wear Materials ASME 229–237, 1981.
60.   R Waterhouse. Fretting wear. Proc Intl Conf Wear Materials ASME 17–22, 1981.
61.   O Vingsbo, S Soderberg. On fretting maps. Proc Intl Conf Wear Materials ASME 885–894, 1987.
62.   R Waterhouse. The role of adhesion and delamination in the fretting wear of metallic materials. Proc Intl Conf Wear Materials ASME 55–59, 1977.
63.   H Ghasemi, M Furey, C Kajdas. Surface temperatures and fretting corrosion of steel under conditions of fretting contact. Wear 162–164:357–369, 1993.
64.   R Bayer, J Gregory. An engineering approach to vibration induced wear concerns of electrical contact systems. Advance in Electronic Packaging, EEP. Vol. 4–1. ASME, 1993, pp 525–536.
65.   P Kennedy, M Peterson, L Stallings. An evaluation of fretting at small slip amplitudes. In: Materials Evaluation under Fretting Conditions. STP 780. ASTM International, 1982, pp 30–48.
66.   P Kennedy, L Stallings, M Peterson. A study of surface damage at low-amplitude slip. ASLE Trans 27(4):305–312, 1984.
67.   S Soderberg, U Bryggman, T McCullough. Frequency effects in fretting wear. Wear 110:19–34, 1986.
68.   K Johnson. Elastically loaded bodies under tangential forces. Proc Royal Soc (London) A 230:531–548, 1955.
69.   A Wayson. A study of fretting on steel. Wear 7:435–450, 1964.
70.   P Blau, ed. Fretting wear. Friction, Lubrication, and Wear Technology, ASM Handbook. Vol. 18. Materials Park, OH: ASM International, 1992, p 244.
71.   R Midlin. Compliance of elastic bodies in contact. J Appl Mech 16:259–268, 1949.
72.   P Blau, ed. Fretting wear. Friction, Lubrication, and Wear Technology, ASM Handbook. Vol. 18. Materials Park, OH: ASM International, 1992, p 242.
73.   J Magee. Wear of stainless steels. In: P Blau, ed. Lubrication, and Wear Technology, ASM Handbook. Vol. 18. Materials Park, OH: ASM International, 1992, pp 710–724.
74.   W Marscher. Wear of pumps. In: P Blau, ed. Lubrication, and Wear Technology, ASM Handbook. Vol. 18. Materials Park, OH: ASM International, 1992, pp 593–601.
75.   R Bayer. The influence of hardness on the resistance to wear by paper. Wear 84:345–351, 1983.
76.   A Rosenfield. Modelling of dry sliding wear. Proc Intl Conf Wear Materials ASME 390–393, 1983.
77.   R Bayer, J Sirico. Some observations concerning the friction and wear characteristics of sliding systems involving cast ceramic. Wear 16:421–430, 1970.
78.   P Engel, R Bayer. The wear process between normally impacting elastic bodies. J Lub Tech Oct:595–604, 1974.
79.   L Mitchell, C Osgood. Prediction of the reliability of mechanisms from friction measurements. Proceedings of the First European Trib. Congress, Inst. of Mech. Eng., 1973, pp 63–70.
80.   D Godfrey. Diagnosis of wear mechanisms. In: M Peterson, W Winer, eds. Wear Control Handbook. Materials Park, OH: ASME, 1980, pp 283–312.
81.   J Martin. IBM Report No. TR07.586, Poughkeepsie, NY: IBM, 1975.
82.   O Ajayi, K Ludema. Surface damage of structural ceramics: Implications for wear modeling. Proc Intl Conf Wear Materials ASME 349–360, 1987.
83.   S Wilson, T Alpas. Thermal effects on mild wear transitions in dry sliding of an aluminum alloy. Wear 225–229:440–449, 1999.
84.   B Bhushan, B Gupta. Nature of solid surfaces. Handbook of Tribology. Section 3.1–3.3. McGraw-Hill, 1991.
85.   R Bill. Fretting wear of iron, nickel, and titanium under varied environmental conditions. Proc Intl Conf Wear Materials ASME 356–370, 1979.

86. V Jain, S Bahadur. Surface topography changes in polymer-metal sliding. Proc Intl Conf Wear Materials ASME 581–588, 1979.

87. T Kosel. Solid particle erosion. In: P Blau, ed. Friction, Lubrication, and Wear Technology, ASM Handbook.Vol. 18. Materials Park, OH: ASM International, 1992, pp 199–213.

88. F Heyman. Liquid impingement erosion. In: P Blau, ed. Lubrication, and Wear Technology, ASM Handbook. Vol. 18. Materials Park, OH: ASM International, 1992, pp 221–232.

89. X Jin, N Kang. A study of rolling bearing contact fatigue failure by macro-observation and micro-analysis. Proc Intl Conf Wear Materials ASME 205–214, 1989.

90. A Graham, A Ball. Particle erosion of candidate materials for hydraulic valves. Proc Intl Conf Wear Materials ASME 155–160, 1989.

91. S Joffe, C Allen. The wear of pump valves in fine particle quartzite slurries. Proc Intl Conf Wear Materials ASME 167–174, 1989.

92. E Iturbe, I Greenfield, T Chou. The wear mechanism obtained in copper by repetitive impacts. Wear 74:123–129, 1981–82.

93. L Ives, A Ruff. Electron microscopy study of erosion damage in copper. In: W Adler, ed. Erosion: Prevention and Useful Applications. STP 664. ASTM, 1979, pp 5–35.

94. R Bayer, E Hsue, J Turner. A motion-induced sub-surface deformation wear mechanism. Wear 154:193–204, 1992.

95. W Glaeser. Light microscopy. In: P Blau, ed. Lubrication, and Wear Technology, ASM Handbook. Vol. 18. Materials Park, OH: ASM International, 1992, pp 370–375.

96. V Venkatesh. Effect of magnetic field on diffusive wear of cutting tools. Proc Intl Conf Wear Materials ASME 242–247, 1977.

97. M Kar, S Bahadur. Micromechanism of wear at polymer-metal sliding interface. Proc Intl Conf Wear Materials ASME 501–509, 1977.

98. D Durkee, H Cheny. Initial scuffing damage studies in simple sliding contacts. Proc Intl Conf Wear Materials ASME 81–88, 1979.

99. C Lutynski, G Simansky, A McEvily. Fretting fatigue of Ti-6Al-4V alloy. In: S Brown, ed. Materials Evaluation Under Fretting Conditions. STP 780. ASTM International, 1982, pp 150–164.

100. R Reinisch. Fretting wear in magnetic memory disk drives. Proc Intl Conf Wear Materials ASME 581–584, 1991.

101. R Bayer, W Clinton, J Sirico. A note on the application of the stress dependency of wear in the wear analysis of an electrical contact. Wear 7:282–289, 1964.

5
Friction

As defined previously, friction is a force, which occurs between two surfaces, is parallel to the interface, and opposes relative motion between the surfaces, as illustrated in Fig. 5.1. There are three general mechanisms, which are proposed as the basis for friction between two solid surfaces (1,2). These are companions to the fundamental wear mechanisms associated with adhesion, single-cycle deformation, and repeated-cycle deformation. With friction the term abrasion in general refers to single-cycle deformation mechanisms. Similarly, hysteresis is used for the friction mechanisms associated with repeated-cycle deformation. When a lubricant is between the two solid surfaces, a fourth mechanism is introduced, namely viscous losses in the fluid. The magnitude of the friction force, F, between two surfaces can be expressed as follows:

$$F = F_{ad} + F_{ab} + F_{hys} + F_{vis} \tag{5.1}$$

where F_{ad} is the friction associated with adhesion; F_{ab}, abrasion; F_{hys}, hysteresis; F_{vis}, fluid viscosity. Dividing this expression by the normal load between the two surfaces, a corresponding expression for the coefficient of friction, μ, can be obtained

$$\mu = \mu_{ad} + \mu_{ab} + \mu_{hys} + \mu_{vis} \tag{5.2}$$

The basic concepts related to these mechanisms are illustrated in Fig. 5.2. The concept for adhesive friction is that a force is required to shear the bonded junctions that are formed between the two surfaces. Similarly for abrasive friction, a force is required to deform the surface, either elastically or plastically, or to cut a groove or chip. Viscous friction is similar in that a force is required to shear a fluid. In these three cases, friction would be the resistive force to such action that the materials exhibit. For hysteretic friction, the concept is somewhat different. As one surface passes over the other, a stress cycle is produced in the material. Generally, materials are not perfectly elastic and there is a hysteresis effect associated with such a cycle. Energy is dissipated in this cycle and can be related to a friction force through the following:

$$E_{hys} = F_{hys}S \tag{5.3}$$

where E_{hys} is the energy dissipated over a distance of sliding, S. This hysteretic effect can be on a micro or macro-scale. In the former, it is associated with asperity deformation and in the latter, with the deformation of the overall or gross geometry of the contacting bodies.

The wear counterpart to each of these friction mechanisms is evident. Adhesive wear occurs with adhesive friction when the shearing of the junctions occurs other than at the original interface. In abrasive friction, wear occurs when there is plastic deformation or

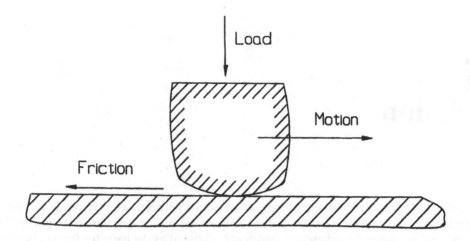

$$\text{Coefficient of Friction} = \frac{\text{Load}}{\text{Friction}}$$

Figure 5.1 Friction between two surfaces.

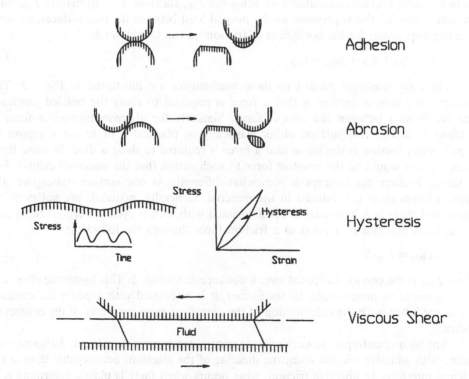

Figure 5.2 Sources of friction.

chip formation. Repeated-cycle deformation wear results from the accumulation of plastic strain associated with the stress cycle, culminating in progressive plastic deformation, crack formation, and crack propagation.

This consideration of hysteretic friction points out a significant aspect of friction, namely that it results in the dissipation of energy. The energy associated with friction is dissipated in two general ways. The vast majority of the energy is dissipated as heat (3–5). A much smaller amount is associated with material loss or deformation, that is, wear. Most estimates indicate that well over 90% of the energy dissipated in friction goes in the form of heat energy.

Models have been proposed for these friction mechanisms but are typically limited in applicability. Generally, all of the models indicate a more complex situation than indicated by the postulations of da Vinci and Amontons (ca. 1500 and 1700, respectively) which frequently are used in engineering. These statements, commonly referred to as Amontons' Laws of Friction, may be summarized as: (1) the friction force is proportional to the normal load, (2) the friction force is independent of the apparent area of contact. The current models, as well as experimental data, indicate that these conclusions should be viewed as approximations with a limited range of applicability. This view can be illustrated by the consideration of some simple models for friction force and the coefficient of friction and some examples of observed behavior. For simplicity, only dry or unlubricated surfaces will be considered at this point. Friction behavior under lubricated conditions will be discussed in the section on lubrication. A model used for paper can be used as a way of illustrating these general models (6).

For adhesion, the general concept is that the F_{ad} is given by

$$F_{ad} = sA_r \tag{5.4}$$

where s is the shear strength of the junctions, and A_r is the real area of contact (7). s is a property of the material system at the interface and is influenced by the same parameters as discussed for adhesive wear (e.g., oxides, cleanliness of the surface, solubility, material strength properties). As was discussed in the sections on wear, A_r can be affected by material properties, asperity distribution, and contact geometry. For example, in the case of a sphere pressed against a plane, a general relationship for A_r is of the following form:

$$A_r = CR^n P^m \tag{5.5}$$

where C is a material parameter, R is the radius of the sphere, and P is the normal load (8,9). The exponents, n and m, are positive and depend on both the nature of the stress system at the junctions and the asperity distribution. For plastic deformation, $n = 0$ and $m = 1$. For elastic deformation, $n > 0$ and m is between 0 and 1 for relatively simple asperity distributions but can be greater than 1 for some complex asperity distributions. The expression for the coefficient of friction then has the following form:

$$\mu_{ad} = K_{ad} R^n P^{m-1} \tag{5.6}$$

where C and s are combined into K_{ad}. Implicit in this relationship is the additional dependencies on load and other parameters as a result of their ability to influence tribosurfaces. These would be contained in K_{ad}.

Abrasive friction can be illustrated by considering a cone of included angle Φ plowing through a softer surface. For plastic deformation, the force required to do this, F', is given by

$$F' = \frac{\cot \Phi P'}{\pi} \tag{5.7}$$

where P' is the load on that asperity. For an array of such asperities supporting a total load, P, F_{ab} can be expressed as

$$F_{ab} = K_{ab}\overline{\cot \Phi}\, P \tag{5.8}$$

where

$$K_{ab}\overline{\cot \Phi}\, P \equiv \pi^{-1} \sum \cot \Phi_i P_i \tag{5.9}$$

In this expression, $\overline{\cot \Phi}$ reflects the average sharpness of the asperities, while K_{ab} is the parameter accounting for the complete description of the asperity distribution. The coefficient of friction is then

$$\mu_{ab} = K_{ab}\overline{\cot \Phi} \tag{5.10}$$

If the cone produced only elastic deformation, there would still be a friction force and the form of the expression for the coefficient of friction would be similar. However, in this case, the friction force would be generated by hysteresis. Equation (5.10) would be modified by a factor ε, which is the ratio of the energy lost to the energy required for the deformation (10). This similarity of form implies that Eq. (5.10) is also appropriate for more realistic material behavior, where the deformation contains both an elastic and plastic portion.

Hysteresis can also be associated with the stress system associated with the macro-geometry of the contact. In this case, the coefficient of friction is proportional to the stress level (11). In the elastic contact between a sphere and a plane, the stress level is proportional to the $P^{1/3}$ and inversely proportional to $R^{2/3}$, where R is radius of the sphere. The friction force, F_{hys}, would be

$$F_{hys} = K_{hys}P^{4/3}R^{2/3} \tag{5.11}$$

and the coefficient of friction would be

$$\mu_{hys} = K_{hys}P^{1/3}R^{-2/3} \tag{5.12}$$

Assume that a fraction of the load, α, is supported by junctions at which adhesion takes place and the remaining fraction by junctions at which deformation occurs, that is by adhesion and single-cycle deformation, respectively. For such a situation, the general expression for the coefficient of friction for a rough sphere sliding on a plane is

$$\mu = K_{ad}\alpha^m P^{m-1} R^n + K_{ab}\overline{\cot \Phi}(1 - \alpha) + K_{hys}P^{1/3}R^{-2/3} \tag{5.13}$$

The observed behavior of the coefficient friction between hardened steel and sheets of paper, supported by a steel platen, shows the same type of dependencies indicated by this equation. These data are shown in Figs. (5.3–5.5). These data show the coefficient of friction to be dependent on the roughness of the sphere, the radius of the sphere, and the normal load. However, the relationships appear more complex than indicated by Eq. (5.13). A dependency on paper thickness is also evident, which modifies the influence of load and radius. This effect can be attributed to the influence that the thickness of the paper layer has on the nature of the stress system developed in the paper. The thinner the paper layer, the greater is its apparent stiffness. This is analogous to the behavior for friction of a layered metallic surface (12,13).

The simple models for the abrasion and adhesion components of friction imply that under rolling their effects should be eliminated (14). Consequently, rolling friction tests

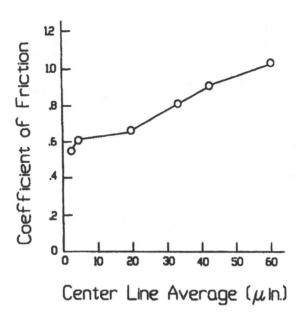

Figure 5.3 Influence of roughness on the coefficient of friction for sliding between steel and paper. (From Ref. 6.)

were conducted to estimate the significance of hysteresis in the overall behavior and the results are shown in Fig. 5.6. It can be seen that while a significant load dependency is indicated that is somewhat stronger than that suggested by Eq. (5.13), actual friction is much lower than for sliding. This suggests that the major contributors to the sliding friction for this system are adhesion and abrasion. Furthermore, the actual data suggest that m and n

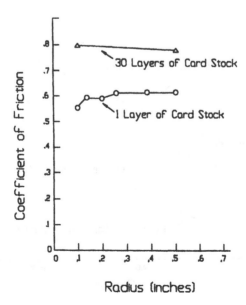

Figure 5.4 Influence of the radius of a steel slider on the coefficient of friction for sliding between steel and paper. (From Ref. 6.)

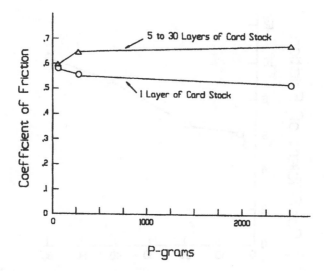

Figure 5.5 Influence of load on the coefficient of friction for sliding between steel and paper. (From Ref. 6.)

of Eq. (5.6) are close to 1 and 0, respectively. The data also show that size is a significant factor.

   While this example of frictional behavior indicates that load and geometry can influence friction, it also shows that Amontons' Laws are also approximately followed, that is, the coefficient of friction is independent of load and geometry. At least over limited ranges, this situation is true for most material systems, especially those that exhibit low hysteresis. This would include metals, ceramics, and the more rigid polymers or plastics.

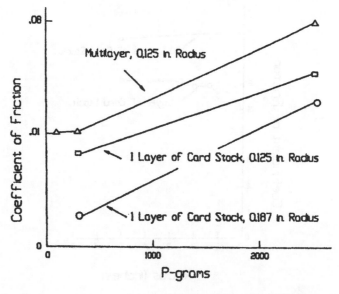

Figure 5.6 Influence of load on the coefficient of friction for rolling between a steel sphere and paper. (From Ref. 6.)

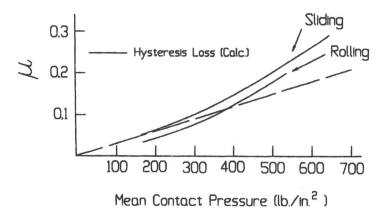

Figure 5.7 Coefficients of friction for sliding and rolling between a steel sphere and a well-lubricated rubber surface. (From Ref. 26.)

Materials, which tend to exhibit high hysteresis, such as rubbers or elastomers, usually exhibit a greater effect of load and geometry on the coefficient of friction. Examples of this behavior are shown in Figs. 5.7 and 5.8.

While the above is generally true, there are examples in which load and geometry significantly influence friction, independent of a hysteretic effect. Sliding over a woven surface provides an example of the influence of geometry and load. Data for such a situation are shown in Fig. 5.9. Metallic and ceramic sliding systems can also show an influence of load on the coefficient of friction. This is generally related to the relationship between oxide formation and sliding parameters (15,16). The coefficient of friction for unlubricated, self-mated copper is shown in Fig. 5.10 as a function of load. Two regimes

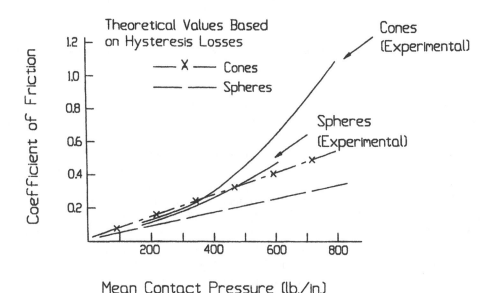

Figure 5.8 Coefficients of friction for spheres and cones sliding on lubricated rubber. (From Ref. 26.)

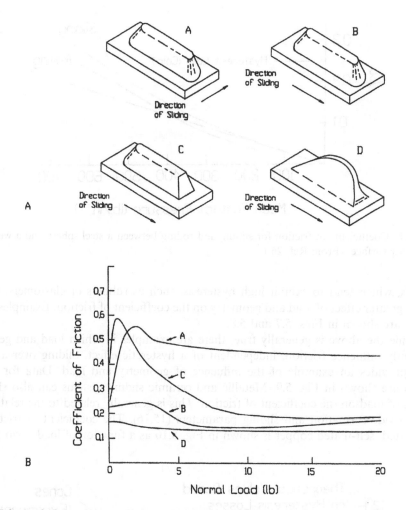

Figure 5.9 Variations of the coefficient of friction for sliding between steel and a lubricated fabric with different steel geometries. (From Ref. 27.)

of friction behavior are evident as function of load. In each, the coefficient of friction is independent of load, indicating that within those regimes Amontons' Law is applicable. In this particular case, the formation of different oxides in the two regions is the explanation for this behavior. This is also an example of a transition in friction behavior as a result of tribosurface modification, similar to the transitions in wear behavior.

As shown by the development of Eq. (5.13), roughness can influence the coefficient of friction through abrasive and hysteresis effects (17). In addition, roughness can affect friction behavior through the adhesive mechanisms. In fact, this can be a stronger or more pronounced effect than those associated with the other two mechanisms, tending to increase the coefficient of friction, as surfaces become smoother. The coefficient of friction for clean, self-mated copper surfaces is shown in Fig. 5.10 as a function of roughness. It can be seen that friction increases much more rapidly for smoother surfaces than it does for rougher surfaces. The explanation for this is that as the surfaces become smoother, the real area increases rapidly and tends to become independent of the load. Perfectly smooth, flat surfaces result in the real area of contact being equal to the apparent area. Under these

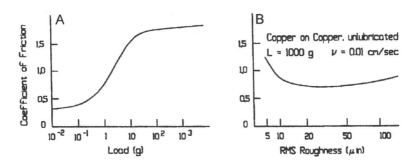

Figure 5.10 Coefficient of friction for unlubricated sliding between Cu surfaces. ("A" from Ref. 28; "B" from Ref. 29.)

conditions, adhesion can be extremely large, with the contact being one large junction. Consequently, there would be a significant increase in the adhesion term, as the surfaces become smoother. This concept is illustrated in Fig. 5.11.

In the case of rubber, smooth surfaces also introduce additional aspects to friction. With rubbers, Schallamach waves or waves of detachment occur and can be the primary contributor to the friction between rubber or elastomers and smooth surfaces (5,18). Fig. 5.12 illustrates this. In this situation, there is strong local adhesion between the rubber surface and the counterface. As sliding occurs, local regions remain attached, at least up to a certain point. Then the bond is broken and the rubber snaps forward. At some point, it again adheres and the process repeats. This process can be viewed as waves, which propagate across the rubber surface. Examinations of sliding contacts and wear scar morphology suggest this behavior as well. This is shown by the examples in Fig. 5.13. In this case, the friction force is associated with adhesion as well as with hysteresis losses in the stretching of the rubber. The following equation has been proposed for this mechanism (5):

$$\mu = \frac{2G^*}{\sigma[(2/\varepsilon_t) + 1]^{1/2}} \tag{5.14}$$

where ε_t is the compressive strain; $G*$, the loss portion of the complex shear modulus; and σ_{ij} the mean pressure.

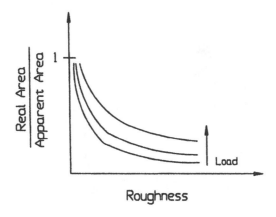

Figure 5.11 Changes in the ratio of the real and apparent areas of contact as a function of load and roughness.

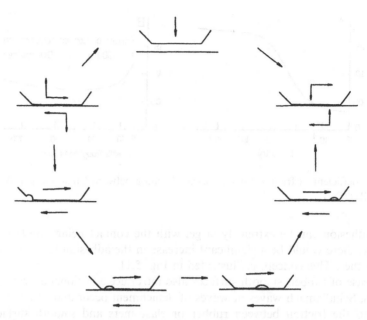

Figure 5.12 Illustration of the propagation of Schallamach waves across a rubber surface (truncated cone in figure) during sliding on a flat surface. (From Ref. 18.)

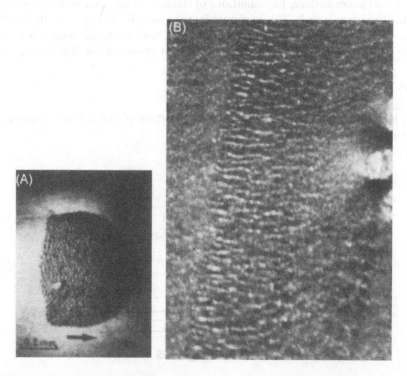

Figure 5.13 Examples of the wave-like wear scar morphology frequently observed on elastomer surfaces as a result of sliding. ("A" from Ref. 30, reprinted with permission from ASTM.)

While the simple models used to describe the mechanisms associated with friction do not explicitly indicate a dependency on sliding velocity, most systems exhibit some velocity effect. The effect is relatively mild for most engineering materials as changes of several orders of magnitude in sliding velocity might result in less than a factor of 2 change in friction. An example is shown in Fig. 5.14. In this figure, the coefficient of friction for unlubricated sliding between steel surfaces is plotted as a function of speed. There can be several reasons for a dependency on speed and the specific reason is usually related to the material or materials involved. For materials, which exhibit creep, such as soft metals and polymers, it is usually associated with viscoelastic behavior of such materials. For such materials, friction usually achieves a maximum value in a particular range of velocity. For other types of materials, friction generally tends to decrease with sliding. Melting and softening at higher speeds can also be a factor, as in the case with polymers, and oxide formation can be a factor in metal systems, as is the case with the data shown in Fig. 5.15. Junction growth phenomena can also contribute to this decrease in friction. This is because junction size tends to increase with time under shear (19). There will be less time for growth at higher sliding speeds. Smaller

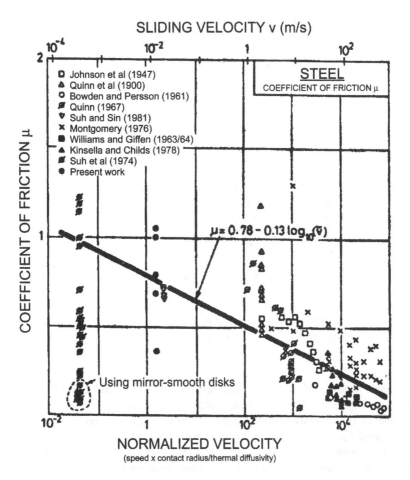

Figure 5.14 Variation in the coefficient of friction with sliding velocity for unlubricated steel. (From Ref. 31, reprinted with permission from Elsevier Science Publishers.)

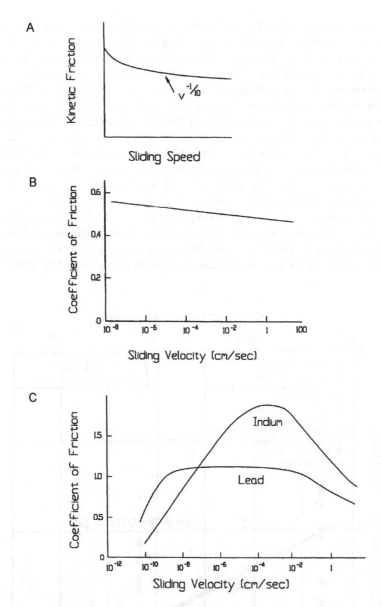

Figure 5.15 Effects of sliding speed on the coefficient of friction for several systems. "A", general trend; "B", Ti/Ti unlubricated; and "C", steel against Pd and In. (From Ref. 31.)

junctions result in less real area of contact and a lower adhesion contribution to friction. The graphs in Fig. 5.15 show some examples of this type of behavior.

Because of the adhesive component of friction, friction behavior is very sensitive to surface film and layers, particularly in unlubricated situations. The effect of oxide formation on the coefficient of friction and the effect of humidity illustrate this. Another example is surface contamination from handling or exposure to contaminating environments. With metals, this can often reduce the coefficient of friction from a value near 1 to 0.3–0.6.

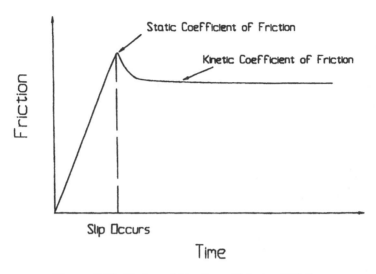

Figure 5.16 Static and kinetic coefficients of friction.

Frequently, the friction associated with the initiation of sliding and the friction associated with maintaining uniform motion are different. In terms of the coefficient of friction, the value associated with the initiation is referred to as the static coefficient of friction; the value for maintaining motion is the kinetic coefficient of friction. Generally, the static coefficient tends to be higher than the kinetic coefficient as a result of increased junction growth that can occur under static conditions. This is the same concept as that associated with the velocity dependency. An example of this type of friction behavior is shown in Fig. 5.16. While this behavior is common, the difference between static and kinetic coefficients can often be negligible and exceptions to this behavior are often encountered.

Before concluding the consideration of friction, the relationship between friction and wear must be discussed. From the previous discussions on the origins of friction, frictional behavior, and the prior treatment of wear, it can be seen that both friction and wear are sensitive to the same parameters and the same general type of phenomena. This frequently is an aid when addressing wear problems. Because of this common dependency, changes to tribosurfaces that result in wear transitions frequently result in frictional changes as well. This is illustrated in Fig. 5.17. As a result, the monitoring of friction behavior during wear tests can aid in the identification of wear transitions. While this is the case, the same trends should not be assumed for both these phenomena. Material systems with higher friction do not necessarily have higher wear. Examples of this can be seen in Table 5.1 and Fig. 5.18. Another example is provided in the case of irradiated PTFE. In this case, increased radiation doses tend to result in increased friction but lower wear rates (20).

One way of understanding this distinction between friction and wear trends is by considering the energy dissipated by the system. Friction can be related to the total energy dissipated. This energy can be considered to consist of two parts, heat energy and wear energy. While the portion of the total energy going into heat generally predominates, the ratio between these two forms can vary between tribosystems and for different wear mechanisms. Consequently the same trends cannot be assumed for both phenomena.

Nonetheless, friction and wear are not independent. Wear can lead to surface modifications, which influence friction, such as film formation and roughness changes.

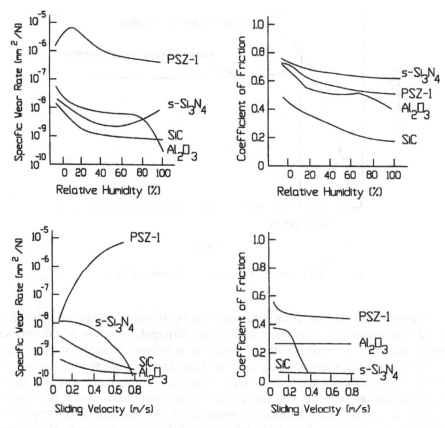

Figure 5.17 Changes in the coefficient of friction and wear rate of several self-mated ceramic couples as a function of humidity and speed. (From Ref. 33.)

Table 5.1 Unlubricated Sliding Friction Coefficients and Wear

Tribosystem	μ	Wear (μ_{in})	
		Sphere	Flat
52100 Sphere			
302 SS flat	1.02	0	8
303 SS flat	0.79	0	5
410 SS flat	0.64	0	20
1018 Steel	0.80	0	4
4150 Steel	0.67	0	2
112 Aluminum	1.08	0	82
220 Aluminum	0.79	0	35
Brass sphere			
410 SS flat	0.62	200	0
440 SS flat	0.72	72	0

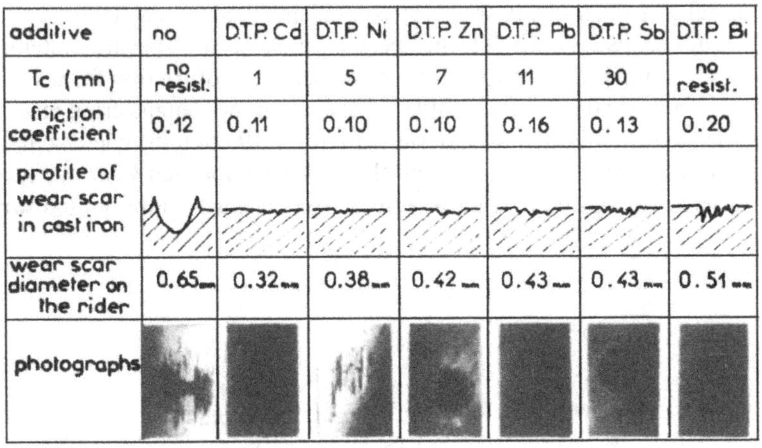

additive	no	D.T.P. Cd	D.T.P. Ni	D.T.P. Zn	D.T.P. Pb	D.T.P. Sb	D.T.P. Bi
Tc (mn)	no resist.	1	5	7	11	30	no resist.
friction coefficient	0.12	0.11	0.10	0.10	0.16	0.13	0.20
profile of wear scar in cast iron							
wear scar diameter on the rider	0.65 mm	0.32 mm	0.38 mm	0.42 mm	0.43 mm	0.43 mm	0.51 mm
photographs							

Figure 5.18 The influence of different lubricant additives on the coefficient of friction and wear of a sliding 52100 steel/cast iron couple lubricated by paraffin oil. (From Ref. 21, reprinted with permission form ASME.)

Friction, through a heating effect, can influence oxide formation and affect material properties, which in turn can influence wear behavior. In addition, friction modifies the contact stress system by introducing a shear or traction component, which can also be a factor in wear behavior (3,4,18,20–24). Because of these aspects, friction and wear must be generally

Table 5.2 Coefficients of Friction

	μ	
Tribosystem	Unlubricated	Lubricated
Sliding		
Steel/steel	0.6–0.8	0.1–0.3
Steel/stainless steel	0.7–1.2	0.1–0.3
Steel/Ni alloys	0.7–1.3	0.1–0.3
Steel/Cu alloys	0.7–1.2	0.15–0.3
Steel/Al alloys	0.8–1.4	0.1–0.3
Stainless steel/stainless steel	0.9–1.5	0.1–0.2
Acetal/steel	≈0.35	≈0.15
PTFE filled acetal	0.2–0.3	
Nylon/steel	0.4–0.6	0.15–0.25
Graphite filled nylon/steel	≈0.6	≈0.25
MoS_2 filled nylon/steel	≈0.6	≈0.25
PTFE filled nylon/steel	0.1–0.2	
PTFE/steel, low speed	0.05–0.08	0.05–0.08
PTFE/steel, high speed	≈0.3	≈0.3
Filled PTFE steel	0.09–0.12	
Polyurethane/nylon	1–1.5	
Isoprene/steel	3–10	2–4
Polyurethane/nylon		0.5–1
Rolling		
Steel/steel	≈0.001	≈0.001

considered as related phenomena, but not equivalent phenomena. However, direct correlation between the two is possible in specific cases or for specific systems.

In Tables 5.2 and 5.3, typical values of the coefficient of friction for a variety of systems are given as a general reference. Table 5.2 is for common engineering materials, while Table 5.3 is for medical and dental materials. Figure 5.19 contains coefficients of friction for different woods. It is interesting to note that while normalized wear coefficients range over many orders of magnitude, friction values cover a much more limited range.

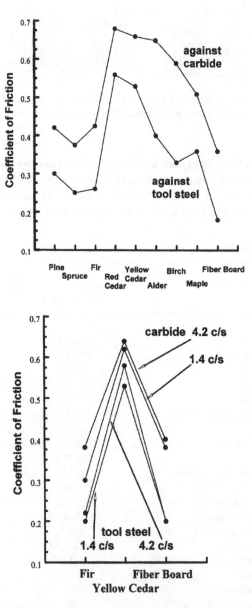

Figure 5.19 Coefficient of friction of various woods sliding against carbide and steel counterfaces. (From Ref. 34.)

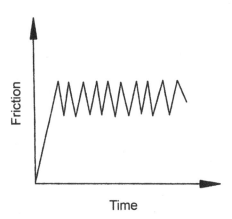

Figure 5.20 Stick-slip behavior.

There is one additional aspect of friction that needs to be considered. This is stick-slip. This is manifested in friction traces of the type illustrated in Fig. 5.20 and frequently as noise in a sliding system. The peaks in these traces associated with stick-slip give the static coefficient of friction. The lower value is dependent on the dynamic characteristics of the system, material properties, and the measurement system used to record the data, and therefore does not provide a measurement of the friction. Two conditions are required for the occurrence of this phenomenon. One is a variable coefficient of friction, the other is

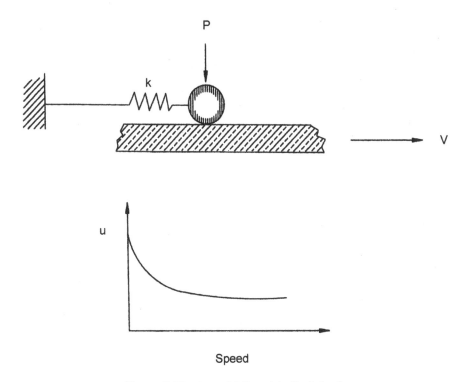

Figure 5.21 A model for stick-slip behavior.

Table 5.3 Coefficient of Friction for Medical, Dental, and Biological Materials

	Coefficient of friction	
Material couples	Dry	Wet
Amalgam on:		
Amalgam	0.19–0.35	
Bovine enamel		0.12–0.28
Composite resin		0.10–0.18
Gold alloy		0.10–0.15
Porcelain	0.06–0.12	0.07–0.15
Bone on:		
Metal (bead-coated)	0.50	
Metal (fiber mesh-coated)	0.60	
Metal (smooth)	0.42	
Bovine enamel on:		
Acrylic resin	0.19–0.65	
Amalgam	0.18–0.22	
Bovine dentin	0.35–0.40	0.45–0.55
Bovine enamel	0.22–0.60	0.50–0.60
Chromium–nickel alloy	0.10–0.12	
Gold	0.12–0.20	
Porelain	0.10–0.12	0.50–0.90
Composite resin on:		
Amalgam	0.13–0.25	0.22–0.34
Bovine enamel		0.30–0.75
Gold alloy on:		
Acrylic	0.6–0.8	
Amalgam	0.15–0.25	
Gold alloy	0.2–0.6	
Porcelain	0.22–0.25	0.16–0.17
Hydrogel-coated latex on:		
Hydrogel		0.054
Laytex on:		
Glass		0.47
Hydrogel		0.095
Metal (bead-coated) on:		
Bone	0.54	
Metal (fibre mesh-coated)		
Bone	0.58	
Metal (smooth) on:		
Bone	0.43	
Prosthetic tooth material		
Acrylic on acrylic	0.21	0.37
Acrylic on porcelain	0.23	0.30
Porcelain on acrylic	0.34	0.32
Porcelain on porcelain	0.14	0.51

Source: Ref. 25

elasticity in the sliding system. The basic concept is that elasticity in the system allows variations in friction to produce oscillations between the two members. If the system were rigid, no oscillations would start, or if the friction were constant, there would be no variable force to initiate the oscillation. This can be illustrated by considering a simple example in which the static coefficient of friction is higher than the dynamic coefficient. Consider the situation shown in Fig. 5.21. As the flat begins to move, the ball follows until the stored elastic energy in the spring overcomes the friction. At this point, the ball becomes free and moves relative to the flat. This results in reduced friction. At some point of time, determined by the stored energy in the spring and the energy dissipated by friction, the ball will stop moving relative to the flat. At this point, the cycle will repeat itself. There can be many reasons for the instability of the coefficient of friction. However, stick-slip is frequently observed under conditions, which favor adhesion, such as clean, dry surfaces, marginal lubrication, etc.

REFERENCES

1. E Rabinowicz. Friction and Wear of Materials. New York: John Wiley and Sons, 1965.
2. F Bowden, D Tabor. The Friction and Lubrictation of Solids. New York: Oxford U. Press. Part I, 1964, and Part II, 1964.
3. M Moore. Energy dissipated in abrasive wear. Proc Intl Conf Wear Materials ASME 636–638, 1979.
4. N Suh, P Sridbaran. Relationship between the coefficient of friction and the wear rate of metals. Wear 34(3):291–300, 1975.
5. D Moore. Proceedings of the Third Leeds–Lyon Symposium on Trib. Guildford, UK: Butterworth Scientific, Ltd., 1976, p 114.
6. R Bayer, J Sirico. The friction characteristics of paper. Wear 17:269–277, 1971.
7. E Rabinowicz. Friction. Friction and Wear of Materials. New York: John Wiley and Sons, 1965, pp 62–64.
8. J Migoard. Proc Phys Soc 79:516, 1962.
9. F Bowden, D Tabor. Polymeric materials. The Friction and Lubrication of Solids. Part II. New York: Oxford University Press, 1964, pp 223–225.
10. F Bowden, D Tabor. The friction of elastic solids. The Friction and Lubrication of Solids. Part II. New York: Oxford University Press, 1964, p 252
11. F Bowden, D Tabor. The friction of elastic solids. The Friction and Lubrication of Solids. Part II. New York: Oxford University Press, 1964, pp 242–254.
12. R Bayer. A general model for sliding wear in electrical contacts. Wear 162–164:913–918, 1993.
13. P Engel, E Hsue, R Bayer. Hardness, friction and wear of multiplated electrical contacts. Wear 162–164:538–551, 1993.
14. F Bowden, D Tabor. The friction of elastic solids. The Friction and Lubrication of Solids. Part II. New York: Oxford University Press, 1964, pp 252.
15. E Rabinowicz. Friction. Friction and Wear of Materials. New York: John Wiley and Sons, 1965, pp 58–59.
16. D Buckley, K Miyoshi. Friction and wear of ceramics. Wear 100:333–339, 1984.
17. M Moore, P Swanson. The effect of particle shape on abrasive wear: A comparison of theory and experiment. Proc Intl Conf Wear Materials ASME 1–11, 1983.
18. G Briggs, B Briscoe. Wear 57:269, 1979.
19. E Rabinowicz. Adhesive wear. Friction and Wear of Materials. New York: John Wiley and Sons, 1965, pp 133–134.
20. B Briscoe, Z Ni. The friction and wear of γ-irradiated polytetrafluoroethylene. Wear 100: 221–242, 1984.

21. J Martin, J Georges, G Meille. Boundary lubrication with dithiophospates: Influence of lubri-cation of wear of the friction interface steel/cast iron. Proc Intl Conf Wear Materials ASME 289–297, 1977.

22. K Muller. Prediction of the occurrence of wear by friction force-displacement curves. Wear 34:439–448, 1975.

23. L Mitchell, C Osgood. Prediction of the reliability of mechanisms from friction measurements. Proceedings of the First European Trib. Congress, Inst. of Mech. Eng., 1973, pp 63–70.

24. H Shimura, Y Tsuyad. Effects of atmosphere on the wear rate of some ceramics and cermets. Proc Intl Conf Wear Materials ASME 452–461, 1977.

25. http://www.lib.umich.edu/libhome/Dentistry.lib/Dental_tables/Coeffric.html, Dentis-try Library, U. of Michigan.

26. F Bowden, D Tabor. The friction of elastic solids. The Friction and Lubrication of Solids. Part II. New York: Chapter XIV Oxford University Press, 1964, pp 242–276.

27. Bayer R, Sirico J. Comments on the frictional behavior between a print character and a carbon ribbon. Wear 11:78–83, 1968.

28. J Whitehead. Proc Roy Soc A 201:109, 1950.

29. E Rabinowicz. Friction. Friction and Wear of Materials. New York: John Wiley and Sons, 1965, p 62.

30. M Barquins. Adherence, friction and wear of rubber-like materials. In: R Denton, M Kesha-van, eds. Wear and Friction of Elastomers. STP 1145, ASTM, 1992, pp 82–113.

31. S Lim, M Ashby. Wear-mechanism maps. Acta Metal 35(1):1–24, 1987.

32. E Rabinowicz. Friction. Friction and Wear of Materials. New York: John Wiley and Sons, 1965, p 60.

33. S Sasaki. The effects of surrounding atmosphere on the friction and wear of alumina, zirconia, silicon carbide, and silicon nitride. Proc Intl Conf Wear Materials ASME 409–418, 1989.

34. P Ko, H Hawthorne, J Andiappan. Tribology in secondary wood machining. In: S Bahadur and J Magee, eds. Wear Process in Manufacturing. STP 1362, ASTM, 1999.

6
Lubrication

A lubricant is any substance, fluid or solid, which when placed between two surfaces reduces either the friction between the two surfaces or the wear of either surface. Consistent with the fact that wear and friction are distinct phenomena, a lubricant does not necessarily have to do both or be effective to the same degree for each of these phenomena. This aspect is demonstrated by the following examples.

The effect of different oils on the coefficient of friction and the wear behavior for several combination of sliding metal pairs is presented in Table 6.1. It can be seen that minimum friction and minimum wear are not necessarily obtained with the same lubricant. It can also be seen that all the lubricants do reduce both friction and wear but that the degree of improvement can be significantly different. Since the coefficient of friction for these systems ranges from approximately 0.5 to 1.0 without the use of a lubricant, the coefficient of friction is reduced by a factor of 1/2 to 1/4 with the use of these lubricants. The reduction in wear is generally far more pronounced, typically being an order of magnitude or more.

While it is generally found that both wear and friction are simultaneously reduced by the use of a lubricant, as illustrated by the data shown in Table 6.1, it is not always the case. It is possible that a lubricant may decrease friction while increasing wear. An example of this is one in which the wear of primary system is controlled by transfer or tribofilm formation film formation. As was discussed in the section on wear phenomena, the addition of a lubricant in such a system can increase the wear by inhibiting the formation of the film. However, the lubricant can still be effective in reducing the adhesive component of friction for the basic pair of materials. Data illustrating this are shown in Table 6.2. For most of these systems, transfer films were observed to form for unlubricated sliding conditions but not under lubricated conditions. In these cases, the data show that the wear increased with the use of the inks while the coefficient of friction generally reduced. For a material pair which did not form a transfer film under the same unlubricated sliding conditions, lubrication by these inks reduced both friction and wear.

Systems, which have low coefficient of friction under unlubricated conditions (< 0.1), can sometimes exhibit the opposite behavior, that is the wear is reduced but the friction increases. In these cases, the viscous losses in the lubricant can significantly contribute to the overall friction. The significance of such a contribution to the coefficient of friction can be illustrated by the case of 302 stainless steel sliding on polytetrafluoroethylene (PTFE). In tests with a ball–plane friction and wear apparatus, the coefficient of friction was measured to be 0.09 without lubrication. With a low viscosity paraffin oil, the coefficient increased to 0.12 and with a higher viscosity paraffin oil, to 0.15 (1). Part of this increase can also be related to the effect that the use of a lubricant has on the formation of transfer films.

Table 6.1 Effect of Three Different Hydrocarbon Lubricants on the Wear and Friction of Different Metal Couples[a]

| | Lubricant | | | | | | Minimum reduction in system wear with lubrication | μ without lubrication |
| | A | | B | | C | | | |
Couple	h (μ in)	μ	h (μ in)	μ	h (μ in)	μ		
52100/*415*	12	0.15	0	0.13	5	0.17	5×10^{-4}	0.97
52100/*440*	6	0.12	8	0.13	5	0.18	5×10^{-4}	0.66
52100/*1060*	77	0.20	26	0.20	38	0.32	0.02	0.73
52100/*phosphor bronze*	11	0.23	0	0.16	0	0.18	3×10^{-3}	0.74
302/*1060*	0	0.15	10	0.15	21	0.16	1×10^{-3}	0.88
302/*220 aluminum*	10	0.18	0	0.17	16	0.25	2.5×10^{-3}	0.92
Brass/*1055*	110	0.20	32	0.19	110	0.25	0.01	0.69
Brass/220 Aluminum	90	0.14	108	0.15	90	0.19	0.5^{b}	0.95
Brass/*Monel C*	21	0.22	21	0.21	108	0.28	5×10^{-3}	0.85

[a]Data from reciprocating ball–plane tests, using different test loads for the different material couples. h is the depth of the wear scar on the wearing member, which is italicized.
[b]With lubrication the brass wears; without lubrication the aluminum wears.

Another example of a tribosystem in which lubrication can reduce wear but increase friction is rolling bearings. Rolling contacts fall into this category of low, unlubricated friction. Coefficients of friction for rolling are typically less than 0.1 (2). With ball and roller bearings, lower friction is usually obtained without the use of oil or grease; however, life and loading capacity are generally increased by the use of a lubricant (3–5). The

Table 6.2 The Effect of Lubrication on Wear and Friction for Several Plastics Against a 302 Stainless Steel Slider

Plastic	Wear depth (cm)	μ	Evidence of film formation
PPS			
Dry	4×10^{-3}	0.5	No
Lubricated	1.5×10^{-4}	0.16	No
PPS + MoS_2 + Sb_2O_3			
Dry	8×10^{-4}	0.5	Yes
Lubricated	1.5×10^{-3}	0.35	No
PPS + Glass + PTFE			
Dry	1×10^{-4}	0.15	Yes
Lubricated	1.5×10^{-4}	0.16	No
Acetal + PTFE			
Dry	1×10^{-4}	0.14	Yes
Lubricated	1.5×10^{-4}	0.12	No
Polyester + Graphite + PTEF			
Dry	4×10^{-5}	0.18	Yes
Lubricated	8×10^{-5}	0.16	No

Lubricant was a non-abrasive aqueous-based electrostatic ink.
Source: Ref. 10.

increase in friction results from the viscous flow of the lubricant. The improvement in wear behavior is related to reduction in sliding wear that lubrication produces. In rolling contacts, there is always some sliding, if only on a micro-scale as a result of deformation that occurs (2). As a result, some sliding wear is involved and adhesion can play a role in rolling wear behavior. A lubricant will tend to inhibit the adhesive contribution to the wear, limit and reduce surface traction (stresses), and provide separation, all of which tend to reduce sliding wear.

The ability of a material to lubricate is a function of its thickness and indirectly the amount. There is a minimum thickness or amount required for maximum effectiveness. This behavior with fluid lubricants is illustrated in Figs. 6.1–6.3. With solid lubricants, the behavior is somewhat different. There is often an optimum thickness range, as illustrated in Fig. 6.4.

Lubricants can be liquids, gases, or solids (6–8). Examples of solid lubricants are PTFE, molydisulfide (MoS_2), graphite, and soft metals, such as lead. Oils are examples of liquid lubricants but this category is not necessarily limited to them. For example, water coolants, refrigerants, and even inks, can provide some lubrication (6,9,10). Also, greases, which are mixtures of oils and thickening agents, are generally considered to be liquid lubricants. Greases are thought to function in two ways. One is as a very viscous fluid and the other is as a reservoir for the oil component. For this latter mode, the concept is that oil leaches out of the grease that surrounds the contact to cover the contact zone (11). The class of liquid lubricants includes a wide range of materials with significantly different rheological properties. As classes, liquid and gaseous lubricants generally have the property of self-healing, which is the tendency to flow back into the region of contact, replenishing any of the lubricant that is displaced during the wearing action. Solid lubricants do not have that ability and, as a result, the durability of the solid lubricant is often a

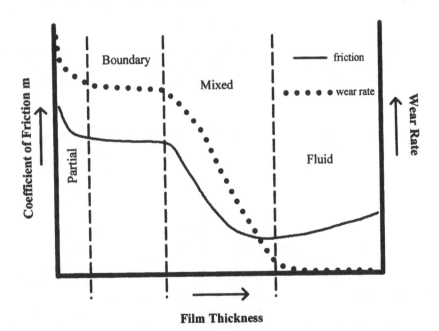

Figure 6.1 Behavior of wear rate and friction coefficient as a function of the thickness of liquid lubricant.

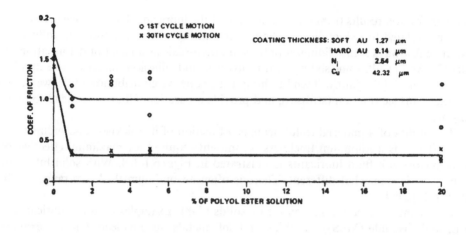

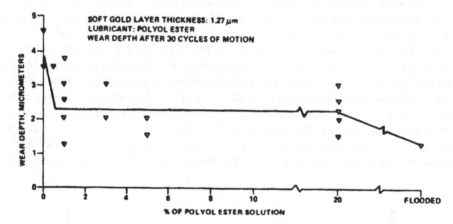

Figure 6.2 Effect of the amount of lubricant on the friction and wear of an electrical contact. The contacts were coated with the lubricant diluted with different amounts of a solvent, which after evaporation of the solvent, resulted in different amounts of residual lubricant on the surfaces. (From Ref. 33.)

factor in engineering situations. Local wear-through of the solid lubricant layer can result in degraded or complete loss of lubrication in the contact region. When this type of lubricant is used, an underlying friction or wear problem is translated into a wear concern with the solid lubricant itself.

This is illustrated by the behavior of a MoS_2 coating used in a band printer application (12). When present, the coating significantly reduced the wear of the interface. A sharp increase in wear occurred with the local depletion of the MoS_2 coating. The durability of this coating was found to be dependent on the processing parameters and the initial wear problem of the interface was converted to optimizing the process for the coating to obtain adequate life. The influence of the MoS_2 coating on wear and the influence of processing parameters on durability of the MoS_2 are shown in Fig. 6.5.

Liquid lubricants can also exhibit "wear-out" characteristics but for different reasons. Liquids can evaporate and spread over available surfaces so that, with time, the amount of lubricant available to the contact interface can decrease. Therefore, if an adequate supply is not maintained, the system will ultimately go dry. Also, the lubricant

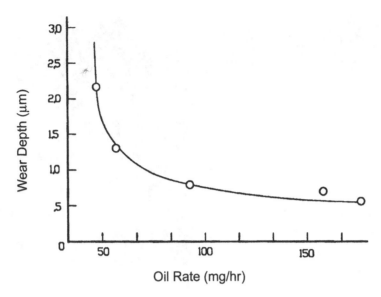

Figure 6.3 Effect of lubricant supply rate on the wear of a type carrier in a high speed line printer. (From Ref. 34.)

may degrade with time as a result of oxidation, polymerization, or some other mechanism, with the consequence that the ability of the fluid to lubricate the contact may degrade (13).

Another aspect of lubrication is that the ability of a material to provide lubrication can change as a result of the conditions surrounding the contact. Pressure, temperature,

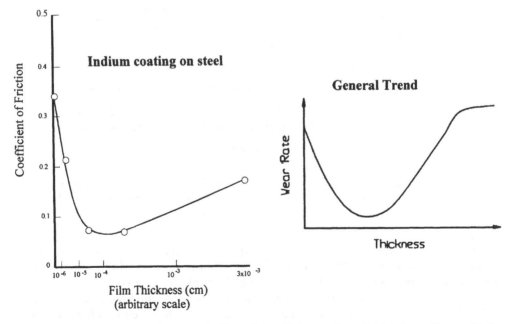

Figure 6.4 Illustration of the general effect of the thickness of a solid lubricant film on friction coefficient and wear rate. (From Ref. 35 reprinted with permission from Oxford University Press.)

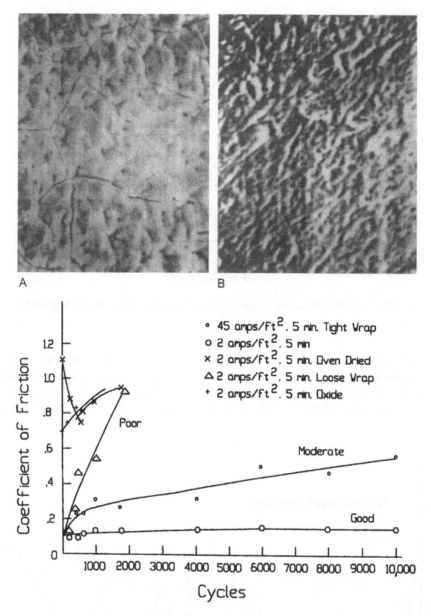

Figure 6.5 Micrographs of MoS_2 conversion coatings resulting from poor ("A") and good ("B") processing conditions. The graph shows the effect of processing conditions on the durability and performance of the coating. (From Ref. 12.)

sliding speed at the interface, and material compatibility can all be factors in determining the degree to which any material can function as a lubricant or, as will be discussed later in this section, the manner by which it provides lubrication. As a consequence, there is generally more distinction in lubricant performance when they are used in harsher, more challenging wear and friction environments, such as in tribosystems with high speeds, pressure and temperature, than in milder wear situations. However, even in the milder situations, there can be significant differences in performance (1). Many of the tests used

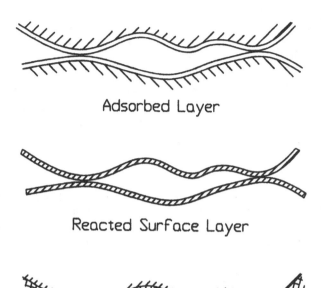

Adsorbed Layer

Reacted Surface Layer

Physical Separation

Figure 6.6 Lubrication mechanisms.

to evaluate, compare, and characterize lubricants focus on their ability to survive as effective lubricants under harsh or challenging conditions (14). These aspects of lubrication, as well as others, such as supply of lubrication to the interface, make the selection of lubricants and lubrication techniques a discipline in itself. Discussions of many of these aspects can be found in references on lubricants and lubrication, such as *Handbook of Lubrication* (15).

The primary way by which a lubricant influences friction and wear is by reducing adhesion (5,9,14,16,17) and there are three general mechanisms for this. One is by absorption on the contact surfaces. The second is by chemically modifying the surface. The third is by physical separation of the surfaces. These three ways are illustrated in Fig. 6.6. The first two mechanisms tend to reduce the strength of the bonds at the junctions, while the third tends to reduce the number of junctions. Secondary effects of lubrication are cooling of the interface, modifications of the stresses associated with the contact, and flushing of wear debris or contamination from the contact region. Since lubricants tend to lower friction, the heat and shear action developed at the contact interface is reduced. For fluid lubricants, additional cooling occurs as a result of the lubricant transporting heat out of the contact area. The lubricant can also influence the distribution of load within the contact by supporting some of the load.

For fluid lubricants, mechanical separation results from the response of the lubricant to being trapped between two surfaces under relative motion. Under such conditions, a fluid can support a normal load, thus providing separation between the two surfaces (5). Two examples of this type of response are shown in Fig. 6.7. In one case, the effect is caused by the constriction of the fluid as a result of tangential relative motion between

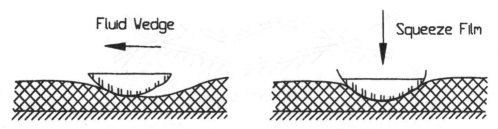

Figure 6.7 Squeeze films and physical wedges in fluid lubrication.

the surfaces, which is referred to as the physical wedge mechanism or simply the wedge mechanism. The second mechanism is referred to as the squeeze film mechanism and results from normal or perpendicular relative motion between the two surfaces. In both cases, the film thickness that results is dependent on the load, geometry, and velocity of the contact and rheological properties of the fluid.

For the wedge mechanism and a simple Newtonian fluid,

$$h \propto \left(\frac{\tau V}{P}\right)^m, \ 0.5 \le m \le 1 \tag{6.1}$$

where h is the minimum thickness; V, the velocity; P, the normal load; and τ, is the viscosity. For more complex fluids, for example non-Newtonian fluids, the relationships are more involved. Temperature gradients within the fluid and a pressure dependency on viscosity can have significant effects on the existence and the thickness of these fluid films. Under these conditions, sufficient pressure can be produced in the fluid so that the surface can be deformed, resulting in local geometry changes (18–21). These changes tend to enhance film formation and the ability of the fluid to support a normal load. Without deformation, lubrication by these types of fluid films is frequently referred to as hydrodynamic lubrication. For deformation of the surface within the elastic range, it is termed elasto-hydrodynamic lubrication. If plastic deformation is involved, it is referred to as elasto-plasto-hydrodynamic lubrication.

Investigations suggest that these additional aspects of deformation, temperature distribution, pressure dependency of viscosity, and non-Newtonian fluid properties, play significant roles in this type of lubrication. The amount of lubricant available is also a major factor. If there is not an adequate supply to the inlet of the contact, the film will not form. As an illustration of the significance of some of these aspects in lubrication, the pressure distributions and wedge shapes for a hydrodynamic model and an elasto-hydrodynamic model for two spheres in sliding contact, are shown in Fig. 6.8 (22,23). The equations for the film thickness are

$$h = 4.9R\left(\frac{\tau_0 V}{P}\right) \tag{6.2}$$

for the hydrodynamic model and

$$h = 2.65 \frac{\alpha^{0.54}(\tau_0 V)^{0.7} R^{0.43}}{P^{0.13} E'^{0.03}} \tag{6.3}$$

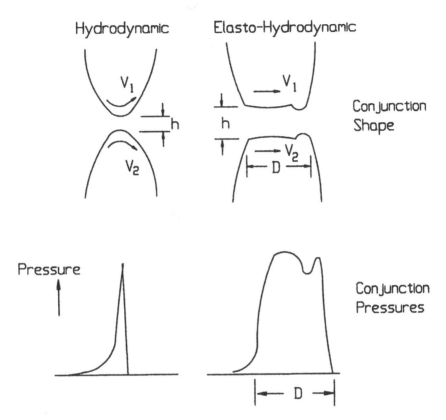

Figure 6.8 Comparison of hydrodynamic and elasto-hydrodynamic models for wedge formation.

for the elasto-hydrodynamic model. τ_0 is the absolute viscosity at the inlet; α is the pressure coefficient of viscosity (allowed for in the elastohydrodynamic model); E', the reduced elastic modulus for Hertzian contacts; and R is the equivalent radius for the contact.

Overall, lubrication as a result of these types of film formation with fluid lubricants is referred to as fluid lubrication. As can be seen, surface separation is the primary way that this type of lubrication effects the amount of adhesion and the degree of separation is directly related to the relative speed between the two surfaces and the geometry. The higher the speed and the flatter the geometries, the thicker the film formed. The other two ways in which a fluid effects adhesive components of friction and wear, the formation of absorption and chemical reacted layers, are generally referred to as boundary lubrication and are not as directly sensitive to these two parameters. The formation, strength, and tenacity of these films are primarily related to the chemical nature of the surfaces and the lubricant (14,16,17,24,25). Boundary lubrication is frequently the key factor in lubrication, especially under extreme conditions. The situation in which both boundary and fluid lubrication occur is normally referred to as mixed lubrication.

In Fig. 6.9, the Streibeck Diagram is used to illustrate these three regions of lubrication (18,26,27). This diagram shows the relationship between the coefficient of friction and wedge formation under sliding conditions. It can be seen that the abscissa of this diagram, the Sommerfeld Number, is related to the thickness of the lubricant layer, based on fluid lubrication concepts [see Fig. 6.1 and Eq. (6.2)]. In the fluid lubrication region, the film thickness is great enough so that the two surfaces do not interact at the asperity

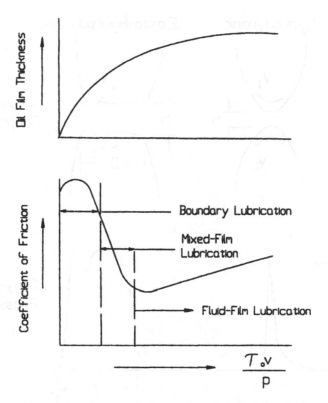

Figure 6.9 Relationship of viscosity, τ_0, speed, V, and load, P, to the coefficient of friction and oil film thickness. ($\tau_0 V/P$) is called the Sommerfeld Number and the lower diagram is referred to as the Streibeck Diagram.

level. In the mixed region, the film thickness allows occasional asperity interaction. In the boundary region, the fluid film thickness is so small that it is ineffective in preventing complete asperity interaction. Generally, wear decreases monotonically as film thickness increases. In the fluid range, wear would be limited to repeated cycle deformation mechanisms associated with the pressure transmitted through the fluid and is generally negligible. As the film thickness decreases and more and more asperity contact occurs, the potential for different wear mechanisms is introduced.

The transition from the mixed region to the fluid region is governed by the ratio of the asperity heights to the film thickness. Conceptually, the film thickness should be greater than the combined asperity heights of the two surfaces for complete separation and to be in the fluid lubrication regime. For two rough surfaces, the film thickness required for full fluid lubrication is given by

$$\Omega = \beta\left(\sigma_1^2 + \sigma_2^2\right)^{1/2} \tag{6.4}$$

where the σ's are the center line average (CLA) roughness of the two surfaces. Studies have indicated that an average value for β is approximately 3 (18,19,28).

The basic concept of solid lubrication or lubrication by solids is that the lubricant or lubricant/counterface junction is easier to shear than the base material or base material/counterface junction (26,29). In the case of solid film lubrication, a relationship between friction and film thickness also exists and is shown in Fig. 6.4. While the shape of the curve is quite similar to the shape of the curve in the Streibeck Diagram for fluids, the reasons are quite different. Initially, for very small thickness of the solid lubricant, the coefficient is reduced by reducing the number of strong adhesive junctions, without increasing the total number of junctions or real area of contact. The real area of contact is controlled by the harder substrate and the real area of contact is thought to be composed of a mixture of junctions, ones without the solid lubricant present and ones with the solid lubricant present. As the percent of junctions with the lubricant increases, the coefficient of friction decreases. Mathematically, the situation is expressed by

$$\mu = \alpha \frac{\tau_s}{\sigma_s} + (1 - \alpha) \frac{\tau_l}{\sigma_s} \tag{6.5}$$

where α is the fraction of non-lubricated junctions; τ_s, the shear strength of the non-lubricated junction; τ_l, the shear strength of the lubricated junction; δ_s, the flow stress of the substrate. Since δ_s is greater than τ_l, μ decreases as α decreases.

In the second region, the film is continuous and all junctions involve the lubricant and the lubricant influences the real area of contact. Because the solid lubricant is softer than the substrate, the real area of contact would tend to increase, leading to increased friction. For very thick films, the coefficient of friction would be the coefficient of friction associated with the solid lubricant and the counterface. Since the coefficient of friction involves the ratio of the shear strength to the compressive strength, the friction between the lubricant layer and the counterface may be higher than that between the substrate and the counterface. Models indicate that the approximate relationship between friction and thickness is

$$\mu \propto \left(\frac{h}{P}\right)^{1/2} \tag{6.6}$$

where h is the thickness of the solid lubricant and P is the normal load (30).

While effective friction behavior as a function of thickness is similar for fluid and solid lubrication, wear behavior is different. Generally for solid lubrication, there is an optimum thickness for wear behavior. If there is not enough solid lubricant, the wear of the surfaces is not significantly effected; if there is a very thick layer, the wear of the system will be the wear of that layer. Since solid lubricants generally have poorer wear properties than the base materials, the effective wear will be initially higher than without lubrication. As the lubricant layer thins, the optimum condition will be produced and improved wear behavior achieved. However, since solid lubricants do not self-heal, there will be a finite time for this period and eventually wear performance will degrade again. Graphically, this behavior is shown in Fig. 6.4.

While the conceptually solid lubrication can be considered simply in terms of thickness and coverage, the actual situation can be more complex. Optimum behavior may involve the formation of a mixed layer on the surface, composed of elements from the lubricant and the base materials. This layer would function, much like a transfer or third-body film in influencing wear and friction. Unlike these films, the solid lubricant equivalent will have a finite life since the lubricant is not replenished.

In addition to the film type of lubrication discussed to this point, lubrication to an interface can also be provided by utilizing materials, which have lubricants "built-in". Examples of this type would be polymers, which have lubricating fillers, like PTFE,

MoS_2, and graphite. An example in metallic systems would be metals, which have lubricating phases in them, such as leaded steels. These generally fall under the category of self-lubricated or self-lubricating materials. In this case, lubrication is usually achieved by transfer and third-body film formation. Hard phases and fillers can also be added to materials to improve wear resistance. Still another example would be porous materials filled with a solid or liquid lubricant, such as oil impregnated sintered bronze journal bearings. In this case lubrication can occur by boundary and fluid lubrication mechanisms (27,31). Circulation of the fluid through the porous media is an additional factor that needs to be considered in the fluid lubrication process of these types of bearings.

Because friction and wear are distinct, the ranking of lubricants in terms of their ability to lubricate can be different for friction and wear. Within the realm of lubricated systems, it is frequently found that while lubricant A gives lower friction than B, there is less wear with B (1,32). While there are these differences in the effectiveness of lubricants, they are generally not as significant as the differences between lubricated and unlubricated conditions. This can be paraphrased by saying that in most cases the biggest improvement obtained in wear and friction performance is associated with the use of any lubricant; a secondary improvement is associated with the selection of a particular lubricant for the system. In terms of wear, the change from unlubricated to lubricated wear generally results in improvement by more than one order of magnitude, with an improvement of 100 times or more being typical. Differences between lubricants are often smaller but can be significant. For lubricated sliding, μ is generally less than 0.3; under unlubricated conditions, μ frequently exceeds 0.6. For rolling, μ is less than 0.01. (see Table 5.2).

REFERENCES

1. R Bayer, T Ku. Handbook of Analytical Design for Wear. New York: Plenum Press, 1964.
2. F Bowden, D Tabor. The mechanism of rolling friction. The Friction and Lubrication of Solids, Part II. New York: Oxford University Press, 1964, pp 277–319.
3. F Bowden, D Tabor. The mechanism of rolling friction. The Friction and Lubrication of Solids, Part II. New York: Oxford University Press, 1964, pp 318.
4. M Todd. Solid lubrication of ball bearings for spacecraft mechanisms. Trib Intl 15(6):331–338, 1982.
5. M Neale, ed. Tribology Handbook. New York: John Wiley and Sons, 1973.
6. E Booser, ed. Handbook of Lubrication, Vol. II. Boca Raton, FL: CRC Press, 1983.
7. F Ling, E Klaus, R Fein, eds. Boundary Lubrication. ASME, 1969.
8. N Soda, T Sasada. Mechanism of lubrication of surrounding gas molecules in adhesive wear. Proc Intl Conf Wear Materials ASME 47–54, 1977.
9. F Bowden, D Tabor. The Friction and Lubrication of Solids. New York: Oxford U. Press, Part I, 1964, and Part II, 1964.
10. R Bayer, J Sirico. Influence of jet printing inks on wear. IBM J R D 22(1):90–93, 1978.
11. I Rugge, E Booser, eds Lubricating greases-characteristics and selection. Handbook of Lubrication. Vol. II. Boca Raton, FL: CRC Press, 1983, pp 255–267.
12. R Bayer, A Trivedi. Molybdenum disulfide conversion coating. Metal Finishing. Nov:47–50, 1977.
13. G Pedroza, C Pettus. NLGI Spokesman. Sept:203, 1972.
14. R. Fein. Presentation at the IRI Conference on Tribology. Warren, MI: GM Research Labs, 22–23 Jan 1985.
15. E Booser, ed. Handbook of Lubrication. Vols. I, II and III. Boca Raton, FL: CRC Press.
16. A Beerbower. Boundary Lubrication, Scientific and Technical Applications Forecast. U.S. Army Research and Development, Contract No. DAHC 19–69-C-0033, 1972.

17. P Kapsa, J Martin. Boundary lubricant films: A review. Trib Intl 15(1):37–41, 1982.
18. Roller Bearings, Vol. 60, Part I and Part II, Lubrication (Jul–Sept and Oct–Dec). Beacon, NY: Texaco, Inc., 1974.
19. H Cheng. Fundamentals of elastohydrodynamic contact phenomena. In: N Suh, N Saka, eds. Fundamentals of Tribology. Cambridge, MA: MIT Press, 1980, pp 1009–1048.
20. A de Gee, A Begelinger, G Salomon. Lubricated wear of steel point contacts – application of the transition diagram. Proc Intl Wear Materials Conf ASME 534–540, 1983.
21. J Dominy. Some aspects of the design of high speed roller bearings. Trib Intl 14(3):139–146, 1981.
22. R Fein, F Villforth. Lubrication. Vol. 51. No. 6. Beacon, NY: Texaco Inc., 1965.
23. D Dowson. In: P Ku, ed., Concentrated Contacts: NASA, , SP-237, 1970.
24. J Martin, J Georges, G Meille. Boundary lubrication with dithiophospates: influence of lubrication of wear of the friction interface steel/cast iron. Proc Intl Conf Wear Materials ASME 289–297, 1977.
25. A Singh, B Rooks, S Tobias. Factors affecting die wear. Wear 25(2):271–280, 1973.
26. E Rabinowicz. Friction and Wear of Materials. New York: John Wiley and Sons, 1965.
27. A Braun. Porous bearings. Trib Intl 15(5):235–242, 1982.
28. T Tallian. On competing failure modes in rolling contact. ASLE Trans 10(4):418–439, 1967.
29. F Bowden, D Tabor. Friction. Malibar, FL: Robert E. Krieger Pub.Co., 1982.
30. E Finkin. A theory for the effects of film thickness and normal load in the friction of thin films. Lub J Tech ASME 7/69(551–556).
31. P Murti. Lubrication of finite porous journal bearings. Wear 26(1):95–104, 1973.
32. K Muller. Prediction of the occurrence of wear by friction force-displacement curves. Wear 34:439–448, 1975.
33. E Hsue, R Bayer. Tribiological properties of edge card connector single/tab interface. IEEE Trans CHMT 12(2):206–214, 1989.
34. R Bayer. The influence of lubrication rate on wear behavior. Wear 35:35–40, 1975.
35. F Bowden, D Tabor. The Friction and Lubrication of Solids. Part I, Chapter V. New York: Oxford Univ. Press, 1964, p 3.

7
Selection and Use of Wear Tests

From an engineering standpoint, the reason for performing a wear test is to provide data that can be applied to a specific application, generally to increase life, reduce cost and maintenance, and provide reliable performance. Frequently in the minds of the engineer or designer, this is translated simply into selecting the best material for the design. However, as will be discussed in *Engineering Design for Wear; Second Edition, Revised and Expanded*, wear tests are used to provide additional engineering information as well. For example, wear tests may be required to help identify the wear mode and wear equation associated with the application; to develop the necessary engineering relationships among various design factors (e.g., shape, roughness, counterface properties, and wear); to determine values of wear parameters associated with models; and to determine and characterize transitions in wear behavior. All this may be summarized by saying that wear tests are done to provide wear data of one type or another, not simply material ranking.

From a designer's standpoint, the primary need is to obtain wear data, preferably without doing a wear test. Frequently as a result, the focus is initially on finding and utilizing available wear data and not on developing or selecting a wear test to generate the needed data. What has to be recognized in such an approach is that implicitly the selection and use of wear data is equivalent to selecting and using a wear test. The data were obtained from some test. As a result, the subject of wear testing is fundamentally equivalent to wear data selection and the points that will be developed regarding wear testing can be applied to the selection of published wear data. Of course with the use of existing data, the cost and the time associated with doing a test are eliminated.

As discussed in Part A Fundamentals, the nature of wear is complex. There are several mechanisms for wear, each of which is sensitive to a wide number of parameters but not necessarily to the same ones nor in the same way. There is no single, unique, universal parameter, which can be used to characterize wear behavior. As a consequence, there is no single, universal test for wear. Rather, this complex nature of wear results in the need for a variety of wear tests, each addressing one particular aspect of wear or wear situations. The large number of wear tests and apparatuses that can be found in the literature serves to illustrate this point (1–6). Another point that needs to be recognized from the information about wear presented in Part A is that wear testing does not define or measure a fundamental or intrinsic material property, like modulus or strength. In that sense, it is not a material's test. Rather, it measures or characterizes a material's response to or behavior in a system environment. Basically, this is because wear is not a materials property but a system property. Materials can behave differently in different wear situations, as has been discussed and illustrated previously. As a consequence, different wear tests tend to provide different rankings of materials.

To the engineer this situation begs the question, "What is the appropriate test for the application at hand?" Hence, wear test selection and use is an appropriate and key aspect in the overall consideration of wear testing. Furthermore, the answer to the question is to select the wear test which best simulates the actual wear situation. The need to simulate the application in the wear test is pointed out again and again in the literature (2–5, 7–13).

The key to the relevance of any wear test to an application lies in the degree to which the application is simulated in the test. There are several levels of simulation, which are significant to the development, selection, and use of wear tests. The most fundamental or basic level of simulation is in terms of the general nature of the wear situation. For example, this level of simulation is concerned with whether both the application and the test represent a rolling, sliding, or impact wear situation; unlubricated or lubricated wear; two- or three-body abrasion; erosion by solid particles or liquids; etc. This level of simulation can be termed as first-order simulation.

The next level of simulation (or second-order simulation) is related to the values of key parameters of the wearing system. Two elements are involved in this: the first is the identification of the significant parameters, and the second is the identification of the appropriate range that is needed for this parameter in the test in order to provide simulation. Examples of elements to be considered in this respect are load, speed, stress, and temperature. Other elements that have to be considered at this level are counterface parameters, nature of the third-bodies involved, amount and type of lubrication, and unidirectional or reversing sliding. However, the list is not limited to these as any aspect or parameter, which can influence wear or friction is a candidate for consideration at this level of simulation.

Third-order simulation, the next level, is essentially replicating the actual wear situation. All parameters and features are similar, if not identical, to those in the application. At this level of simulation, the wear tester is often very similar to the actual device and may be an instrumented version of the device or a replica of a portion of the overall machine or mechanism. Wear testers at this level of simulation may be called wear robots, to contrast them to the type of apparatuses used in first- and second-order simulation, which are generally laboratory type devices. The differences between third-order simulation and actual machine testing or field-testing generally lie in the area of control and data acquisition. At this level of simulation, testing conditions are generally more controlled and wear measurements are more frequent and refined than in field-testing.

The level of simulation that is required in a wear test depends on the purpose of the wear test. If the intent of the test is to provide only general type of information, then first-order simulation is adequate. Tests to understand the general nature of wear occurring in a given type of situation, to provide broad ranking of material groups, to identify major factors effecting wear and to identify general trends, are examples of this type of purpose. When more specific information is required, such as the need to rank or select materials for a given application, to project wear performance of a given design, or to determine the value of a specific design parameter required for optimum performance, second- or third-order simulation is required. The need for specific information of this type is generally characteristic of engineering applications and consequently wear tests generally done for engineering purposes will require this higher degree of simulation. Tests associated with more fundamental or research studies generally have only first-order simulation, when compared to applications. Tests used by material developers tend to provide first-order simulation for most applications.

Providing second-order simulation assumes that the major factors influencing the wear have been identified. That identification might in itself require some testing, possibly involving first-order simulation, or might be available from experience or published information. The thoroughness to which this is done influences the degree of risk associated with the use of the data from the test. Another way of stating this is that the correlation expected between the test and actual performance is controlled by this element. The less thorough this is done, the higher the risk associated with projecting the actual performance or the lower the anticipated correlation. With the use of third-order simulation, risk is minimized and improved correlation with actual performance can be obtained.

For most engineering situations, third-order simulation is not required to provide the useful and specific information desired. Second-order simulation is usually adequate, provided the parameters influencing the wear are correctly identified and understood. That is the key. Frequently though, tests that are basically representative of third-order simulation are used as a result of pragmatic considerations. In certain cases, it may not be practical or desirable to spend the time to identify the major factors in the wear situation or to develop an apparatus that provides the adequate simulation and control over these. It may be easier to instrument the device itself (or a replica of the device) and use it as a wear robot to provide data under actual use conditions. Because this approach tends to include all interactions, it reduces risk and enhances correlation. This type of test does have some negative aspects, though. While time and effort are usually saved by avoiding tests to identify significant parameters, these robot-type tests tend to be more lengthy and involved than those associated with second-order simulation. Also, robot tests generally do not directly provide information about fundamental relationships. However, robot tests do provide information regarding parameters, which, while not basic, may be more relevant and significant to the application.

The choice of the apparatus used is a key part in any simulation. While this is the case, there are other elements, which are equally as important to the simulation and have to be considered. For example, the environment in which the wear test is done, the properties of the counterface(s), and the characteristics of the wearing media (particularly in erosion and abrasion testing) are equally as important. In addition to simulation, there are other testing and tester aspects which are also important to the proper conduction of a wear test. Sample preparation, data recording, wear measurement technique, and analysis of the data are examples. Variations in these elements are generally sources for the scatter in test data. While procedures for these elements are often specified for standard tests, they may not be adequate. It is also necessary to recognize the primary purpose of the standard test. It may not be wear but friction or lubricant evaluation. As a result, it is necessary to review these procedures and perhaps modify them for use as a wear test. These elements, along with simulation, will be discussed in greater detail in subsequent sections.

Because of the need to simulate and the complex nature of wear, most laboratories associated with wear testing have a variety of test apparatus and procedures that are used, often with modifications, to address specific problems (2,10,14–17). The particular complement of test apparatus that a laboratory has and the procedures used generally reflect the nature of the industry that the laboratory supports and the purpose for which the testing is done. For example, a laboratory associated with the wear of office and data processing equipment typically utilizes different apparatuses than a laboratory associated with the wear of airframes (18,19). Similarly, both will likely have different tests and procedures than a laboratory supporting a light manufacturing operation (20–24).

Laboratories associated with material suppliers and developers tend to form a unique category that tends to be somewhat different than laboratories associated with design. Generally, laboratories associated with material development have testers and procedures, which allow them to differentiate material behavior quickly for some broad area of application and which are appropriate for one particular class or type of material. For example, tests used in laboratories concerned with the development of hard, bulk materials, such as tool steels or ceramics, are generally not the same as those used in laboratories associated with the development of coatings or plastics (3–5,25,26). High speeds, high stress, and, in the case of ceramics, high temperature, are typical features of tests used for the former; for the latter, milder tests conditions and different durations are generally required to differentiate between materials. The harsher conditions used for tool steels or ceramics would result in such large and more severe wear for the other two types of materials that differences in performance would be less apparent. Conceptually this is illustrated in Fig. 7.1, where wear rate is plotted as a function of test severity or harshness. Above the mild/severe wear transition, there is less difference in rate than below the transition. Transition points can also vary with material. As a result, movement of the transition point can also confound the comparison as well, as is illustrated in the same figure. The situation with coatings is shown in Fig. 7.2, where wear depth is plotted as a function of test duration. As can be seen, if the test results in wear-through of the coating, the ability to differentiate is again reduced.

The milder conditions required for plastics and coatings evaluations, as compared to tool steels and ceramics evaluations, also reflect the differences between the typical applications for these types of materials. In effect, this demonstrates the requirement of simulation. The situations illustrated in Figs. 7.1 and 7.2 indicate the source of some of the problems that can occur as a result of lack of adequate simulation, namely improper ranking and selection of materials.

A common feature of most of the tests used by materials-oriented laboratories is the tendency to focus simply on providing material rankings, rather than on the determination of parameters needed for wear prediction or selection of an over-all design (25–34). These latter aspects tend to be found in the tests used by the laboratories associated with the design and development of new equipment and the development of design information. Examples of this type of data might be specific values of wear parameters to be used in conjunction with a model (35), the determination of transition points (36,37) and the influence of design parameters other than material selection on system wear (38,39). In tests used for material ranking purposes, it is often the practice to use the amount of wear generated after a particular amount of time, number of revolutions, abrasive consumed, etc.,

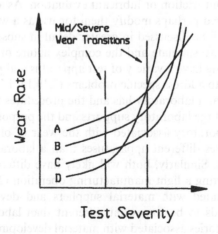

Figure 7.1 The effect of test severity on relative wear behavior.

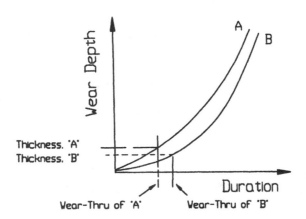

Figure 7.2 The effect of coating wear-through on relative wear performance.

to provide the ranking. In tests used to provide more design-oriented data, tests involving the generation of a wear curve, that is, a plot of wear or wear rate vs. usage or exposure, are frequently desirable or needed. In general, the wear curve provides more information than a single point and may be needed to differentiate behavior, particularly when the possibility of different wear modes exists with different materials.

The need to simulate the application in the wear test and the confounding influences of purpose and materials on that simulation can have significant effect on the testing procedures and equipment used. One way of illustrating this is to consider the wear tests and approaches associated with three different laboratories that have been published in the literature. The first laboratory is associated with the wear of components found in business machines and peripheral computer equipment (18,40–42). The next is associated with the wear of components of light manufacturing equipment in a chemically oriented industry (20–24). The third is concerned with the wear of airframe elements (19). Figures 7.3, 7.4, and 7.5 contain illustrations of the testers used by these laboratories, along with a description of the data generated in the tests and the purpose of the test. An examination of these figures shows that the apparatus and procedures are quite different for each of the laboratories. This is a consequence of the need to simulate the significantly different applications as well as difference in the purposes of the wear tests.

In the first laboratory, the focus was to select a design which would achieve a given life and therefore the tests were used to provide more general engineering information, not simple material selection. The tests were used to develop engineering models for wear, determine values of parameters associated with those models (including material parameters), and investigate the influence of other design parameters on the wear, such as radius or shape, thickness of coatings or layers, roughness, edge conditions, and alignment. Once a model was developed, an appropriate test to evaluate and compare materials was usually identified, since material selection is always a part of a design approach.

The wear situations encountered in this laboratory included: sliding, rolling, impact, and mixtures of these motions; metal/metal, metal/polymer, polymer/polymer interfaces; wear both by and of paper, inks, ribbons, and magnetic media; some form of boundary or dry lubrication; generally mild environmental conditions, e.g., room temperature or near room temperature and normal atmosphere. Normally, only mild wear behavior could be accepted in these applications. While in most of the applications loads tended to be small (e.g., order of pounds or less), stress levels could be high because of

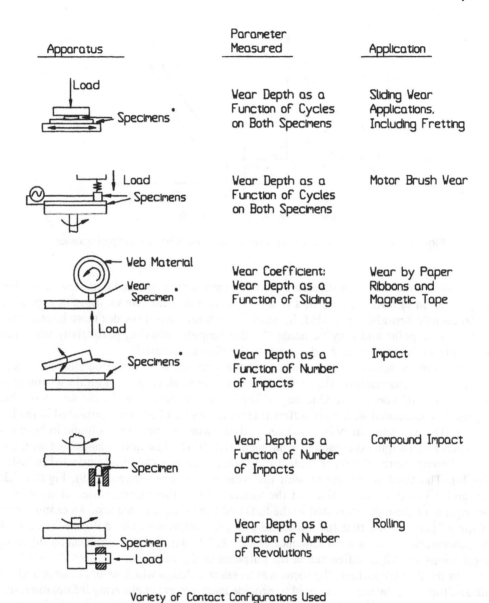

Variety of Contact Configurations Used

Figure 7.3 Wear tests used for computer peripheral applications. The tests were used to develop wear models, to determine wear coefficients, to investigate the effects of different parameters, the selection of design parameters, and to rank materials.

small contact areas in the applications. However, since long life generally requires the stress level to be well below the elastic limits of the materials, stresses in the applications were generally a fraction of the elastic limits of the materials used.

Often the loads in these applications were generated from kinematic conditions or were time varying, rather than a constant load supplied by a dead-weight or spring. Parts were relatively small and contacts were generally nonconforming. Performance was typically affected by small amounts of wear. Changes in the range from 0.001 to 0.010 in.

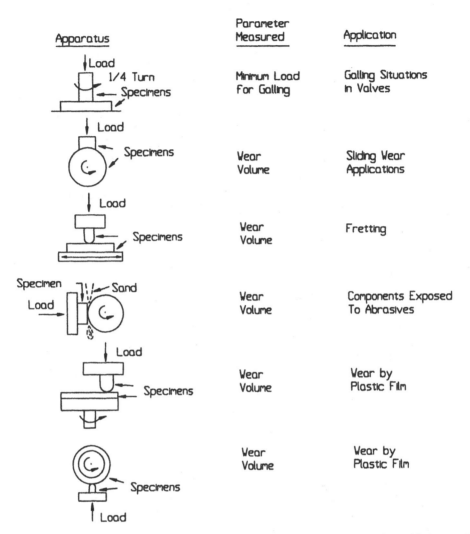

Apparatus	Parameter Measured	Application
Load / 1/4 Turn / Specimens	Minimum Load for Galling	Galling Situations in Valves
Load / Specimens	Wear Volume	Sliding Wear Applications
Load / Specimens	Wear Volume	Fretting
Specimen / Sand / Load	Wear Volume	Components Exposed To Abrasives
Load / Specimens	Wear Volume	Wear by Plastic Film
Specimens / Load	Wear Volume	Wear by Plastic Film

Figure 7.4 Wear tests used for manufacturing equipment in a chemically oriented industry. The tests were used to rank materials in terms of their resistance to different types of wear. Often, several tests were combined into a screening procedure of the selection of materials for a given application.

(sometimes less) of a critical dimension frequently resulted in functional failure in the applications.

A review of the apparatuses used by this laboratory and their features, shown in Fig. 7.3, indicate that these apparatuses have the same general features of the applications. The apparatuses accommodate small specimens, provide light loads and different motions, accommodate different materials, and generally involve nonconforming contacts. The nature of these wear situations has typically resulted in the development of unique apparatuses and test methods in order to simulate these situations and to provide the needed data. The impact wear apparatuses, the drum tester, the C-ring configuration, and the configuration used for elastomer drive rolls (Fig. 7.3), are examples of some of the unique test configurations used. Since initial wear cannot be ignored in applications which are sensitive to small amounts of wear, many of the tests involve the development

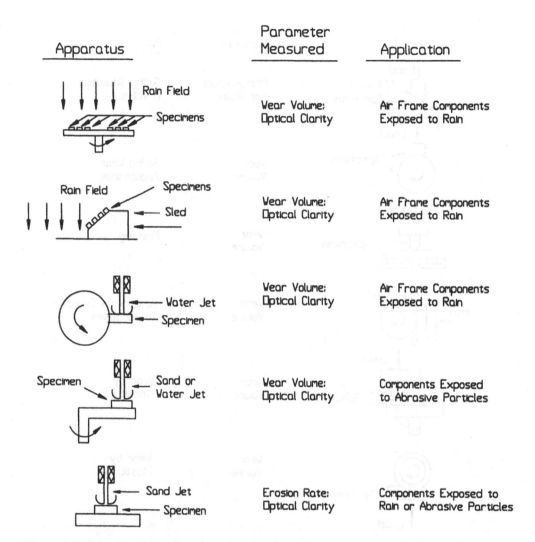

Apparatus	Parameter Measured	Application
Rain Field / Specimens	Wear Volume: Optical Clarity	Air Frame Components Exposed to Rain
Rain Field / Specimens / Sled	Wear Volume: Optical Clarity	Air Frame Components Exposed to Rain
Water Jet / Specimen	Wear Volume: Optical Clarity	Air Frame Components Exposed to Rain
Specimen / Sand or Water Jet	Wear Volume: Optical Clarity	Components Exposed to Abrasive Particles
Sand Jet / Specimen	Erosion Rate: Optical Clarity	Components Exposed to Rain or Abrasive Particles

Figure 7.5 Tests used by a laboratory concerned with the selection and development of materials for use on airframes, including components that had to be optically transparent. The tests were used to rank materials, to determine the effects of different parameters, and to investigate wear mechanisms.

of wear curves rather than simply utilizing data after a stable wear situation is achieved. An illustration of this is the procedure employed with the ball/plane tester used for sliding wear. In this case, a wear curve was developed to determine the exponent associated with different wear modes as well as the determination of a material wear factor (43,44). This method is illustrated in Fig. 7.6.

In order to establish a more complete engineering approach, many of the tests and testers were developed or selected so that the design parameters other than material selection could be evaluated and to provide the basis for the development of engineering models. Examples of this are the approaches used for impact wear (42,45,46), rolling/sliding wear (41), the abrasive wear of a magnetic sensor (47), and C-ring wear (48). In these cases testers were developed in which the effects of geometry, loading, and other design

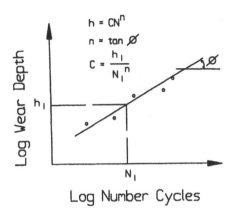

$$h = CN^n$$

$$n = \tan \phi$$

$$C = \frac{h_1}{N_1{}^n}$$

Figure 7.6 Example of a wear curve and the data obtained from the ball-plane test for sliding wear used in addressing wear concerns in computer peripheral applications.

factors could be evaluated as well providing a capability for evaluating different materials. Because of the complexity of simulating some of the situations, many studies were performed utilizing robots rather than developing specific testers; in some cases, both were used to varying degrees. The wear of electroerosion print elements (49), type carriers (50), print cartridges (51), and band/platen interfaces (Fig. 4.46) are examples of situations for which robots were used extensively in this laboratory.

In contrast to this situation, the focus in the other two laboratories was to maximize the machine or component life by selection of the optimum material or material pairs. Consequently, testing was primarily associated with material ranking. The thrust was to develop a test procedure that simulated the application and allowed differentiation of materials in a reasonable length of time. The test was then used to evaluate a matrix of materials or material pairs for the application. A higher degree of simulation was employed than is typical of simple material testing. Both laboratories generally establish a second-order simulation in their tests.

In the second laboratory, that is the laboratory supporting a chemically oriented light manufacturing operation, many of the wear situations were more representative of classical contact situations, found in bearings, gears, and cams, than those encountered in the first laboratory. Since hostile environments frequently limit the choice of materials in chemical environments, the approach in this laboratory was generally to modify a standard tester and test methods to account for the specific conditions of the application, rather than to develop unique testers or test methods. This point is evident from examination of Fig. 7.4, which contains a summary of the tests used and the applications to which they are applied. In complex wear situations in which several distinct wear modes are present, perhaps in different regions of the part, this laboratory tended to utilize a series of tests, each focused on a particular mode, to provide a full evaluation (52). This is an approach used in many laboratories (14,53,54).

While the third laboratory, that is the laboratory supporting airframe applications, also focused on material selection, significantly different test apparatuses were required as a result of the differences in the wear situations encountered. The primary concerns were with the wear produced by high-speed motion through the atmosphere. Solid particle erosion due to air-borne dirt, sand, etc. was one concern, others were the effects of rain drop impingement and cavitation. The apparatuses developed and used for these types

Table 7.1 Levels of Simulation in Tests

Level of test simulation	Use	Correlation with applications
First-order (general nature of wear situation replicated)	To obtain general information	Generally poor except in terms of general trends; often not adequate for engineering
Second-order (key parameters replicated)	To obtain engineering information when the influences of parameters understood	Generally good; correlation tends to improve with the level of understanding regarding the effects of key parameters; often adequate for engineering
Third-order (most or all parameters replicated)	To obtain engineering information when the influences of parameters not understood	Good; correlation tends to decrease with lack of replication

of situations are shown in Fig. 7.5. Again, simulation is evident. In this laboratory, the need was to select a single material for a range of environmental and operating conditions. This is different from many of the situations encountered by the second laboratory in that the wear performance of a material pair was of importance. The testers and methods reflect this aspect as well. In the third laboratory, the wearing media is changed to match the conditions of the environment associated with the situation and wear evaluations simply focus on the wear of the target material in the test. In many of the tests employed by the second laboratory, material pairs were evaluated and counterface wear was an important element in the evaluation. In this respect, the first and second laboratory are similar in that for many of their wear situations they have the ability to control both members of the wear system and wear of both is important.

All three laboratories have reported successful application of their approaches. Fundamentally they employ the same simulation strategy. The first differs from the second and third primarily in that it strives to optimize the entire design to achieve a specific life or performance target, while the other two aim only to insure adequate life and to achieve as long a life as possible. The combined influence of the need to simulate and the purpose for which the test is done on the selection of the wear test is clearly indicated by this comparison of these three laboratories.

Further examples of simulation and the impact of purpose and materials on wear testing can be found in a series of books published by the American Society for Testing and Materials on the subject of wear testing (2–6). A summary of first-, second-, and third-order simulation and their relationships to the nature of the test, their use, and correlation is given in Table 7.1.

REFERENCES

1. R Benzing, I Goldblatt, V Hopkins, W Jamison, L Mecklenburg, M Peterson. Friction and Wear Devices. Park Ridge, IL: ASLE, 1976.
2. R Bayer, ed. Selection and use of Wear Tests for Metals, STP 615. West Conshohocken, PA: ASTM, 1976.
3. R Bayer, ed. Wear Tests or Plastics: Selection and use, STP 701. West Conshohocken, PA: ASTM, 1979.

4. R Bayer, ed. Selection and use of Wear Tests for Coatings, STP 769. West Conshohocken, PA: ASTM, 1982.
5. C Yust, R Bayer, eds. Selection and use of Wear Tests for Ceramics, STP 1010. West Conshohocken, PA: ASTM, 1988.
6. R Denton, K Keshavan, eds., Wear and Friction of Elastomers, STP 1145. West Conshohocken, PA: ASTM, 1992.
7. R Bayer. Wear Testing; Mechanical Testing. In: J Newby, ed. Metals Handbook. Vol. 8. 9. Materials Park, OH: ASM, 1985, pp 601–608.
8. M Moore. Laboratory simulation testing for service abrasive wear environments. Proc Intl Conf Wear Materials. ASME 673–688, 1987.
9. M Olson et al., Sliding wear of hard materials - The importance of a fresh countermaterial surface. Proc Intl Conf Wear Materials ASME 505–516, 1987.
10. D Rosenblatt. Factors involved in liner wear. Proc Intl Conf Wear Materials. ASME 158–166, 1977.
11. S Bhattacharyya, F Dock. Abrasive wear of metals by mineral and industrial wastes. Proc Intl Conf Wear Materials. ASME 167–176, 1977.
12. D Gawne, U Ma. Wear mechanisms in electroless nickel coatings. Proc Intl Conf Wear Materials. ASME 517–534, 1987.
13. M Ruscoe. A predictive test for coin wear in circulation. Proc Intl Conf Wear Materials. ASME 1–12, 1987.
14. A Begelinger, A de Gee. Wear in lubricated journal bearings. Proc Intl Conf Wear Materials. ASME 298–305, 1977.
15. H Avery. Classification and precision of abrasive tests. Proc Intl Conf Wear Materials. ASME 148–157, 1977.
16. C Young, S Rhee. Wear process of TiN coated drills. Proc Intl Conf Wear Materials. ASME 543–550, 1987.
17. H Hawthorne. Wear debris induced friction anomalies of organic brake materials in vacuo. Proc Intl Conf Wear Materials. ASME 381–388, 1987.
18. R Bayer, A Trivedi. Wear Testing for Office and Data Processing Equipment. In: R Bayer, ed. Selection and Use of Wear Tests for Metals, STP 615. West Conshohocken, PA: ASTM, 1976, pp 91–101.
19. G Schmitt. Influence of Materials Construction Variables on the Rain Erosion Performance of Carbon-Carbon Composites. In: W Adler ed. Erosion: Prevention and Useful Applications, STP 664. West Conshohocken, PA: ASTM, 1979, pp 376–405.
20. K Budinski. Wear of tool steels. Proc Intl Conf Wear Materials. ASME 100–109, 1977.
21. T Groove, K Budinski. Predicting Polymer Serviceability for Wear Applications. In: R Bayer ed. Wear Tests for Plastics: Selection and Use, STP 701. West Conshohocken, PA: ASTM, 1979, pp 59–74.
22. K Budinski. Wear Characteristics of Industrial Platings. In: R Bayer ed. Selection and Use of Wear Tests for Coatings, STP 769. West Conshohocken, PA: ASTM, 1982, pp 118–133.
23. K Budinski. Incipient galling of metals. Proc Intl Conf Wear Materials. ASME 171–178, 1981.
24. K Budinski. Tribonetic characteristics of copper alloys. Proc Intl Conf Wear Materials. ASME 97–104, 1979.
25. Standard Test Method for Wear Testing with a Crossed-Cylinder Apparatus. West Conshohocken, PA: ASTM, G83.
26. Standard Test Method for Measuring Abrasion Using the Dry Sand/Rubber Wheel Apparatus. West Conshohocken, PA: ASTM, G65.
27. Standard Test Method for Conducting Wet Sand/Rubber Wheel Abrasion Tests. West Conshohocken, PA: ASTM, G105.
28. Standard Test Method for Ranking Resistance of Materials to Sliding Wear Using Block-on-Ring Wear Test. West Conshohocken, PA: ASTM, G77.

29. Standard Test Method for Determination of Slurry Abrasivity (Miller Number) and Slurry Abrasion Response of Materials (SAR Number). West Conshohocken, PA: ASTM, G75.
30. Standard Test Method for Conducting Erosion Tests by Solid Particle Impingement Using Gas Jet. West Conshohocken, PA: ASTM, G76.
31. Standard Test Method for Cavitation Erosion Using Vibratory Apparatus. West Conshohocken, PA: ASTM, G32.
32. Standard Test Method for Jaw Crusher Gouging Abrasion Test. West Conshohocken, PA: ASTM, G81.
33. Standard Practice for Liquid Impingement Erosion Testing. West Conshohocken, PA: ASTM, G73.
34. Standard Test Method for Abrasinvess of Ink-Impregnated Fabric Printer Ribbons. West Conshohocken, PA: ASTM, G56.
35. R Bayer. Tribological Approaches for Elastomer Applications in Computer Peripherals. In: R Denton, K Keshavan, eds. Wear and Friction of Elastomers, STP 1145. West Conshohocken, PA: ASTM, 1992, pp 114–126.
36. R Lewis. Paper No. 69AM 5C-2. Proceedings of 24th ASLE Annual Meeting, Philadelphia, 1969.
37. S Lim, M Ashby. Wear-mechanism maps. Acta Metal 35(1):1–24, 1987.
38. R Bayer, J Sirico. The Influence of surface roughness on wear. Wear 35:251–260, 1975.
39. R Bayer. Design for wear of lightly loaded surfaces. Stand. News 2(9):29–32; 57, 1974.
40. R Bayer, P Engel, J Sirico. Impact wear testing machine. Wear 19:343–354, 1972.
41. P Engel, C Adams. Rolling Wear Study of Misaligned Cylindrical Contacts. Proc Intl Conf Wear Materials. ASME 181–191, 1979.
42. R Bayer. Impact wear of elastomers. Wear 112:105–120, 1986.
43. R Bayer. Predicting wear in a sliding System. Wear 11:319–332, 1968.
44. R Bayer, T Ku. . Handbook of Analytical Design for Wear. New York: Plenum Press, 1964.
45. P Engel, R Bayer. The Impacting Wear Process Between Normally Impacting Elastic Bodies. J Lub Tech Oct:595–604, 1974.
46. P Engel, T Lyons, J Sirico. Impact wear model for steel specimens. Wear 23:185–201, 1973.
47. R Bayer. A Model for Wear in an abrasive environment as applied to a magnetic sensor. Wear 70:93–117, 1981.
48. R Bayer. Wear of a C Ring seal. Wear 74:339–351, 1981–1982.
49. R Bayer. Wear in electroerosion printing. Wear 92:197–212, 1983.
50. R Bayer. The influence of lubrication rate on wear behavior. Wear 35:35–40, 1975.
51. R Bayer, J Wilson. Paper No. 71-DE-39. Design Engineering Conference and Show, 4/71. New York: ASME, 1971.
52. K Budinski. Wear of tool steels. Proc Intl Conf Wear Materials. ASME 100:109, 1977.
53. S Calabrese, S Murray. Methods of Evaluating Materials for Icebreaker Hull Coatings. In: R Bayer, ed. Selection and Use of Wear Tests for Coatings, STP 769, West Conshohocken, PA: ASTM, 1982, pp 157–173.
54. N Payne, R Bayer. Friction and wear tests for elastomers. Wear 130:67–77, 1991.

8
Testing Methodology

8.1. GENERAL

While wear testing may not be an exact science, it is also not a black art. There is a general methodology that can lead to the successful selection, development, and implementation of wear tests for engineering applications. As should be evident from the preceding section, the cornerstone of this methodology is simulation. However, several other elements also have to be contained in the methodology if useful engineering information is to be obtained. The methodology requires that the appropriate degree of control be used, that appropriate measurement and analysis techniques be used, that the appropriate information and observations be recorded, and that a suitable degree of acceleration be associated with the test. If any of these elements are not addressed or inadequately addressed in the test strategy, correlation with the application is generally reduced and, in the extreme, may be completely missing. On the other hand, if these elements are correctly addressed, excellent correlation can be obtained. The individual elements of this test methodology are treated in the following sections on Simulation, Control, Acceleration, and Data Acquisition, Analysis, and Reporting.

8.2. SIMULATION

A minimum of second-order simulation is required a priori to insure good correlation. That is to say, the test must simulate the application in all key aspects. A good way to establish that degree of simulation is to start with two assumptions. One is that all attributes of the application are key and that specific values of all the parameters should be the same in the test as in the application. The second assumption is that any deviation from complete replication, unless justified, will tend to reduce or even eliminate correlation. Basically, this amounts to assuming that third-order simulation in the test is required for good correlation. A test representing second-order simulation evolves by accepting modification of only those parameters or attributes that can be shown or expected to have negligible influence on the wear behavior.

Generally the attributes surrounding a wear situation that should be considered for simulation can be grouped into seven broad categories:

1. materials,
2. geometry,
3. motion,
4. loading,
5. lubrication,

6. environment, and
7. heat dissipation/generation.

There are many elements in each of these categories which have to be considered and these can vary with the situation. Table 8.1 provides a summary of some of the typical elements considered in these categories. The significance of some of these elements is discussed in the following paragraphs.

Since wear testing frequently has the element of material selection associated with it, the need to consider materials in the simulation process appears obvious. However, the degree to which this should be considered may not be as obvious. First of all, it has to be recognized that wear is influenced by both bulk and surface properties and that these properties are not solely controlled by composition. Consequently, simulation simply in terms of bulk composition is frequently not sufficient for second-order simulation. Beyond composition, material and surface preparation has to be considered. A wrought specimen of the same alloy may exhibit different wear characteristics than a cast version of the same alloy. Differences in machining techniques also have to be considered; a ground surface is not necessarily equivalent to one prepared by milling or polishing. In addition to differences in surface topography that might be associated with these methods, there may be difference in such things as residual stresses, degree of work hardening or microcracking that can be very significant to wear behavior. In the case of polymers, skin effects can be significant. Testing with a machined surface, where the skin is removed, may not provide a valid simulation where the same material is to be used in molded form, since the wear in the skin may influence behavior in the application. With polymers and possibly with other materials, some environmental preconditioning of the specimens might be required to simulate behavior in service, since such things as absorbed and adsorbed moisture have been observed to influence wear rate. In the case of coatings or platings, consideration must also be given to the substrate, not just to the coating. The wear can be influenced by deformation and thermal characteristics of the substrate in addition to adhesion aspects.

When associated with the wearing member of a device, many of these aspects are almost automatically recognized and addressed by the design engineer or wear test developer. The significant point that has to be brought out is that the same issues and concerns apply to the counterface or wearing media in the test and in application. Its surface and

Table 8.1 Simulation Categories

Category	Typical Aspects to Be Considered
Materials	Composition, processing conditions, cleaning, surface preparation, sources, coating thickness
Geometry	Line, point or area contact, size, roughness conditions
Motion	Rolling, sliding or impact, unidirectional, oscillatory
Loading	Constant, fluxuating, impact, contact stresses, uniformity
Lubrication	Lubricated or not, solid, grease or fluid, composition and properties, amount, supply, aging, breakdown, boundary
Environment	Temperature, relative humidity, gaseous and particulate composition, abrasive, corrosive, fluid flow characteristics
Heat dissipation and generation	Heat conduction paths, source of heat, cooling, specimen thickness, surface temperatures, flash temperatures

bulk properties can influence wear behavior as well. In the case of a solid counterface, the elements to be considered are identical. In the case where the wear is the result of fluid or slurry encounter, the composition of the fluid, its pH, its viscosity, and the particles contained in it, as well as their hardness and number, are examples of additional material aspects which have to be considered.

Geometry or shape is also an element that has to be considered in simulating a wear situation. One element of this consideration is simply the nature of the immediate contact situation (e.g., point contact, line contact, or conformal contact). A prime difference between these contacts is the stress distribution. Since stress can be a factor in some wear situations, it may not be appropriate to simulate a wear situation that is basically a conformal contact with a point or line contact test configuration. An illustration of this would be the evaluation of coatings for such an application. A point contact which concentrates loading might result in immediate break-through of the coating, while in the conformal application, the coating will fail by gradual wear. Another aspect about these different contact configurations is that stress levels change with wear for both the point and line contact configuration. In a conformal contact, such as a flat on a flat, they may not. In the former cases, wear modes can change as wear progresses as a result of this changing stress condition, while a conformal situation may not exhibit a similar change. Phenomena, such as temperature rise, transfer film formation, and hydrodynamic lubrication, can also be influenced by the nature of the contact as well.

A further element that has to be considered in terms of shape or geometry is the general or over-all nature of the contact (e.g., a journal or a thrust-bearing configuration, a roller bearing, piston ring, etc.). This general nature can influence such aspects as heat dissipation, debris entrapment or removal, tribofilm formation, as well as lubrication effects. For example, while a line-contact test geometry, such as a rotating cylinder on a flat, might simulate some aspects of a journal bearing, it does not provide a complete simulation. Any effects of clearance between shaft and bearing on wear would not be simulated with this simple configuration. Another example of this sensitivity to the over-all nature of the contact is in an erosive wear situation, like an airfoil moving through a fluid. Differences in geometry between the actual case and a proposed test configuration could result in differences in the type of flow across the wearing surface. For example, in the actual device, the flow might be turbulent, while in the test, the flow is streamline. Since the nature of the flow can have significant impact on wear in these situations, significant differences in wear behavior might result. As a consequence, this would not provide good simulation.

Geometrical considerations can also be coupled with surface roughness or topography. Consideration should be given to whether the test configuration provides the same orientation as the application with respect to any lay that the surface might have. A companion consideration of geometry is motion. The basic level of this consideration is in terms of the general nature of the motion (e.g., rolling, sliding, impact, or fluid flow). For example, a rolling wear test is generally needed to simulate a rolling wear situation; a sliding test, to simulate sliding wear; etc. However, the consideration cannot be limited to that level and it is necessary to consider several other aspects of the motions. For example, is the motion unidirectional or reciprocal? Are stop/starts involved? If the motion is reciprocating, what is the length of the stroke? In a rolling situation, is there slip or sliding involved and if so, how much? Is it on a micro- or macro-level? Does the motion involve a combination of impact and sliding, impact and rolling, etc.? The magnitude of the velocities and any acceleration also need to be considered for effective simulation, as well as the repetition rate in the case of cyclic motion. In the case of erosion, the angle of impingement and the fluid velocity are also included in this category. These aspects can influence the mixture of the basic wear modes involved, material response, temperature rise, formation of surface films and tribofilms, and the influence of debris.

Some of the elements discussed so far in the motion category are relatively obvious. Perhaps a less evident element is the relative amount of wearing action that each element of the couple experience. For example, in the case of a cam/follower system, the follower will generally experience more sliding or rolling than the cam. Second-order simulation would typically require that this feature be in the test configuration. Another element that may not be immediately evident is the possible presence or absence of vibrations superimposed on the general motion. In sliding, impact, and rolling situations, this additional fretting component can be a significant factor in total wear behavior. Differences in this aspect can frequently be related to differences between the mechanical stiffness of the device and the tester. This can also be a factor in data scatter and the differences in results obtained with similar but different test equipment (1).

Like the previous categories that need to be considered for simulation, loading must be examined from several different perspectives. Perhaps the most obvious is the nature of the force between the wearing bodies. Magnitude and direction of the force must be part of that consideration but it should not be limited to these. For example, is the force constant in the wear situation, and if it does vary, what is the loading rate? The latter aspect can be significant with materials, which are strain-rate sensitive. In impact wear situations, the loading is in the form of a pulse. In this case, not only should the magnitude of the pulse be considered but also its shape and duration. These features can also influence wear behavior. Beyond the consideration of the force, the stress systems that result in the application and the test need to be considered as well. This should not be done only in terms of contact pressure, but it should also be done in terms of the entire stress system, including the subsurface stress, which can influence wear behavior. For example, these subsurface stress systems influence subsurface deformation, crack formation, and crack propagation. Consideration should also be given to the general stress/strain level in the test and the application so that the material is being tested in the same region of behavior as in the application. If only elastic deformation is present in the application, the test should have the same feature; conversely if the application involves plastic deformation, then so should the test.

The stress systems associated with point and line contacts are significantly different than those in conformal contacts. In the former, maximum shear can be below the surface, while in the latter, maximum shear always occurs on the surface. As was mentioned in prior considerations, the stress system can change with wear, as a result of changing geometry. Such changes can be different for different geometries and can have different effects on the total wear behavior of the system.

The elements that influence loading between a fluid and a surface need to be considered in a similar fashion. These would include nature of the flow, velocity of the fluid, angle of impingement, and abrasive content of the fluid.

It can be seen in these discussions that individual parameters associated with wear may enter into several of the considerations for simulation. For example, the nature of the contact (e.g., point, line, or conforming), has both shape and loading considerations. This is also true in two of the remaining aspects, heat generation/dissipation, lubrication, and the environment. The primary concern with heat generation/dissipation is to insure that there is no significant difference in the temperature of the wearing surface in the test and in application. Geometry, loading, and motion parameters are involved in this consideration. The temperature of the environment and the ambient temperature of the components are also factors to be considered in this respect. The considerations regarding lubrication has similar features. Since the intent is to insure that lubrication in the test is the same in the test and in application, the type of lubrication is significant. The mixture or degree of boundary and hydrodynamic lubrication are important and; speed, geometry, and load influence this. In addition to using the same lubricant in the test and in application, the supply, quantity, possible aging, and contamination are elements that also need to be considered. All could have an effect on wear performance. With the environment, the general concern is to insure that the temperature of the surfaces, surface films, and chemical interactions are similar for the test and application. Typical considerations with this element are the temperature, humidity, and chemical composition of the atmosphere surrounding the contact, but other elements could be involved as well. For example, motion and geometry can be factors as the test geometry might allow the formation of a stagnant region around the wear spot. This would tend to inhibit or reduce chemical effects. In the application, this may not occur and the wear would be modified.

While these discussions of the seven attributes of a wear test illustrated elements that need to considered for simulation, they do not indicate how one goes about establishing simulation in practice. As was mentioned previously, the starting point should be from the standpoint that the actual device or wear situation must be replicated for simulation. Then, by considering the various elements, a judgement can be made as to whether or not certain features need to be replicated or how close the replication should be. This is usually done on a hierarchical basis. Those elements, which constitute first-order simulation and basically define the basic wear situation, need to be replicated. What this means is that for a rolling wear situation, the test should be a rolling test; if erosion, erosion; etc. Furthermore, the relative amount of wearing action that each member experiences should be similar in the test and in application. In a cam-follower application for example, the follower surface tends to experience more rubbing than the cam. Consequently, a ball-plane test configuration, where the cam material is the ball and the follower materials is the flat, would not be an appropriate simulation of the situation. The material for the cam should be used for the slider or ball for simulation, as illustrated in Fig. 8.1. As a rule, it is also generally necessary to replicate the nature of the contact configuration (e.g., flat-on-flat, thrush washer, point contact, etc.).

For adequate simulation, there is generally more latitude in the selection of the specific values of the parameters associated with these features, as well as for secondary elements, than there is with the selection of the basic elements. Values of velocities, loads, sizes, repetition rate, etc., typically do not have to be identical in the test and in application. This is also true of such aspects as the use of unidirectional or reciprocating motion, degree of vibration present in the test, method of applying or developing the load, as well as others. However, they should be in appropriate ranges. This is also appropriate for the considerations of lubrication, environment, and thermal aspects. To a large extent, what defines these ranges are the natures of the materials involved, including known sensitivities to different wear situations. These ranges are also defined by the sensitivity of relevant wear phenomena to these elements and parameters. The intention is to insure that the same relative mixtures of wear phenomena and mechanisms occur in the test as in the application. For example, the general sensitivity of most materials to high temperature would suggest that tests for engine components (e.g., piston rings or valves) should simulate the high temperature of that application. For plastics, greater consideration needs to be given to frictional heating elements, such as speed and heat conduction paths, in wear evaluation for nominally low or room temperature applications than for most metals and ceramics. This is because the temperature

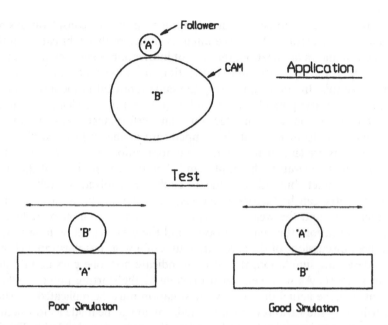

Figure 8.1 Illustration of good and poor simulation for a cam-follower application in a ball-plane test.

sensitivity and poor heat conduction properties of plastics generally make them more sensitive to these elements.

Some additional examples may help to illustrate this as well. Consider the simulation of an erosion or cavitation wear situation. If the materials involved in the evaluations are inert with respect to the chemical make-up of the fluids involved in the application, there is little need to replicate that element in the test. However, if the materials are sensitive, then this would become a significant element in the simulation. Such things as temperature and depletion of the corrosive element would now have to be considered. Another example would be in the case of simulating a wear situation, which is primarily a normal impact but does involve a small amount of sliding. Since studies have shown that a small sliding component has a small effect on the overall wear of metals in such situations, the small amount of sliding can most likely be ignored in the simulation if only metals are to be evaluated. However, similar studies have shown that elastomers are much more sensitive to the sliding component. As a result, if elastomers are to be investigated for the application, then the sliding element must be replicated in the simulation. Another illustration of these types of consideration involves stress levels. Since many materials have both an elastic and plastic range of behavior, a primary requirement for simulation is that the loads and stresses in the test and in application be in the same range. However, the actual values in the range may be relatively unimportant, provided they are representative. Testing under elastic conditions for a wear situation where plasticity is a factor will result in missing those mechanisms, which are related to plasticity.

In these considerations for simulation, it must not be forgotten that surfaces are modified as a result of wear. The influence of load, geometry, etc., on such things as transfer film, third-body film, and oxide layer formation must be considered. In a material-wear system where transfer film formation is known to be a major factor, care has to taken to replicate those elements, which influence the formation and durability of such films. Such elements as unidirectional vs. reciprocating motion, amplitude of motion cycle, which member experience the greater amount of rubbing, and relative size of the components are major factors in the considerations involved with simulation in these cases. On the other hand, if transfer film formation is not a factor, these considerations may be relatively unimportant and need not be replicated.

Having gone through these considerations, an apparatus is selected or designed and built. Given that there is the ability to perform some tests, it is desirable to investigate the influence of some parameters and features before being satisfied with the degree of simulation. If this testing indicates strong sensitivity to an element of the test or the test apparatus, this would suggest that it is an important element to simulate. The degree of replication of that element between the application and the test should be reviewed at that point. It may be necessary to modify the test or apparatus to replicate completely that element.

The litmus test of simulation is the comparison of the wear-scar morphology that occurs in the test to that which occurs in the application. The primary consideration is the nature of the damage and the appearance of the worn surfaces. These give indications of the wear mechanisms and phenomena involved. Differences in morphology would indicate that simulation is poor. Since wear is a system property, this comparison should be done for both of the surfaces involved; not just one. In addition to comparison of wear-scar morphology, comparison of wear debris and the condition of abrasives, lubricants and fluids after the wear exposure can provide additional insight into the adequacy of the simulation. Difference should always be viewed as an indication of lack of simulation. Two examples of wear-scar morphology comparisons are shown in Fig. 8.2, illustrating good and poor simulation.

Many of the considerations addressed for simulation are somewhat intuitive from a materials perspective. While this is the case, it is important to keep in mind that for wear, the considerations for simulation must be not be limited to that element, since wear is a system property. One key element of this is that materials concerns must focus on both members of the wearing couple; it cannot be limited to simply the wearing member or the material being evaluated.

These general areas of considerations involved with simulation should be used as a checklist. The selection of parameters associated with these considerations depends on the degree of simulation desired and the purpose for which the test is being done. In addition to using the list to design or develop a test, it can be used to assess the relevance of existing data to an application or the applicability of an existing test or test apparatus. The general concepts of this chapter are illustrated in the case histories covered in *Engineering Design for Wear: Second Edition, Revised and Expanded* (EDW2E).

Before concluding this treatment of simulation, there are two additional points that require some specific comments. One is concerned with the simulation of break-in. Since break-in can have a significant influence on ultimate wear behavior, specific attention should be given to this element in considering simulation. This can be extremely important in providing successful correlation to field performance. For good simulation, it is generally desirable that the test replicates the break-in and wear-in of the application. A second point is in connection with wear situations that involve multiple wearing actions. In cases where these multiple actions are independent, each action can be simulated in separate tests and still provide correlation with the field. For example, two modes of wear or wear actions were identified for a roller used in a check-sorting machine. One was an abrasive wear action as a result of slip on the surface of a check. The other was a tearing or fatigue type of wear as a result of engagement with the edge of a check. In this case, two wear tests were developed to characterize wear behavior and used in the resolution of the wear problem; one simulated the abrasive action, and the other, the edge engagement (2,3). Another example of this approach is associated with the development and selection of

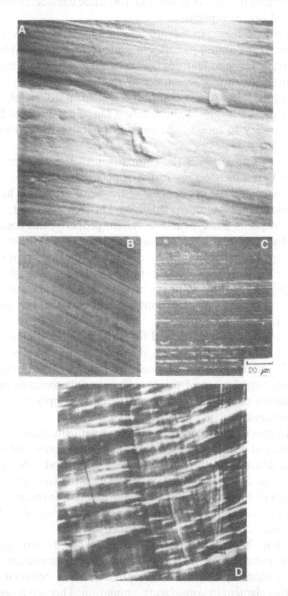

Figure 8.2 Example of test and field wear scar morphology. "A" is the wear surface of UHMW polyethylene from an application; "B" through "D", from laboratory tests. The magnification of the micrographs is similar. "B" shows an example of good simulation; "C" and "D", poor simulation. "E" through "H" show examples of good simulation in the case of abrasive wear of metals. "E" and "F" show examples of the range of wear-scar morphology found in a field test on steels. "G" and "H" show the morphology on these same steels in a laboratory test. ("B" from Ref. 35; "C" from Ref. 36; "D" from Ref. 37; and "E"–"H" from Ref. 38; reprinted with permission from ASME.)

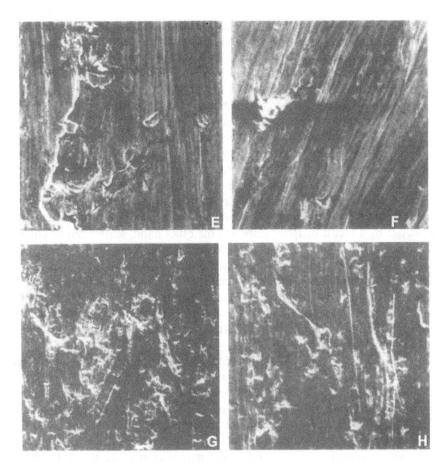

Figure 8.2 (*continued*)

coatings for icebreaker hulls, where several tests were developed to simulate different wear actions (4). In both of these cases, examination of actual wear scars and theoretical considerations indicated that the underlying assumption of independence was valid. This type of approach is sometimes more convenient or practical than trying to develop a test combining all the elements. However, the assumption of independence has to be verified.

A system for characterizing a wear situation in terms of mechanical features has been proposed for use in selecting tests for evaluating materials for different applications. In this system, a Tribological Aspect Number (TAN) describes both the test and application. Simulation is obtained by matching TAN values (5). This system is described in Sec. 8.7.

8.3. CONTROL

Wear testing and wear tests generally have poor reputations. A common impression about the general characteristics of wear tests and wear testing is that there is: large scatter in the data, poor reproducibility, and that correlation between laboratories is poor. Unfortunately, this is an accurate description of many wear tests and evaluations that have been done. While it might be correct to conclude from this that wear testing has a definite tendency for such a behavior, it is not correct to conclude that such a behavior is inherent to wear testing. Many studies and standardization activities have shown that such a behavior is not an intrinsic feature (6–14). In standardized tests, repeatability of 25% or less has been shown to be achievable with wear tests, both within a single laboratory and between laboratories. Reproducibility of results within a laboratory tends to be higher than that between laboratories. Within a laboratory, repeatability of 10% or less has been demonstrated in some cases. There are many non-standard tests reported in the literature, which show similar repeatability within a laboratory.

The tendency for large scatter and poor repeatability in wear testing can be understood in terms of the complex nature of wear and wear phenomena. A wide range of factors influences wear behavior and parameters, as has been discussed in Part A, which provides a summary of wear and friction behavior. Lack of control on any of these parameters during a wear evaluation may result in large scatter. Since wear coefficients can vary several orders of magnitude, lack of adequate control of key parameters can easily result in scatter by a factor of 10 or more. On the other hand, scatter can be considerably reduced with adequate control, with repeatability in the range of 10% being possible. Likewise, differences between laboratories in the values of these influencing parameters or the degree of control of these parameters can result in poor inter-laboratory correlation, while adequate control and the use of the same parameter values result in good agreement. Without proper control, completely different results are possible between laboratories (e.g., different rankings or large differences in absolute values), but with proper control agreement within 10% is possible.

The considerations involved with control are very similar to the considerations involved with simulation; in fact a checklist that could be used for control considerations is the same as that used for simulation. However, there would be a difference in the focus. For simulation, the focus is the correspondence between test and application. While for control, the focus is the consistency within the test itself. Another way of considering the control aspect of wear testing is to focus on four general areas of the test or testing. One is the test apparatus. The second is the materials used in the test. The third is the environment surrounding the test. The fourth is the procedures associated with the conduction of the test. The subject of control in wear tests will be addressed from the standpoint of these four areas with emphasis placed on the more common contributors to scatter and poor inter-laboratory agreement in these areas.

With respect to the test apparatus, the primary concern is associated with the ability of the testing apparatus to consistently provide the proper wear situation (e.g., load, motion, lubrication, etc.). Generally, the scatter of wear test data is inversely related to the ability of the wear tester to repeatedly provide the same wear conditions, such as alignments, loads, and motions; hence, the design, construction, maintenance, and calibration of these apparatuses are significant factors. For good control, wear test apparatuses should have the characteristics of precision equipment and, in addition, be durable. Typically, repeatable positioning and alignment capabilities within a few 0.001 in. or less is required, as well as load and speed control of better than 10%. In tests which involve the feeding of abrasive materials or fluids in the wear zone, a similar precision with respect to such aspects as flow rates, pressures, particle velocities, and orientation of the stream is generally required. The ASTM's wear tests referred to previously offer good examples of this requirement. In these tests, tolerances on key dimensions of the test apparatus are specified, as well as tolerances on loads, alignments, and speeds. In erosion and sand abrasion tests, tolerances on velocities and sand feed rates are also specified. In addition to being concerned with this intrinsic nature of the wear apparatus, it is also necessary to be concerned with the continued performance of the apparatus. This means that it is also necessary to establish procedures and techniques to monitor the performance and calibration of the apparatus. For many of these aspects, typical engineering techniques can be used for checking loads, dimensions, speeds, etc., but because of the unique nature of certain tests, it might be necessary to develop a technique to measure key parameters in the test as well. An example of this is the special techniques used for measuring particle velocity in the standard test for erosion (11).

This need for checking or monitoring the condition of a wear apparatus is of particular significance in cases where the counterface is part of the wear tester. An example of this is the rubber wheels used in the dry and wet sand wear tests, where a rubber wheel is used to press and rub abrasives against a wear specimen (7,8). The purpose of these tests is to measure the wear resistance of the specimen to low stress cutting abrasion and wear or damage to the wheels is not of interest. However, while the wheel materials used are quite resistant to wear or damage in this situation, they do degrade and wear, which can affect the wear of the specimen. Consequently, it is necessary to monitor the condition of the wheels and to either change or dress them to insure repeatability. As part of the development of these test methods, standard techniques for dressing the wheels and guidelines for wheel replacement were developed. Another example of this type of concern is testing with the Taber Abraser, which uses standard abrasive wheels to evaluate materials. Their state must also be monitored. An example of such a test is ASTM F1978, which is used for the evaluation of abrasion resistance of metallic thermal spray coatings. Beyond these unique types of concerns, there has to be a general concern with the overall wear of the test apparatus as well. Bearings and reference surfaces of the apparatus can wear with use; nozzles used in erosion and cavitation tests can wear and change dimensions. Hence, it is desirable to continually monitor the status of the apparatus so that tolerances stay within the appropriate limits and performance is maintained.

The next area to consider in terms of control is the materials involved in the wear test. This consideration is of equal importance to the concern with the apparatus and is not limited to the wear specimens. Control of the other materials associated with a wear test, such as lubricants, abrasives, slurries, or counter face materials, is of importance as well. The particular aspects that need to be controlled vary with the materials and the test. Lack of consistency and uniformity in composition and purity are common materials aspects that generally increase scatter in wear results. Variations in hardness, cure, heat treatment, as well as surface finishing techniques, are also common contributors to scatter. In the dry sand/rubber wheel test, for example, the dressing, cure, composition, and durometer (hardness) of the rubber wheel all need to be controlled. The abrasive sand used in that test also must be controlled in terms of composition, size, and angularity. Moisture content of the sand must also be controlled. A drying procedure is specified to insure consistency. For fluids used in wear tests, other aspects, such as viscosity, viscosity index, and pH, may need to be controlled as well. In certain cases, this same level of control needs to be extended to materials, which are used to prepare the wear specimens or counterface surfaces. For example, a high level of purity for solvents or cleaning agents used to prepare wear specimens and counterfaces may be required (e.g., reagent grade).

For wear specimens and non-specimen counterfaces, such as the wheels in the referred to dry and wet sand/rubber wheel tests or the Taber Abraser type test, control of dimensions is also important. These might be in the form of tolerances on sizes and shape, such as length, width, thickness, or radius, or they can be in the form of concentricity, flatness, and parallelism requirements. These aspects obviously are significant factors in achieving reproducibility since they directly influence the geometry of wear contact and can also influence stress level, load distributions, and shape of the wear scar. This element generally has to be addressed if a test is to be well controlled. However, what parameters have to be specified and the degree to which they have to be controlled vary with the test and materials involved. For example, point contact wear configurations are generally more forgiving then those involving area or line contact. However, tight tolerance might be required for the radii involved in the point contact configuration. Materials also influence this as well. In tests involving elastomer and less rigid polymers, alignment is less critical than with more rigid materials, like metals or ceramics. Certain test techniques and methods of analysis may be developed to minimize some of these sensitivities as well.

Another aspect that is frequently a concern with wear specimens is surface preparation or cleanliness. Wear specimens can be produced or need to be produced by a variety of means. Some are machined from wrought stock; others are molded and still others may be weldments or castings. In addition, the specimens can be handled, stored, or packaged in a variety of ways. The net result is that the surface of the specimens can be contaminated by a variety of organic and inorganic materials in an uncontrolled manner as a result of these processing and handling techniques. Since wear is influenced by absorbed and adsorbed layers and by the nature of surface films, the presence of these uncontrolled layers can result in scatter in wear performance. Consequently, it is desirable to clean or prepare the specimens in a prescribed and controlled manner to insure consistent surface conditions. Again, the particular procedures and techniques vary with type of materials and the nature of the test and may involve several steps, each addressing a particular aspect. For example, a solvent may be needed to remove organics, such as oils from a machining process or handling. An abrasive action might be required to remove oxide and scale from a surface, followed by a rinse and drying procedure to remove the abrasive. This can be a very difficult area to address because of the wide range of contaminants possible, the need to use techniques and solvents which do not effect the base material, and the fact that some solvents and cleaning techniques can leave residues. The need to control surface contamination is especially important in unlubricated wear tests and the challenge is greatest in these cases. Lubricated tests tend to be less sensitive to contamination. Even in such cases where there is not as strong a sensitivity to contamination, some form of surface preparation and cleaning generally needs to be established.

The third general area that has to be considered for control is the environment surrounding the test or wear couple. In wear tests in which unique and specific environments are established, the need to provide a stable and repeatable environment is self-evident. In such a case, the uniqueness of the environment has been recognized as a major factor in the wear situation and the significant parameters identified. Examples in this category are wear tests done for space applications, engine components, and devices operating in atmospheres other than air (15–17). Temperature, composition of the environment, uniformity of gas mixture, and relative humidity are examples of specific parameters, which have to be controlled in these cases. However, the majority of wear tests are done for room-ambient applications and therefore, there is not an a priori sensitivity to environmental conditions. In these cases, the parameters normally of interest are temperature and relative humidity. Both can have significant effects on wear behavior and may need to be controlled to a few degrees and a few percent of relative humidity. The tolerances on both depend on the wear situation and the materials involved. It is possible that the tolerance on either of these parameters may be tighter than that provided by normal air conditioning systems. In such a case, special air conditioning may be required to minimize scatter and improve repeatability. Alternatively, an environmental chamber may be required for the tester to provide the needed control.

A case history might serve to illustrate this point. In determining the abrasivity of papers, a test was developed and found to be quite repeatable over a three-month period of extensive testing. Tests generally repeated within 10–15%. The need for testing declined for a period of time and resumed about six months later. Increased scatter was observed in the second series of tests and differences of an order of magnitude were found between the two testing periods. After some investigation, it was finally concluded that the abrasivity of paper was extremely sensitive to moisture. Both test series were conducted in the same air-conditioned laboratory. The first occurred during winter months and the second in late spring and early summer. While the room in which the testing was done was air-conditioned, the relative humidity varied with the season. The relative humidity was low during winter months, possibly as low as 10–20%, while in the spring it typically was 60%. It was concluded that for adequate control it was necessary to conduct further tests in specially controlled test chambers.

In conducting tests in ambient environments, which are not specifically controlled, it is generally a good rule to monitor and record temperature and humidity. Such records might be useful in sorting out problems with test scatter.

Identification of the parameters and estimates of the tolerances that are required for the different parameters associated with the apparatus, the materials, and the environment may be made from theoretical considerations regarding possible wear mechanisms and material behavior considerations, published data, and prior experience. While these considerations should always be involved in test development or selection, they are often not adequate. Systematic testing is frequently required and generally desirable to characterize the significance of these parameters on the results and, in some cases, to determine the required tolerances. It is often desirable to do some initial experiments with a jury-rigged or preliminary apparatus to address some of these aspects prior to building a final apparatus. Alternatively, a design might be developed which allows for modification as this information is developed. While it might be concluded from these activities that a particular parameter need not be controlled tightly, it is generally a good rule of thumb to build-in as much control as possible.

In addition to using specific methods and techniques to monitor individual features of the apparatus, materials, and the environment, a frequently used technique in wear testing is to develop a reference wear test with the apparatus. This test would be done utilizing controlled reference materials and a fixed set of other test parameters. First, a database for this condition and representative of stable performance of the apparatus is established. This test is repeated from time to time and the results compared to the database. If the results fall outside the range that is typical for the test, investigation into possible reasons for the change is then done, including a review of the calibration and current status of the apparatus, the procedures used in preparing the specimens and controlling the materials, and the environmental controls.

Many standardized tests specify this approach for controlling the test and provide reference conditions for such tests. In this comparison, not only can the magnitude of the wear be used as a check, but also the shape, location, and morphology of the wear. These individual elements can highlight different problems. For example, variations in the shape or location of the scar could point to an alignment problem. If the magnitude of the wear has changed, this could point to a loading problem. Changes in the morphology of the scar and the presence or absence of films could indicate loss of control in environmental conditions or preparation procedures.

Surrounding the entire issue of control is the establishment of procedures for conducting the test and insuring routine monitoring of the critical parameters involved.

As stated, wear testing is prone to large scatter. One element in minimizing this is to exercise good discipline in conducting the tests and paying attention to details. It is a good practice to include the use of the reference test as part of this procedure to help insure this discipline. For example, the reference test could be done at the start and end of a testing sequence to insure that nothing has changed. If the test or apparatus is particularly sensitive or is prone to large scatter, interspersing this reference test during a sequence may be desirable as well. When this is done, these reference tests can often be used to provide or develop a correction or scaling factor for the data obtained between the reference tests. This is often an effective way of accounting for variations in environmental conditions or some of the materials used in the tests, like abrasives, that cannot be controlled as well as desired (18,19).

The routine used in performing the test is as important an element as control of the apparatus, materials, and environment. For example, such details as maintaining the same sequence of tightening screws in positioning a wear specimen, length of time between preparing a specimen and starting the test, method of stopping the test or performing the measurements, can be important in minimizing scatter. These details should be covered in the procedures established for tests.

If a wear apparatus has the capability of measuring friction, this can also be used as a monitor for control. Wear and friction are sensitive to many of the same parameters. If changes in the friction behavior are seen in repeated tests, this would tend to imply that one or more of the parameters involved are varying. The reasons for these variations should be investigated and proper controls established. This would be both for the reference test as well as replicates of other tests.

8.4. ACCELERATION

There is a desire to minimize or reduce test time in most wear testing situations. This can be for a variety of reasons but it is usually so that the testing does not delay the achieving of the primary goal, such as material or design optimization or problem resolution. In any event, this desire brings an element of acceleration into wear testing. To achieve acceleration, one or more of the parameters associated with the wear needs to be more severe in the test than in the application. This directly conflicts with the primary rule in wear testing, namely, to simulate the application in the test. However, effective acceleration is possible in some cases, but in general it is risky and should be approached in a careful and deliberate manner. There are several reasons for this. One is that there are many mechanisms for wear and in any wear situation one or more of them are typically present. Since the different mechanisms do not necessarily depend on the same parameters in the same manner, the relative contribution of these mechanisms can change as a result of changing the value of a parameter. Consequently, the effect of the acceleration on each mechanism needs to be understood if simulation is to be maintained. Another aspect is that changes in the parameters can also eliminate or introduce different mechanisms, phenomena, and material changes that can significantly alter the situation. An example of this would be thermal softening of a polymer or the formation of different oxides on a metal surface as a result of the use of too high a speed in the test. Increased loading might introduce plastic deformation that is beyond the level found in the application. In general, the sharp and dramatic transitions that have been found in wear behavior make acceleration in wear testing an element that needs to be approached with caution.

Acceleration in wear testing can be approached by considering physical parameters, such as load, speed, temperature, or environmental composition. Depending on the situation, altering these parameters may provide some degree of acceleration and still maintain adequate simulation. However, there is not a general rule of thumb associated with these factors like there is for the acceleration of chemical reactions by increasing the temperature (i.e., a factor of $2\times$ for every $10°C$). When relationships between wear measures and these parameters are approximated by a power relationship, x^n, there is considerable variation in the value of n that is obtained. In some cases, it is less than 1. In which case, the amount of acceleration that may be achieved may be minor. Double of the parameter would result in less than doubling the wear or wear rate. In other cases, it can be much larger than 1 and significant acceleration can be obtained with relatively small changes in the parameter. For example, with fatigue wear, wear is related to some high power of the stress, for example, $n > 5$. In cases where this mechanism is dominant, a small increase in stress level may provide significant acceleration. Because of this wide range of possibilities that exist with wear, it is not possible to identify a universal rule of thumb for estimating acceleration factors.

Effective acceleration can also be achieved in other ways. One way is by increasing the amount of wearing action that takes place in a unit time, in the case of abrasive wear, for example, the amount of abrasive applied to the interface could be increased beyond the quantity that is present in the application (18). Another example would be in the case of rolling with some slip (20–22). The amount of slip in the test could be made larger than that in the application. For applications in which the wearing action is intermittent, acceleration can be achieved by shortening the time between wearing actions. Approaches based on increasing the wearing action per unit time are generally less risky than significant modifications of the physical parameters and can provide significant acceleration and good simulation. However, problems can occur with this type of acceleration as well. For example, too short a time period between wearing actions could result in increased surface temperature or decreased healing of a lubricant film. In the rolling case, too much slip could be introduced and overshadow the rolling aspects of the contact. In addition, there might be saturation effects associated with this type of approach, limiting the amount of acceleration that can be achieved. An example of this is in the case of abrasive wear. Abrasive action tends to increase with the amount of abrasive present up to a certain level but beyond that level, any increase in the amount of abrasive does not increase the amount of wear (17). Similarly, there might be a maximum amount of sliding that can be introduced in the rolling situation. In addition, duty cycles can only be increased to 100%.

A third way of approaching acceleration is by refining the amount of wear that can be measured and using smaller amounts of wear to project performance or to base decisions. This approach does not effect simulation, since it does not involve changing parameters. Because this approach implies that no changes in wear-behavior occurs as the magnitude of the wear increases, the test duration should be long enough so that significant wear is produced and stable wear-behavior is apparent. With this approach, special care has to be taken in situations in which materials properties vary with distance from the surface. This is a situation frequently encountered in practical wear situations, such as the situation where platings or other coatings are used. Situations involving casehardened steels and molded plastics are other examples. In cases like these, the wear test has to be carried to a depth representative of the depth to which it will be allowed in the application. If not, erroneous projections or assessments will be made since the wear properties of the underlying layers will not be measured in the test. In many printer applications, for example a hard Cr plating is used on the surface of case- or through-hardened steel parts. The thickness of the plating is typically in the range of 0.001 in., while end-of-life wear depth on these components can be in the range of 0.005 in. This means that the useful life of the part involves the wear resistance of the Cr plating and the wear resistance of the hardened steel. This is in addition to any role that the hardened steel has as a substrate for the Cr layer. A wear test which results in wear less than 0.001 in. would only provide information regarding the Cr plating, supported by the particular substrate. It could not be used to project or assess the performance in an application where the wear would be allowed to progress into the substrate. A test resulting in a wear depth of greater than 0.001 in. would be required for that assessment.

In practice, wear tests frequently involve all three ways of providing acceleration. Values of individual parameters will be used which are higher or more severe than in the application, wearing action per unit time is increased by using saturated amounts of abrasives or by insuring continuing wearing action, and small amounts of wear are measured. The amount of acceleration that any or all of these elements provide or the effect of these on simulation has to be evaluated on a case-by-case basis. The amount of acceleration that the parameters or wearing action intensity can provide can be estimated from models associated with different wear mechanisms and phenomena. While these same models can be used to estimate the effect on simulation, the effect of the acceleration on simulation still needs to be addressed empirically. One way of doing this is to conduct wear tests at accelerated and non-accelerated conditions and compare the results. This comparison should focus on the physical appearance of the wear scars and debris. The intention is to verify that the same mechanisms and wear phenomena occur under accelerated conditions as they do under normal conditions. Comparison to actual wear scars from applications would also be of benefit. The amount of acceleration that is provided can be verified or established by comparison of the quantitative results of these tests as well. Because of the complex nature of wear behavior, any claim to simulation is suspect if these comparisons are not done.

An additional point to be recognized in both the considerations of simulation and acceleration is that conditions, which are acceptable for one type of material may not be acceptable for another type. For example, acceleration by increasing speed or repetition rate may be acceptable for metals, but the same acceleration may result in significant temperature rise with plastics and invalidate simulation. Similarly, test duration and loading that are developed to evaluate hard bulk materials might be too coarse to evaluate softer materials or thin coatings. Several of the ASTM standard tests have different practices for the different types of materials because of these considerations.

8.5. DATA ACQUISITION, ANALYSIS, AND REPORTING

Obviously, one piece of data from a wear test that is to be obtained, used for analysis, and reported is a measure of the wear on the wear specimen. This can be a direct measure, such as volume loss or change in a dimension, or an indirect measure, such as time to seizure or malfunction of a device. A common presumption is that this measurement provides an adequate or complete summary of the wear behavior. This is usually more an idealistic desire from a testing standpoint than a realistic assumption in wear testing. Generally, a simple measurement of the wear of one member by itself is not adequate. It is a good practice in most wear testing and, in certain cases necessary, to gather, analyze, and report other pieces of data as well. In addition, it may not only be useful but necessary to develop

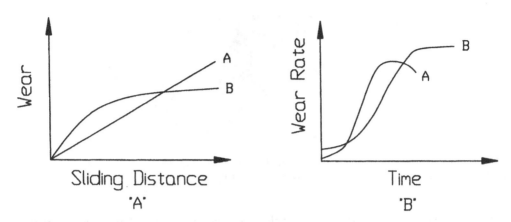

Figure 8.3 The effect of possible non-linear behavior on the interpretation of wear test results. "A" represents possible wear behavior of two materials in the same sliding wear test. "B" represents wear behavior often found in cavitation tests. In these cases, the ranking of the materials, using a single end point value, depends on the duration of the tests.

a wear curve with the test, rather than use a single end-point value. A wear curve is a plot of wear or wear rate values obtained for different amounts of wearing action, such as test time, amount of sliding, or amount of abrasive supplied. Typical wear curves that might be obtained with different materials are illustrated in Fig. 8.3. As can be seen, end-point values do not necessarily provide sufficient information when variations in behavior and non-linear behavior are involved. As can be seen in this illustration, even ranking of the materials can change depending on the duration of the test.

The amount, type, and value of additional data that is obtained can vary with the state of development and application of a test and also with the use of the test. It is generally the greatest during the initial stages of the development of a test and the establishment of simulation. Simple material ranking or selection tends to require less than the development of a complete model or the establishment of correlation with the field. However, even in simple material ranking, it is desirable to obtain more data than the amount of wear on one member. As a minimum, data regarding the degree, magnitude, or state of the wear of the counterface are generally significant and should be obtained.

The following examples indicate the need for additional and various types of data obtained from a wear test. While it is often a practical necessity to summarize wear in terms of a single parameter, such as a scar volume or depth, this single parameter does not necessarily describe all the significant attributes of the wear. For example, morphological aspects of the worn surfaces, such as evidence of transfer, cracking, and plastic flow, are aspects that are needed to identify mechanisms and provide a basis for modeling and simulation. This is also true for transfer and third-body film formation. Do they occur or not? The nature and extent of these films are additional attributes that might be noted. Also, the general shape and uniformity of the scars often can be significant in terms of control.

Finally, the relevance of the primary measure can change, depending on the nature and occurrence of some wear phenomena. For example, if mass loss is being used as the measure of wear, adhesive transfer of material to the surface can result in misleading values. In fact, when this measure is used, negative wear or mass gain can occur. This condition is illustrated in Fig. 8.4. In this case, mass loss would not provide a good

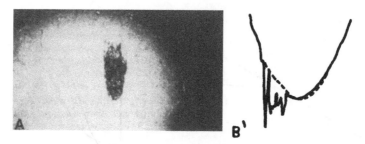

Figure 8.4 "A" is a micrograph of sphere, showing mass gain in a sliding wear test. "B" is a profilometer trace across the region, showing the build-up of material on the sphere.

measure of wear volume. Another example would be in the case of the measurement of wear of a coating. If there is no deformation of the substrate, the depth of the wear scar can be directly related to the thinning of the coating. However, if deformation of the substrate occurs, there may be no thinning of the coating but a depression or wear depth can still be measured. The micrographs in Fig. 8.5 show the wear of printed circuit contact tabs. In one, no noticeable wear of the Au plating has occurred, while there is a deep wear scar, caused by deformation of the substrate. In the other, there is no deformation of the substrate and the Au layer is severely worn. These two situations are extremely different in terms of the corrosion protection and contact resistance that is provided by

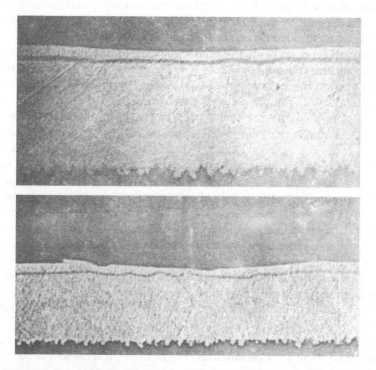

Figure 8.5 Cross-section through the wear scar produced on Au plated printed circuit card tabs. "A" shows the situation when there is little or no substrate deformation involved; "B", when there is significant deformation.

the Au layer. In this case, wear scar depth is not always a measure of the residual thickness of the Au layer.

Consequently, it is a good practice to collect data regarding the general nature and appearance of the wear scar produced in the test. This usually means observing and recording information regarding the nature of the scar to supplement the primary measurement. The most basic methods for this are: visual and low power optical microscopy, followed by scanning electron microscopy (SEM), and higher power optical microscopy. Some additional techniques that may be appropriate include electron dispersive x-ray analysis (EDX), Fourier Transform infra-red analysis (FTIR), electron scanning for chemical analysis (ESCA), and Auger analysis (AUGER), as well as profilometer measurements and cross-section examinations. In the extreme, it may be necessary to routinely take SEM or optical micrographs of the wear scars, along with the primary quantitative measurement. In cases where profilometer traces of the wear scars do not provide the primary wear measure, consideration should be given to using them as a means of providing additional information. In situations like the electrical contact situation shown in Fig. 8.5, routine cross-sectioning on the scar might be needed or desirable. These additional pieces of data have often proved to be very valuable and in some cases essential in either improving control, modeling, or establishing correlation.

In sliding and rolling wear situations, it is generally desirable to measure or monitor frictional behavior during the test. The correlation of frictional changes with wear behavior frequently provides useful information regarding mechanisms and modeling. This aspect will be treated further in a following section.

The minimum information that should be recorded and associated with each wear test is the values of the major parameters that influence the wear situation. These would be those parameters, which are specifically controlled in the test. Beginning and ending values of these parameters are desirable. It is also desirable to note and record information regarding additional parameters and factors, which could influence the wear. Basically, a fundamental set of auxiliary information that should be kept is information regarding specimen preparation, operational parameters of the test, and the environment surrounding the test. When lubrication is involved, additional data and observations regarding the lubricant and the method of supplying or application are also needed. Noting the state of lubrication at the start, during, and end of the test is generally a good practice. Characterization of the lubricant by Fourier Transform infra-red analysis and chromatography techniques may be needed where degradation of the lubricant is a factor. Within a laboratory, it is sometimes useful to note which engineer or technician conducted the test as well. This might be of use in resolving problems associated with technique.

Several all-inclusive lists or forms for this purpose have been proposed (23,24), one such form or list is shown in Fig. 8.6. Examination of this form in Fig. 8.6 shows that the listing is very extensive and identifies all the additional information and data referred to in the preceding paragraphs. These lists include data and information regarding the nature of the wear and wear scars for both members of a wearing interface. It is also recommended that such information be provided with the results reported in the literature. This enables the reader of such literature to assess the relevance of the published data to their specific situations and can alert them to specific problems and sensitivities.

While it may not be practical or necessary to record all the information requested by such a form, it is frequently desirable to review such a form when developing a wear test. This helps to identify aspects that should be controlled in the specific case or that should be recorded to aid in the resolution of potential problems with the test or with correlation to the field. Experience with the development and utilization of wear tests indicates that what initially appears to be unnecessary attention to detail often turns out to be valuable information needed for the improvement of a test and establishing correlation. Once a test is well-established, the amount of information that is needed can often be reduced to the minimum set of major parameters or factors and attributes of wear scars.

One of the initial points made in this chapter was that it is frequently desirable to generate wear curves in a test, rather than simply determine a single end-point value (see Fig. 8.3). In the former, a curve of wear vs. amount of wearing action is developed in the test. For example, this could be a curve relating depth of wear scar to number of cycles, amount of time, or amount of abrasive used. In the latter, the wear is simply measured after a specific amount of wearing action or exposure has taken place. There are a few reasons for developing a wear curve. A fundamental one is that, in using a single value, an underlying assumption is made that all the systems evaluated follow the same relationship with duration or amount of wearing action. Consider a situation for

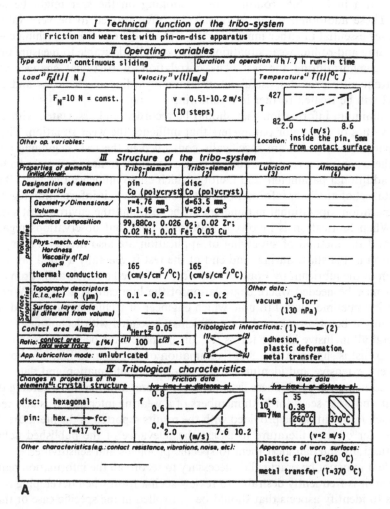

Figure 8.6 Examples of wear test data sheets. "A", pin-on-disc test; "B", ball-on-cylinder test. (From Ref. 23, reprinted with permission from Elsevier Science Publishers.)

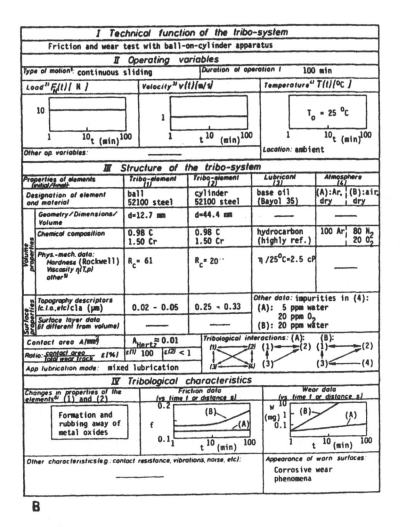

I Technical function of the tribo-system

Friction and wear test with ball-on-cylinder apparatus

II Operating variables

Type of motion[1]: continuous sliding   Duration of operation t   100 min

Load[2] $F_N(t)$ [N]   Velocity[3] $v(t)$ [m/s]   Temperature[4] $T(t)$ [°C]

$T_0 = 25$ °C

Other op. variables: ———   Location: ambient

III Structure of the tribo-system

Properties of elements (initial/final)	Tribo-element (1)	Tribo-element (2)	Lubricant (3)	Atmosphere (4)	
Designation of element and material	ball 52100 steel	cylinder 52100 steel	base oil (Bayol 35)	(A):Ar, dry	(B):air, dry
Geometry/Dimensions/Volume	d=12.7 mm	d=44.4 mm	———	———	
Chemical composition	0.98 C 1.50 Cr	0.98 C 1.50 Cr	hydrocarbon (highly ref.)	100 Ar	80 N₂ 20 O₂
Phys.-mech. data: Hardness (Rockwell) Viscosity η(T,p) other[5]	R_c= 61	R_c= 20··	η /25°C=2.5 cP	———	
Topography descriptors (c.l.a.,etc)/c.l.a. (μm)	0.02 - 0.05	0.25 - 0.33	Other data: impurities in (4): (A): 5 ppm water 20 ppm O₂ (B): 20 ppm water		
Surface layer data (if different from volume)	———	———			
Contact area A [mm²]	$A_{Hertz} \approx 0.01$	Tribological interactions: (A): (B):			
Ratio: contact area/total wear track ε[%]	ε[1] 100 ε[2] < 1				
App lubrication mode: mixed lubrication					

IV Tribological characteristics

Changes in properties of the elements[6] (1) and (2)	Friction data (vs time t or distance s)	Wear data (vs time t or distance s)
Formation and rubbing away of metal oxides	0.2 (B) f (A) 0.1	w (mg) 10 (B) 1 (A) 0.1

Other characteristics(e.g.: contact resistance, vibrations, noise, etc): ———

Appearance of worn surfaces: Corrosive wear phenomena

B

Figure 8.6 *(continued)*

which the depth of wear, h, is related to the distance of sliding, S, by a relationship of the following form:

$$h = KS^n \tag{8.1}$$

The use of a single value implies that the wear situation in the test is characterized by K. This is valid if n is the same for all tribosystems involved in the evaluation. If n varies with the system, which is often encountered in practice, a value for K no longer provides an adequate characterization. In this case, both K and n are needed for complete characterization. The use of a single value approach in such a case can lead to errors, as illustrated in Fig. 8.7. This figure shows wear curves for two different materials obtained in the same test. In this case, the wear curve is plotted on log–log paper. The linear behavior on the log–log plot of these materials indicates that the behavior of both materials can be described by Eq. (8.1) but with different values for n. As a result, it can be seen that if an end-value of a short-term test is used to compare materials, material A would be

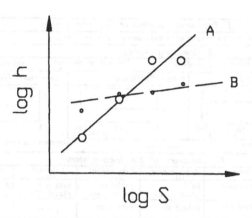

Figure 8.7 Hypothetical wear data for two materials in the same wear test, illustrating the significance of wear curves in evaluations and the need for more than one parameter in characterizing wear behavior.

selected since it has less initial wear than material B. If, on the other hand, a longer test time is used, material B would be selected. In this case, a value for both K and n is needed to compare the behavior of these two materials.

The example shown is not an isolated case. There are many examples in the literature which show wear curves intersecting, indicating that it is not appropriate to simply assume that all systems will have all the same relationships, even in the same test (25–27). At the very least, some testing should be done initially to demonstrate that such an assumption is valid. If it is valid, then the test can be reduced to a single point; if not, a curve or multiple point approach needs to be used to characterize a system. For a relationship of the type shown, values of K and n as determined from the curve could be used to characterize the system.

Transitions in wear behavior as the amount of wearing action or test duration increase can occur and is another reason for utilizing wear curves. Break-in and the introduction of additional modes of wear due to debris action are examples. In addition, there can be incubation periods, such as can occur with fatigue wear, or there might be two distinct modes of wear possible with one being more significant early in the test and other becoming more significant in the longer term. An example of this behavior is the impact wear of elastomers as shown in Fig. 8.8 (28). Initially, the impact wear of elastomers is associated with a compression set (or creep type) behavior, which asymptotically approaches a limiting value. After an incubation period, the elastomer begins to wear by a fatigue process. A single measurement cannot provide adequate information about such behavior and can often be misleading, as can be seen in this example. A short-term test (e.g., less than 10^5 impacts) provides only comparison of materials in terms of their initial creep behavior. It would, therefore, not be a good indicator for long-term performance and in fact the wrong material could be selected for an application which involves a few million or more impacts. On the other hand, if simply an end-point value is obtained in the long-term behavior region, the tendency would be to explain the wear behavior and relative wear performance simply in terms of a fatigue mechanism. The result is at best an incomplete, or more likely, an erroneous model or understanding. This would be a problem in an application where the creep behavior is a significant contributor to the dimensional change caused by the impact.

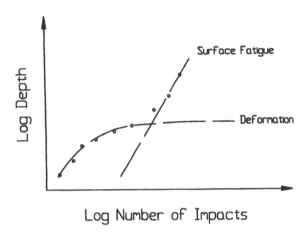

Figure 8.8 Impact wear behavior of elastomers, showing the effect of the presence of two wear modes on the shape of the wear curve. In the initial region, a deformation mode predominates. In the later stages, a surface fatigue mode predominates.

Another reason for the use of wear curves in wear tests is associated with the more sophisticated uses of such tests, that is, projecting field performance. Generally, a relationship between time, cycles, distance, etc., and wear is needed. This relationship can be established from a wear-curve approach but not from a single value. From these considerations, it is evident that the use of wear curves in wear testing not only provides more data and information about the wearing system but helps to avoid errors in wear testing. Because of these aspects, wear curves should be used during the development of wear tests. When a system or wear situation is sufficiently understood, then the reduction to a single value may be appropriate. Also, if the relationship between wear and duration is established, the wear measurements obtained at the individual data points used to establish the wear curve can be normalized to provide additional values of the wear coefficient. For example, in a case where the wear can be described by Eq. (8.1), the depth of wear, h_I, determined at different numbers of cycles or time, S_I, can be divided by S_i^n to provide values of K.

Having addressed the merits and need for the development of wear curves in wear testing, it is worthwhile to consider how such a curve can be developed. There are two general ways of doing it. One is to interrupt the test at intervals for measurement and characterization of the wear. The other is to use several different specimens or individual tests, each being performed for different duration. Both approaches have been used successfully. However, there is an inherent concern with the former. Since that method involves interruption of the test, the wear situation could be altered. For example, behavior of the type shown in Fig. 8.9 has been reported (29). This behavior suggests a break-in type phenomenon occurring at each continuation point. Possible perturbations, which can result in this type of behavior include: changes in alignment or positioning, oxide growth or contamination of surfaces, and disturbance of wear debris. The sensitivity to these elements or even relevance of these elements is not the same for all wear situations. Consequently, provided proper care is taken, the perturbations can have a negligible effect and the technique is acceptable in some cases. The second method, which utilizes several specimens, eliminates this type of concern but in addition to requiring more specimens, generally requires somewhat longer time. In addition, each data point obtained has

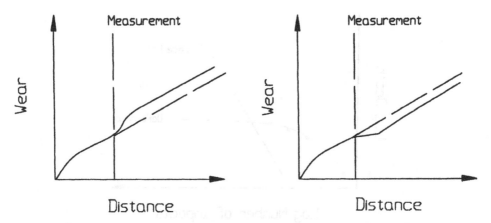

Figure 8.9 Illustration of the possible effects of intermediate measurements and test interruptions on wear behavior.

included its own scatter due to specimen variations. Therefore, specimen control is extremely important in this approach since specimen-to-specimen variations can mask other dependencies. An additional advantage of this approach is that worn samples from each data point can be retained for more intensive analysis. With the single specimen approach only samples from the end point are available for such type of analyses. Because of this element, most testing programs should involve some testing using the multiple specimen approach so that detailed analysis of the state of wear in various regions can be analyzed. This sort of analysis is generally required in addressing simulation, acceleration, and correlation concerns.

Another element with the development of a wear curve is the selection of the intervals for measurements. A good initial approach to use is to obtain data at an order of magnitude difference (e.g., 1, 10, 100 cycles), or on a logarithmic scale rather than on a linear one. This provides a broad enough perspective that can be used to identify the general dependency on duration, while minimizing concerns with measurement accuracy. This approach is generally useful since for stable wear situations, the relationship between wear and duration is at most linear and, if the measure is other than volume, typically less than linear. This means, for example, that doubling of duration typically does not result in doubling the wear measure. In mild wear situations that are appropriate for many engineering applications, the wear often has a very low dependency on duration, such that an order of magnitude increase in duration might only result in a 20–30% increase in the wear measure (30,31). For this type of behavior, it is difficult to discern differences in wear behavior over a small increment of duration.

These large increments also help to identify transitions and help in the development of models and projections. Figure 4.46 is an example of a wear curve that shows a transition in wear behavior that is very significant for the application. In addition, it can be seen that because of the relationship between wear and duration prior to the transition, small increments of duration are not an appropriate way of gathering data in that region. In the refinement of a test, smaller intervals can be used and may be desirable with certain types of behavior. In the application associated with Fig. 4.46, for example, finer intervals near the transition region are appropriate since the value of this transition point is quite significant.

At each of the data points associated with the development of a wear curve, the data gathering should not be limited to the wear measure. The additional observations discussed at the beginning of this section should be done and it is desirable to record information regarding operational parameters, environment conditions, etc., at these points. With either approach, that is the single point or wear curve approaches, the investigator should review all the data and observations, and look for consistencies or discrepancies between the various observations. From these detailed examinations, one can develop information regarding possible wear mechanisms and phenomena to aid in the formulation and selection of models. By coupling these types of observations with similar ones from actual applications, simulation can be refined and verified. Such examinations also aid in the establishment of adequate control. For example, alignment problems or the appearance of different surface films or oxide layers could be identified. On the basis of these observations, the need for improvement in fixturing or environmental control could be assessed. In performing these types of analyses, it is generally very helpful and informative to use several levels of examination of the wear scars. Each level can provide a different perspective about the wear situation. The use and coupling of visual examination of the wear scar, low power stereo optical microscopy, metallographic microscopy, and SEM is frequently desirable. Figure 8.10 shows an example of such a sequence. The visual and stereo microscopic views provide information about the overall nature of the scars (e.g., shape, gross features, debris formation, etc.).The higher levels of magnification associated with metallographic microscopic and SEM examination provide information regarding the microscopic details of the wear (e.g., transfer, cracks, etc.).

It is generally prudent not to rely on a single test; in wear testing, replicates are generally necessary. Statistical design of experiments and statistical analysis can be used to establish the proper number of replicates. However, as a rule of thumb, it is generally desirable to use a minimum of 3. One is simply not sufficient. If there is a big difference in the results, it is difficult to differentiate between simply a poor result, or inherent scatter. A third test can help to resolve that issue. More replicates might be required to provide a good determination of the mean and standard deviation associated with the test. However, for many engineering cases, estimates of these based on three tests are often sufficient.

8.6. FRICTION MEASUREMENTS IN WEAR TESTING

In wear tests involving sliding or rolling, it is useful to incorporate friction measurements, so that the coefficient of friction can be determined and to monitor tribosystem behavior. There are several reasons for this. One is that in many engineering applications tribological concern is not limited to wear. For example, in cases where traction, power consumption, noise, and heat generation is of concern, knowledge of the coefficient of friction and its behavior with wear is of direct interest and needs to be determined. Another reason is that the coefficient of friction can be a factor in wear behavior. This is the case with wear mechanisms that are sensitive to stress distribution and shear stresses, since contact stress distributions and shear stresses can be modified by friction. The maximum shear stress is a function of the contact pressure and the coefficient of friction as well. For example, the maximum shear stress of a flat on a flat is given by

$$\tau = q_0(1 + \mu^2)^{1/2} \tag{8.2}$$

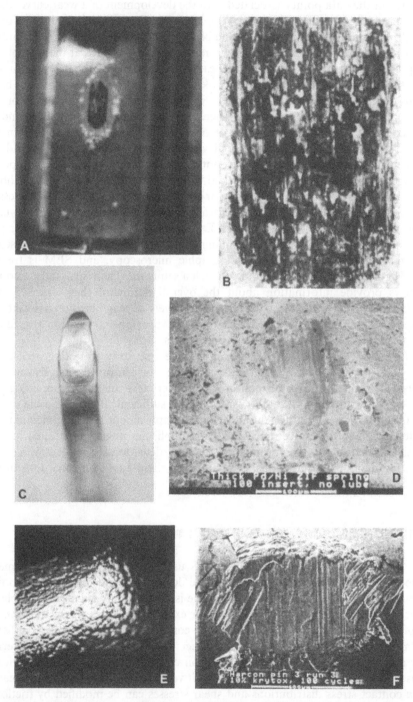

Figure 8.10 Example of the use of several examination techniques. "A", wear scar observed with a stereo microscope and "B", the same wear scar observed in a metallographic microscope; "C", another wear scar observed with a stereo microscope and "D", the same wear scar in an SEM; "G" and "H", another wear scar observed in an SEM, using two different detection modes.

Figure 8.10 (*continued*)

where τ is the maximum shear stress, q_o, is the average pressure and μ is the coefficient of friction. Other contact situations have similar relationships (32). The location of maximum shear stress and the distribution of shear stresses in a contact can also change with friction. For example, for a point or line contact, the distribution changes with the coefficient of friction. For coefficients less than 0.3 the maximum shear is below the surface. As the coefficient increases the surface shear increases and above 0.3 exceeds the subsurface shear stress (32). An additional reason for including friction measurements in a wear test is that friction and wear are often sensitive to the same parameters. Changes in friction are often associated with and are an indication of changes in wear-behavior.

In summary, knowledge of the coefficient of friction and frictional behavior is generally needed for a complete approach to the wear. Beyond this as indicated previously, friction measurements during the course of a wear test can often be of use in analyzing wear behavior and insuring proper control in a wear test. This is because wear and friction, while distinct, are sensitive to the same general range of parameters. Transitions in wear behavior, which are associated with surface modifications, are frequently associated with changes in friction. For example, in the case of unlubricated sliding between metals, the

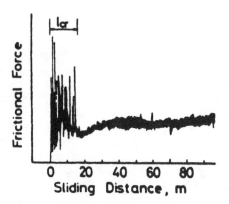

Figure 8.11 Example of a change in friction during a wear test as a result of oxide formation. (From Ref. 39, reprinted with permission from ASME.)

occurrences of different oxides have been shown to influence both wear and friction, as shown in Figs. 3.46 and 8.11, respectively. Also, many systems show a decrease in friction with break-in, as shown in Fig. 8.12. Similarly, increases in friction are also often observed with changes from mild to severe wear. Therefore, monitoring friction during a test can help point to changes either in wear behavior or other changes that will influence wear-behavior. This is of particular significance since this monitoring can be done without interrupting the test.

Friction measurements can also be used to establish adequate control. This is of particular use in terms of specimen preparation and environment control. The initial friction behavior (i.e., coefficient of friction) is often a good indicator for whether or not there is sufficient control in these areas. For example, inadequate cleaning or variations in cleaning often result in different values of initial friction. In the case of dry sliding with metals, clean surfaces generally have a high coefficient of friction (e.g., 0.6 or greater). But if oils from machining or from handling contaminate the surfaces, initial values are in the range of 0.2. In the case of polymer specimens, which tend to have skins, variations with the presence or absence of the skin frequently can be observed as well (33). In a testing procedure, the initial friction can be used as an

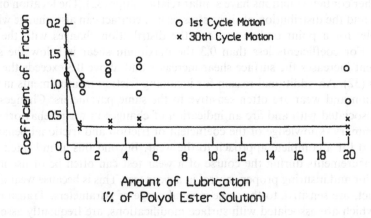

Figure 8.12 Example of a change in friction with break-in. (From Ref. 40)

indicator as to whether to continue with the test or not. This would avoid wasting time and effort on a test, which would ultimately be considered invalid. In fact, being able to obtain accepted and consistent values of the coefficient can be used as a training technique for specimen preparation with those conducting friction and wear tests. The procedures can be practiced until the accepted value is consistently obtained. Also, such measurements can be used as a monitor of laboratory environment. If the standard values are not achieved, this could point to contamination in the atmosphere, such as hydrocarbon vapors, or drifting of temperature or relative humidity.

Long-term friction behavior can be of significance from a control standpoint as well. For example, if friction is monitored through the test and the same behavior and values are not obtained from test to test, then lack of adequate control should be suspected. A thorough investigation into the reasons for these types of variations and an assessment of their possible significance to wear behavior needs to be done.

With these uses of friction measurements in wear testing, it must be recognized that stable or consistent friction behavior does not automatically imply stable and consistent wear behavior. Similarly, variation in friction does not necessarily mean changes in wear. However, when changes are seen, an understanding of why they occur and their possible significance to wear and performance should be determined. It should be kept in mind that sensitivity plays a role in this as well. For example, some changes associated with wear might be associated with only a 10% change in the coefficient of friction. If the instrumentation used to monitor friction force does not allow that resolution, a change in friction would not be observed; similarly the wear measurement might be too coarse to detect a change in wear behavior.

The incorporation of friction measurements into a wear test generally requires providing a means to measure friction force. Sometimes, this can be done indirectly, such as monitoring the torque of a motor. At other times, a direct measurement of the friction force can be done such by the use of strain gages or force transducers. An important point is that in providing for this capability care has to be taken so that it is done in a way which does not perturb the wear situation and therefore might result in a reduced or poor simulation.

Another use of friction measurements in wear tests is to use friction as the wear measure. For example, a rise in the coefficient of friction or a friction force to a certain level can indicate that a coating has been worn-through. The time it takes for this to occur can be a measure of wear resistance (34). Also, a rise in friction could be correlated to a change into a catastrophic wear mode (e.g., very high wear rate), that is associated with the end of useful life. Wear testing of brake liners is an example of this type of use.

8.7. TRIBOLOGICAL ASPECT NUMBER

The Tribological Aspect Number or TAN is a method for characterizing the general nature of the basic mechanical parameters of a contact situation, namely, motion, loading, and geometry (5). TAN characterization of a wear test and a wear situation can be a useful tool for looking at simulation. However, it does not address all aspects that need to be considered for simulation nor is it equally applicable to all wear situations that can be encountered. For example, it does not address magnitudes, environment, impact, erosive, and abrasive conditions; however, the method could be expanded upon to include these and other aspects.

With the TAN system, a sequence of four numbers are used to characterize a contact. The first number in the sequence describes contact velocity characteristics: the second number contact area characteristics; the third number contact pressure characteristics; and the fourth and final number the entry angle characteristics. The system uses four general relative motion conditions to characterize the contact velocity. Eight contact area conditions are identified in the system. Three contact pressure conditions and nine entry angle conditions are used. This classification system is shown in Table 8.2.

As can be seen from table, the code primarily relates to sliding motion. Additional conditions could be added to cover pure rolling and different impact conditions. These are shown in Fig. 8.13. The contact area classification provides a classification of contact geometry in terms of how contact area changes with time. The first six conditions describe wear situations in which contact is maintained between the same regions on the surfaces of the two contacting bodies. The last two conditions, described as "open", apply to situations where contact is always with new areas of one of the surface, such as pin following a spiral path on a flat surface. If the pin had a spherical tip in this case, it would be classified as 8; if flat, 7. If the sliding path formed a ring on the disk, the

Table 8.2 Tribological Aspect Number (TAN)

TAN: <u>A</u> <u>B</u> <u>C</u> <u>D</u>
TAN Code

(A) Contact velocity characteristic number, 1–4   (C) Contact pressure characteristic number, 1–3

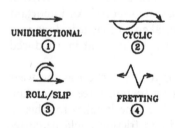

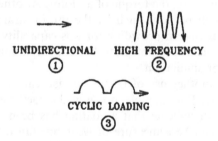

(B) Contact area characteristic number, 1–8   (D) Interfacial entry angle number, 0–9

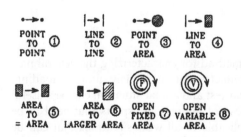

Source: Ref. 5.

Contact Velocity Characteristics

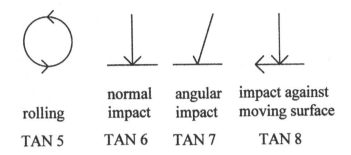

rolling	normal impact	angular impact	impact against moving surface
TAN 5	TAN 6	TAN 7	TAN 8

Contact Pressure Characteristic

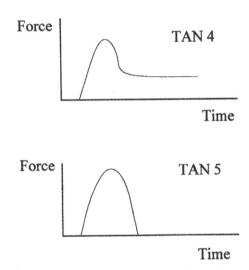

Figure 8.13 Additional categories for contact velocity and contact pressure characteristics for use with the TAN methodology of characterizing wear situations. These extend the concept to rolling and impact wear situations.

classification would be 3 and 5, respectively, and referred to as "closed". This distinction is illustrated in Fig. 8.13. Conceptually, the "open" area categories could be expanded to allow the same degree of differentiation used for "closed" contacts, that is 1–6.

The third number characterizes contact pressure behavior over one second or less time intervals. Three different conditions are identified. One is when the contact pressure is constant. This condition is referred to as "unidirectional". If the pressure slowly varies, the condition is classified as "cyclic". If there are rapid variations or if the pressure goes negative, it is classified as "high frequency". The difference between these three conditions is graphically illustrated in Table 8.2. If the TAN concept were extended to included impact wear situations, additional contact pressure behavior categories, 4 and 5, would be needed to differentiate different impact conditions. One would be where there is an

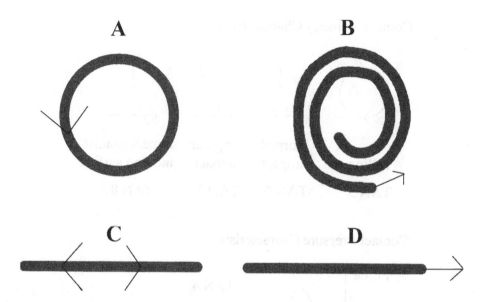

Figure 8.14 Examples of wear paths on the larger surface for TAN closed and open contact areas. "A", where a body is sliding is sliding over the same circular path, is an example of a closed area situation. "C", where a body is sliding back and forth over the same track, is also an example of a closed area. "B", where a body is sliding in a spiral-like manner, always encountering unworn regions of the surface, is an example of an open area situation. "D", unidirectional sliding, is also an example of the wear path for an open contact area.

impact load, such as when a cam follower is driven onto a cam, 4; the other would be a ballistic impact where the momentum of the impactor generates the pressure, 5. These pressure profiles are shown in Fig. 8.14.

The fourth number characterizes the entrance angle or the angle formed between the leading edge of one surface and the contacting surface, as illustrated in Table 8.2. This aspect of the contact can affect lubrication, debris entrapment, and tribofilm formation and needs to be considered in simulation.

This method of characterization has been used to select test methods for the evaluation of materials for specific applications. The approach is to select a test method, which has the same TAN value as the application. Values for the various test parameters and environmental conditions are then chosen to provide further simulation (5).

REFERENCES

1. R Bayer ed. Effects of Mechanical Stiffness and Vibration on Wear, STP 1247. West Conshohocken, PA: ASTM, 1995.
2. R Bayer. Tribological Approaches for Elastomer Applications in Computer Peripherals. In: R Denton, K Keshavan, eds. Wear and Friction of Elastomers, STP 1145. West Conshohocken, PA: ASTM, 1992, pp 114–126.
3. N Payne, R Bayer. Friction and wear tests for elastomers. Wear 130:67–77, 1991.
4. S Calabrese, S Murray. Methods of Evaluating Materials for Icebreaker Hull Coatings. In: R Bayer, ed. Selection and Use of Wear Tests for Coatings, STP 769. West Conshohocken, PA: ASTM, 1982, pp 157–173.

5. R Voitik. Realizing Bench Test Solutions to Field Tribology Problems by Utilizing Tribological Aspect Numbers. In: A Ruff and R Bayer, eds. Tribology: Wear Test Selection for Design and Application, STP 1199. West Conshohocken, PA: ASTM, 1993, p 45.

6. Standard Test Method for Wear Testing with a Crossed-Cylinder Apparatus. West Conshohocken, PA: ASTM, G83.

7. Standard Test Method for Measuring Abrasion Using the Dry Sand/Rubber Wheel Apparatus. West Conshohocken, PA: ASTM, G65.

8. Standard Test Method for Conducting Wet Sand/Rubber Wheel Abrasion Tests. West Conshohocken, PA: ASTM, G105.

9. Standard Test Method for Ranking Resistance of Materials to Sliding Wear Using Block-on-Ring Wear Test. West Conshohocken, PA: ASTM, G77.

10. Standard Test Method for Determination of Slurry Abrasivity (Miller Number) and Slurry Abrasion Response of Materials (SAR Number). West Conshohocken, PA: ASTM, G75.

11. Standard Test Method for Conducting Erosion Tests by Solid Particle Impingement Using Gas Jet. West Conshohocken, PA: ASTM, G76.

12. Standard Test Method for Cavitation Erosion Using Vibratory Apparatus. West Conshohocken, PA: ASTM, G32.

13. Standard Practice for Liquid Impingement Erosion Testing. West Conshohocken, PA: ASTM, G73.

14. Standard Test Method for Abrasinvess of Ink-Impregnated Fabric Printer Ribbons. West Conshohocken, PA: ASTM, G56.

15. R Bayer. Influence of oxygen on the wear of silicon. Wear 69:235–239, 1981.

16. W Wei, K Beaty, S Vinyard, J Lankford. Friction and Wear Testing of Ion Beam Modified Ceramics for High temperature Low Heat Rejection Diesel Engines. In: C Yust, R Bayer, eds. Selection and Use of Wear Tests for Ceramics, STP 1010. West Conshohocken, PA: ASTM, 1988, pp 74–87.

17. G Lundholm. Comparison of seal materials for use in stirling engines. Proc Intl Conf Wear Materials. ASME 250–255, 1983.

18. R Bayer. A model for wear in an abrasive environment as applied to a magnetic sensor. Wear 70:93–117, 1981.

19. J Miller. American Institute of Mining, Metallurgical, and Petroleum Engineers (AIME). Paper 73-B-300. Society for Mining, Metallurgy, and Exploration (SME) Meeting Pittsburgh, 1973.

20. P Engel, C Adams. Rolling wear study of misaligned cylindrical contacts. Proc Intl Conf Wear Materials. ASME 181–191, 1979.

21. R Morrison. Test data let you develop your own Load/life curves for gear and cam materials. Machine Design. 8/68. Cleveland, OH: Penton Publishing Co, 1968, pp 102–108.

22. E Buckingham, G Talbourdet. Roll Tests on Endurance Limits of Materials. New York, NY: ASME, 1950.

23. H Czichos. Tribology. Tribology Series. New York: Elsevier Science Publishing Co, 1978.

24. M Godet, Y Berthier, J Lancaster, L Vincent. Wear Modeling: How Far Can We Get With Principles? In: K Ludema, R Bayer, eds. Tribological Modeling for Mechanical Designers, STP 1105. West Conshohocken, PA: ASTM, 1991, pp 173–179.

25. R Barkalow, I Goebel, F Pettit. Erosion-Corrosion of Coatings and Superalloys in High-Velocity Hot Gases. In: Erosion: Prevention and Useful Applications. W Adler, ed. STP 664. West Conshohocken, PA: ASTM, 1979, pp 163–192.

26. C Yang, S Bahadur. Friction and wear behavior of alumina-based ceramics in dry and lubricated sliding against tool steel. Proc Intl Conf Wear Materials. ASME 383–392, 1991.

27. C Young, S Rhee. Wear process of TiN coated drills. Proc Intl Conf Wear Materials. ASME 543–550, 1987.

28. R Bayer. Impact wear of elastomers. Wear 112:105–120, 1986.

29. P Blau, C Olson. An application of thermal wave microscopy to research on the sliding wear break-in behavior of a tarnished Cu-15 wt%Zn alloy. Proc Intl Conf Wear Materials. ASME 424–431, 1985.

30. R Bayer. Predicting Wear in a Sliding System. Wear 11:319–332, 1968.
31. R Bayer, J Wilson. Paper No. 71-DE-39. Design Engineering Conference and Show, 4/71. ASME, New York, 1971.
32. R Bayer, T Ku. Handbook of Analytical Design for Wear. New York: Plenum Press, 1964.
33. R Bayer, N Payne. Wear Evaluation of Molded Plastics. Lub Eng May:290–293, 1985.
34. R Bayer, A Trivedi. Metal Finishing Nov:47, 1977.
35. K Tanaka, Y Yamamuda. Effect of counterface roughness on the friction and wear of polyethylene under a sliding condition involving surface melting. Proc Intl Conf Wear Materials. ASME 407–414, 1987.
36. M Watanabe. The friction and wear of high density polyethylene in aqueous solutions. Proc Intl Conf Wear Materials. ASME 573–580, 1979.
37. M Kar, S Bahadur. Micromechanism of wear at polymer-metal sliding interface. Proc Intl Conf Wear Materials. ASME 501–509, 1977.
38. P Swanson. Comparison of laboratory and field abrasion tests. Proc Intl Conf Wear Materials. ASME 519–525, 1985.
39. A Iwabuchi, K Hori, H Kudo. The effects of temperature, pre-oxidation and pre-sliding on the transition from severe wear to mild wear for S45C carbon steel and SUS304 stainless steel. Proc Intl Conf Wear Materials ASME 211, 1987.
40. E Hsue, R Bayer. Tribological properties of edge card connector spring/tab interface. IEEE Trans CHMT 12(2):206–214, 1989.

9
Wear Tests

9.1. OVERVIEW

A number of different wear tests will be reviewed in this section. There are several purposes associated with these reviews. One is to provide examples of the methodology; another is to use these tests to point out deficiencies and areas of concerns that are frequently encountered in wear testing, and a third is to illustrate some of the approaches used to resolve these deficiencies and to address these concerns. However, the application and use of wear tests to address design concerns and to resolve problems will not be specifically treated. These aspects will be illustrated in *Engineering Design for Wear: Second Edition, Revised and Expanded* (EDW2E). In addition to discussing these tests in terms of such aspects as simulation, acceleration, and control (which were addressed in Chapter 8 Methodology), the correlation of these tests to field performance will also be reviewed.

It is useful to consider and classify wear tests in terms of two general categories, phenomenological wear tests and operational wear tests. These two categories tend to have different degrees of simulation associated with them, be used in different ways, and require different levels of knowledge about the wear situation. Phenomenological wear tests are those tests which tend to focus on some broad or general type of wear situation or phenomenon, such as sliding wear, erosion, or low-stress cutting abrasion. Such tests tend to provide generic information. In terms of simulation, these tests provide either first- or second-order simulation. Frequently, there is an underlying element that a particular mode or mechanism of wear is induced, and this element is often key in the development of the test. Operational wear tests, on the other hand, focus more on specific applications or situations. The names of these types of tests often indicate this focus, such as a wear test for brake liners, journal bearings, or gears. The operational parameters associated with the situation or application tend to be duplicated in the test, and the test is generally representative of second- or third-order simulation. In designing the test, the simulation of the loading, motion, geometry, and environment of the application is the primary consideration. Consideration of wear mechanisms is secondary. Wear mechanisms are identified by analysis of the results and frequently may change with the materials or particular values of the parameters.

In many engineering environments, wear tests tend to migrate from one category to the other. Frequently, initial forms of wear tests developed to address engineering concerns can be classified as operational; however, with experience, increased knowledge and further development, these tests tend to be modified and used in a manner more consistent with the phenomenological category than the operational. The converse is also true. In engineering situations, phenomenological tests are often modified to provide better simulation of an application and evolve into operational tests.

Tests used by material developers tend to fall into the phenomenological category. Many of the standardized tests are also representative of this category and are used to support material development activities.

Phenomenological tests are useful in understanding and characterizing basic wear behavior, either of materials or wear mechanisms. On the other hand, operational tests are useful in characterizing and understanding wear behavior of devices and accounting for subtle difference in applications. Less is presumed about the wear situation with an operational test approach than with a phenomenological approach. Achieving correlation is in general more direct and less involved with the operational approach than with a phenomenological approach, and the degree of simulation is high enough that little has to be done to establish correlation. With a phenomenological approach, additional information is required (e.g., failure analysis of worn components, demonstration of simulation, or identification of limits of applicability) to establish correlation between the test and application.

The tests that will be discussed are grouped into these two general categories, phenomenological and operational. The reasons for the particular classification in each case will be identified and will further illustrate the difference in focus and relevance associated with these two categories. It will be seen that in many cases a test is composed of both phenomenological and operational elements, but is biased in one direction or the other.

9.2. PHENOMENOLOGICAL WEAR TESTS

9.2.1. Dry Sand-Rubber Wheel Abrasion Test

This is a good example of a phenomenological wear test. The test was developed to simulate wear situations in which low-stress scratching abrasion is the primary mode of wear. This is the mode of wear associated with loose abrasive grains being dragged across a surface under loading conditions which do not induce fracture of the abrasive particles. The test has been used to investigate the influence of various parameters on this mode of wear, such as abrasive particle size and shape and material parameters. In addition, since the test generally correlates with field conditions, it has been used effectively to rank materials in terms of their resistance to this type of wear and to select materials for applications in which this type of wear occurs (1). For example, good correlation in rankings was demonstrated between the test and the tilling of sandy soils (2).

Low-stress scratching abrasion is simulated in the test by trapping a stream of free-falling abrasives between a wear specimen and a rotating rubber coated wheel. The test configuration is shown in Fig. 9.1. The rim of wheel is coated with rubber to avoid crushing the grains in the nip between the specimen and the wheel. A typical wear scar is shown in Fig. 9.2. The wear is usually determined by weight loss. However, since wear volume is generally the preferred way for describing the magnitude of wear, weight loss is converted to wear volume by dividing by the density. A standard version of this test for material ranking has been developed by the ASTM (ASTM G65). In this standard key design elements of the apparatus are specified, along with load, test duration, specimen shape and tolerances, cleaning procedures, abrasive definition, method of analysis, and data reporting. When the test is conducted in the manner outlined in the standard, a coefficient of variation of 5% or less is achievable.

The low variation associated with the standard method results from the degree of control that this version of the test insures. The overall design of the apparatus and the associated tolerances insures good stability and alignment, as well as control of the normal

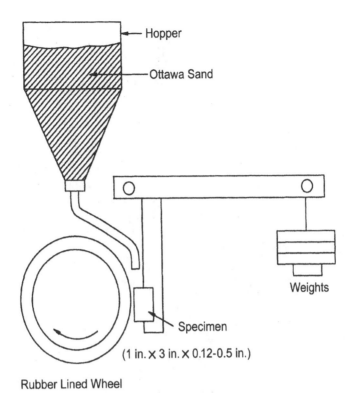

Rubber Lined Wheel

Figure 9.1 Sketch of the dry sand–rubber wheel test.

load and speed. In addition, a fundamental element in the repeatability of this test is control of the amount and type of abrasive used. The test concept is to operate in a region in which the results are insensitive to feed rate of the abrasive. This is done by using a nozzle that consistently provides a uniform curtain of abrasive wider than the wheel and by using a feed rate, which insures saturation in the nip. The shape and dimensions of the nozzle are critical to provide this control. In addition, the abrasive is controlled both in terms of size, type, and moisture content. This last element is needed because variation in moisture content can influence feed rate, as well as directly influencing the wear. Since the test causes some wear and damage to the abrasives, it is recommended that it be used only once.

The properties of the rubber can also influence the wear. Therefore the composition, cure, and durometer are specified. Since the rubber also wears and heats up in the test, intervals between tests are controlled, a dressing procedure defined, and an interval for dressing defined. The amount of rubbing that the wheel provides also influences the amount of wear that is generated. A standard amount of rubbing is used as a reference in order to provide a valid comparison of materials. While rotational speed of the wheel is specified with tolerances, a fixed time is not used. Specifying the number of revolutions for the test controls the amount of rubbing. Since wheel diameter changes as a result of wear and dressing, a correction or scaling factor is also introduced to account for such changes.

The actual speed, load, and number of revolutions used in the standard test were determined empirically to produce a level of wear that can be measured with sufficient

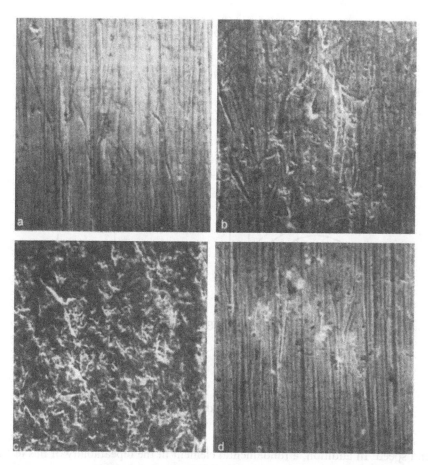

Figure 9.2 Examples of wear scar morphology from the dry sand–rubber wheel test. (From Ref. 113.)

accuracy to provide material ranking in a reasonable length of time. In addition, they were also chosen to avoid complicating elements, such as excessive heating of the rubber or degradation of the abrasive. In the standard test, different practices are defined for different classes of materials to achieve these goals. Milder conditions are specified for less abrasive resistant materials and coatings.

The ASTM test procedure also advises the visual examination of the wear scar to insure that there is good alignment and no problem with the flow of abrasives. In addition, the test method also uses tests with reference materials to insure consistency. The ASTM procedure also calls for the reporting and measuring of a number of parameters associated with the test. The recommended data sheet is shown in Fig. 9.3. This test configuration can be used for other purposes other than ranking materials, such as evaluating the abrasivity of different materials. When used in this fashion, the standard test procedure can be used as a guide in establishing proper control, etc.

One point should be made regarding the weight loss method for wear determination used in the procedure. To provide valid comparisons, the volume of wear should be used; this requires knowledge of the specimen's density. If this is not done, ranking might be in error because of density differences. This is of particular concern when

TABLE 3 Data Sheet

Dry Sand/Rubber Wheel Test

ASTM G-65 Procedure __

Qualification of Apparatus (114)					Date __	
Reference Materials __					Quantity __	
Adjusted Volume Loss (avg) __ mm³ Coefficient of Variation __						__ %

Test Data

Material Description __					Wheel diameter: __	
Heat Treatment __					Wheel width: __	
Hardness __					Wheel hardness: __	
Surface Preparation __						

Test No.					
Test load					
Wheel revolutions					
Sand flow, g/m					
Initial mass, g					
Final mass, g					
Mass loss, g					
Density, g/cm³					
Volume loss, mm³ (mass loss/density) x 1000					
Adjusted volume loss, mm³					

Comments __

Company Name __	Tested by __	Date __

Figure 9.3 Data sheet for the dry sand–rubber wheel test. (From Ref. 114, reprinted with permission from ASTM.)

coatings are evaluated, since the densities of the coatings may be unknown. Therefore other techniques, such as computation of the wear volume from wear scar dimensions, might yield better results.

An important element to recognize in this test is that while it simulates a wear mode and correlates to field performance, it does not directly provide a wear constant or parameter that can be applied to different situations. To determine absolute performance in an application, a reference material for which there is field and test data must be used and a scaling factor established for that application. This is because the test does not address the question of relative severity in the test and in application, either on an empirical or a theoretical basis. This is also the case in several of the other tests that will be discussed. This implies that there is an unknown amount of acceleration provided by the test and that the acceleration could be different for different applications. The amount of acceleration that exists can be significantly dependent on differences in the amounts and type of abrasives in the test and in application.

9.2.2. Wet Sand-Rubber Wheel Abrasion Test

This test also simulates low stress scratching abrasion and is very similar to the dry sand-rubber wheel test. A diagram of the wet sand test is shown in Fig. 9.4. It can be seen that the basic test concept is very similar to the dry sand test, namely using a rubber wheel to drag abrasives across the face of a wear specimen. In the wet sand test, however, the abrasive is in the form of a water slurry. The wear in both tests is determined by weight loss and converted to volume for material rankings. Both tests have been used extensively to investigate scratching abrasion and both have been found to correlate well with many practical applications (1–3). While there is the potential for differences, similar material behavior is generally found with both tests. The wet sand

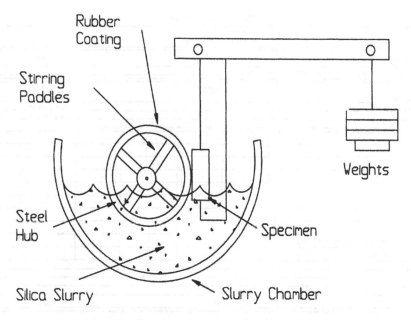

Figure 9.4 Sketch of the wet sand–rubber wheel test.

test or test configuration offers an advantage in that the normal procedure can be modified to utilize a slurry more representative of an application. In this way, chemical effects associated with an application can be addressed.

As with the dry sand test, a standard version for material comparisons has been developed by the ASTM (ASTM G105). Similar repeatability is found with both the wet and dry sand tests when the procedures are followed. A coefficient of variation of 5% is characteristic of the wet sand test. To achieve this, the test procedure controls many of the same factors that are controlled in the dry sand test. These include design and construction of apparatus, measuring techniques, specimen tolerances and preparation, and rubber wheel and abrasive specifications. A procedure and interval for dressing the rubber wheel are also given. However, some of the approaches used are different.

As with the dry sand test, the feeding and control of the abrasive is a significant factor in the test. Composition of the slurry is defined in terms of abrasive size, type and source, and the type and amount of water. A uniform supply to the interface is achieved by having the contact below the surface of the slurry. To maintain uniformity of the slurry during the test, paddles on the side of the wheel provide agitation (Fig. 9.4). The test procedure also requires that the wear specimens be demagnetized prior to the test. This is to avoid problems with magnetic wear debris adhering to the wear surface or clustering in the slurry.

In the wet sand test, three rubber wheels of different durometer (A50, A60, and A70 durometer) are used to minimize the effect of wheel variation. This is done by using the wear obtained with these different durometer wheels to define a best-fit linear relationship between durometer and log of the wear. The value of wear reported for the test is that obtained for A60 durometer from that curve. Graphically, this is illustrated in Fig. 9.5.

Another difference between the two tests is that the wet sand procedure requires a break-in cycle with the A50 durometer wheel. This is to minimize the effects of surface

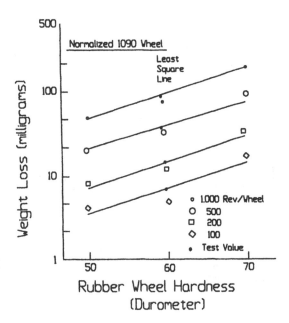

Figure 9.5 Illustration of the method of analyzing wear data in the wet sand–rubber wheel test. (From Ref. 3.)

defects and variations. After each of these tests, including the break-in, it is specified that the slurry chamber be cleaned and new slurry used. The technique also requires repositioning of the specimen, which is controlled by the fixturing and associated tolerance. Visual inspections and statistical criteria are associated with the procedure as well. For example, tolerance on the correlation coefficient for the linear fit is specified. Also, if the coefficient of variation is beyond 7%, the test is considered out-of-control and the test procedures and apparatus should be examined to determine the reason for this condition.

Like the dry sand test, the wet sand provides a ranking of materials but does not provide absolute values. Different test conditions are specified for different classes of materials to provide suitable resolution and separation in wear resistance.

9.2.3. Slurry Abrasivity

This test is also associated with low stress scratching abrasion. However, it was primarily developed to address slurry abrasion problems in pipelines, which handle slurries. It has been used extensively for these applications with good correlation (4,5). The initial focus for this test was to rank the abrasivity of the slurries encountered, but the test method was expanded to provide a ranking capability for wear resistance against a particular slurry. For the former, the test results in what is termed a Miller Number; for the latter, a SAR Number. SAR stands for slurry abrasion response. Both are based on a weight loss technique for measurement of wear, although a wear rate is used for ranking. This test is also an ASTM test procedure (ASTM G75).

The basic configuration of the test is a flat rectangular wear specimen sliding back and forth across a rubber lap, flooded by a slurry; both the wear specimen and the lap are submerged in the slurry (Fig. 9.6). The flat wear specimen is mounted on an arm, which moves it back and forth at a controlled speed. The arm lifts the specimen at the end of each

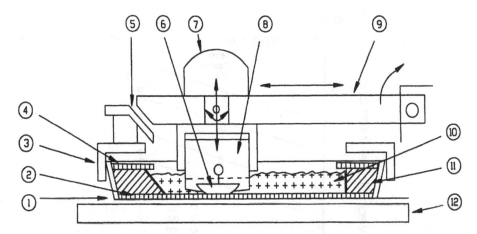

Figure 9.6 Sketch of the slurry abrasivity test apparatus. 1, Molded plastic tray; 2, neoprene lap; 3, tray clamp; 4, splash guard; 5, block lifting cam; 6, standard wear block—27% Cr-iron; 7, dead weight; 8, adjustable plastic wear block holder; 9, pivoted reciprocating arm; 10, sand slurry; 11, molded plastic filler "V" channel; 12, tray plate.

stroke to allow the slurry to come between the specimen and the lap. The tray that holds the slurry has tapered sides so that with the motion of the wear specimen, mixing and circulation of the slurry occurs. The wear specimen is removed periodically and weight loss determined. By varying the slurry and keeping the wear specimen constant, the abrasivity of slurries can be ranked. This is essentially the Miller Number. By keeping the slurry constant and varying the wear specimen, wear resistance of different materials to that slurry can be ranked. This is used to generate the SAR Number.

The reproducibility of the test is good, as the coefficient of variation is approximately 6% for the Miller Number and 12% for the SAR Number. The larger value for the SAR Number reflects the decrease in control that this version of the test has. A carefully controlled reference wear specimen is used in the Miller Number test, whereas in the SAR version the wear specimen is simply a material of interest. One element contributing to this increased variation would be lack of precision in the value for specific gravity of this specimen; other factors are variation and nonuniformity in other material properties, such as composition and grain structure.

The test procedure associated with each of these tests involves a well-designed and specified test apparatus, cleaning and specimen preparation steps, specifications and controls on the materials involved, and carefully outlined steps for conducting the test. The test method calls for replacement of the rubber lap with each test to avoid problems associated with degradation and contamination of the lap. In particular, extensive details are provided to insure proper alignment of the wear specimen to the lap, including a method of checking alignment prior to the start of the test. Procedures for preparing the slurry are also given.

The analysis method involves the developing of a wear curve and using a wear rate to rank either the abrasivity of the slurry or the abrasion resistance of a material to a given slurry. Measurements are made after 4, 8, 12, and 16 hr. Using a least square method, the wear data obtained are fitted to the following equation:

$$\text{mass loss (mg)} = A \times t^B \tag{9.1}$$

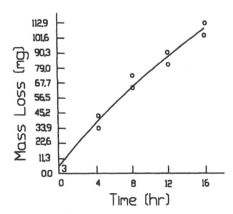

Figure 9.7 Example of the method of analysis used in the slurry abrasivity test. The curve represents the least squares fit of the data, using Eq. (9.1). Best fit values of A and B are, respectively, 13.4262 and 0.749182, which result in a wear rate of 8.47 mg/hr at 2 hr and a Miller Number of 154. (From Ref. 6.)

where t is time. Using the value of A and B obtained in this manner, the wear rate (mg/hr) at the 2 hr point is determined. This method is illustrated in Fig. 9.7. The wear rate is then multiplied by a scaling factor of 18.18 hr/mg to provide the dimensionless Miller Number. The higher this number, the more abrasive the slurry is. The scaling factor was determined so that the Miller Number of a reference slurry of sulfur is 1 and of a reference slurry of corundum is 1000.

The method of analysis is the same for the SAR Number. However, the wear rate at the 2 hr point is multiplied by the scaling factor and the ratio of the specific gravity of the reference material to that of the test specimen

$$\text{SAR Number} = (\text{wear rate at 2 hr}) \times 18.18\ (\text{h/mg}) \times \frac{7.58}{p} \qquad (9.2)$$

where p is the specific gravity of the test specimen. To reduce scatter, the method of analysis uses the average value of two runs for the fit. The test apparatus is usually built with two separate trays so that two tests can be performed at the same time. In addition, the test procedure calls for the samples to be rotated 180° between each measurement to reduce effects of misalignment and orientation. Since separation of the slurry can occur while these measurements are made, the procedure requires stirring of the slurry prior to continuing the test.

The format for reporting the results of the test is shown in Fig. 9.8, which shows that more data and observations regarding the test are recorded than simply the mass loss of the wear specimen(s). Included in this are the temperature and pH of the slurry, observations regarding the degree of wear seen on the lap, and a detailed identification of the slurry used in the test, including the preparation technique of the slurry. The recommended slurry for the test is a 50% mixture of the solid extracts from the slurry in question and water. However, the procedure allows the use of other types of slurries, such as the slurry in question or the mixing of the dry extracts with other media.

The test parameters and procedures were selected and developed with the use of additional studies, aimed at insuring simulation with the intended application, minimum scatter, and a convenient test (4–6). One focus of the studies was the influence of

Project   —Test Number_____

               —Date_____

               —Description_____

Slurry   —Description_____

               —Concentration(S)_____

               —Temperature_____°C

Wear Specimen—Description(S)_____

               —Specific Gravity_____

               —Hardness_____Rc

Lap Material   —Description (S)_____

               —Wear   Trace Light Moderate Heavy Severe

Tray   —   (_____)   (Wear Specimen_____)   (Corr Specimen_____)

               (pH)   (Mass   Mass Loss)   (Mass   Mass Loss)

Initial   — _____   _____ _____   _____ _____

1st 4 Hours — _____   _____ _____   _____ _____

2nd 4 Hours — _____   _____ _____   _____ _____

3rd 4 Hours — _____   _____ _____   _____ _____

4th 4 Hours — _____   _____ _____   _____ _____

Total   —   _____   _____

Notes   — _____

Notes   — _____

Tray   —   (_____)   (Wear Specimen_____)   (Corr Specimen_____)

               (pH)   (Mass   Mass Loss)   (Mass   Mass Loss)

Initial   — _____   _____ _____   _____ _____

1st 4 Hours — _____   _____ _____   _____ _____

2nd 4 Hours — _____   _____ _____   _____ _____

3rd 4 Hours — _____   _____ _____   _____ _____

4th 4 Hours — _____   _____ _____   _____ _____

Total   —   _____   _____

Notes   — _____

Notes   — _____

Figure 9.8 Format for reporting wear data in the slurry abrasivity test. (From Ref. 6, reprinted with permission from ASTM.)

concentration and slurry media; a nonlinear effect was found (Fig. 9.9). The Miller Number becomes less sensitive to concentration as the concentration increases. This means that performing the test at higher concentration can minimize variation. The use of higher concentration also enhances simulation in that many of the applications involve high concentration (e.g., in the 50% range).

These same studies also showed effects associated with the liquid phase of the slurry. For example, oil-based slurries tend to result in lower wear than water-based slurries. Some of this behavior can be associated with the different fluids providing different lubrication and dispersion. In addition, this type of behavior can also be the result of differences in chemical or corrosion effects that occur with different fluids. Several things were done to control this element in the test. One was to use a design, which employed the use of nonconducting materials to eliminate electrochemical effects; another was to use a chromium iron, which is used in some applications and is somewhat corrosion

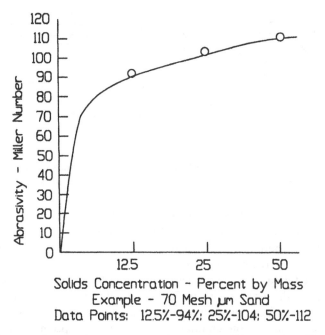

Figure 9.9 The effect of slurry concentration on the Miller Number. (From Ref. 6.)

resistant, as the standard material for the Miller Number. These effects also influenced the selection of a water/dry solids mixture as the preferred slurry for the test. However, because of the varied and uncontrolled ionic nature of the dried solids from various applications, these do not necessarily eliminate corrosion effects from tests. To help separate and assess the significance of corrosion on the test results, it is suggested that a second test be done with a slurry specifically buffered to eliminate corrosion. This concern is reflected in the pH data requested in the data sheet. It also underscores the desirability to test with the actual slurry from an application to enhance simulation and the use of the SAR Number test. Since the material of interest is tested in the SAR Number test, the specific corrosion effects on that material are included in a test with the actual slurry. It can also be seen that both the SAR and Miller Numbers must be referenced to a particular slurry.

The use of several different test conditions suggested with the slurry abrasion test procedure illustrates the need in some cases to modify standard wear tests to improve correlation with an application.

The primary intention of both the dry and wet sand test and the slurry test is to rank materials in terms of their abrasion resistance. Neither the wet nor the dry sand test is as sensitive to the corrosion element as the slurry test. In the dry sand test, it is eliminated since a fluid is not used. In the wet sand test, corrosion effects are limited by the use of a nonionic abrasive in water. From a test for pure abrasion resistance standpoint, the dry and wet sand tests have some advantage in that the wear situation is less complex. However, from an application standpoint, they may be poorer since the actual situation may involve corrosion. Consequently, in performing evaluations for an application, which involves fluids, the slurry test provides the better simulation in that the confounding effects of corrosion can be assessed. This may or may not be significant, depending on the relative degree of pure abrasion and corrosion, in any given situation.

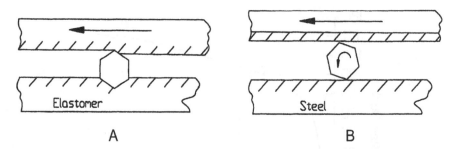

Figure 9.10 Behavior of abrasive particles trapped between two surfaces, which are sliding relative to one another. "A", when one surface is an elastomer; "B", when both surfaces are metals.

These three tests, dry and wet sand, and slurry abrasivity, have a common feature in that they use a rubber lap as a counterface. In simulating low stress scratching abrasion, the use of rubber as a counterface has an advantage in the way that it helps to reduce the stress on the abrasive particle and thereby to reduce fracture. Also, since they tend to provide large contact area with the particles, they tend to preferentially hold the particles and drag them across the wear specimen. Figure 9.10 illustrates this for both a rubber and rigid (e.g., steel), counterface. This action plus the use of a much larger surface area for the rubber counterface results in preferential wear of the wear specimen. These two effects account for the apparent superior abrasion resistance of the rubber laps in comparison with the much harder materials evaluated in these tests. However, in a situation where the abrasive would be dragged across the rubber surface, the rubber materials would show a large reduction in wear resistance and inferior behavior to many of the materials evaluated in the test.

The slurry abrasion test allows for different procedures for different material categories. It does not provide absolute values of wear performance but rather a material ranking.

9.2.4. Erosion by Solid Particle Impingement Using Gas Jets

Gas jets have been used to investigate solid particle erosion and to rank materials in terms of resistance to this mode of wear (Fig. 9.11). It has been found to correlate well with erosion situations characterized by near normal particle impacts, such as erosion of valves in coal gasification and similar equipment (7,8), but not with situations which involve grazing impacts. Examples of these latter situations would be erosion by wind-blown dust and erosion of airfoil surfaces. Differences in the impingement conditions for these cases are illustrated in Fig. 9.12. The effect of incident angle on wear scar morphology is shown in Fig. 9.13.

Using the conditions and procedures for this type of test outlined in ASTM G76, coefficients of variation in the range 5–20% are achievable. The standard conditions for the test are shown in Table 9.1, which lists the significant parameters to be controlled in the test and tolerances are specified for each. The test procedure also requires the routine use of a reference material and the monitoring of the nozzle for signs of wear (erosion). If the diameter of the nozzle increases by 10% or more, the nozzle is to be replaced. Procedures for specimen preparation, cleaning, and repeatability are also addressed in the standard.

Weight loss is the method used for determining the amount of wear that occurs. However, the resistance to erosion is measured in terms of the wear volume per gram

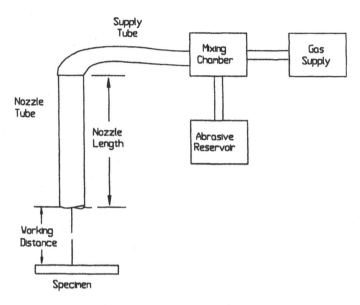

Figure 9.11 Schematic of the solid particle erosion test.

of abrasive. This is obtained through the use of a wear curve that is generated by measuring the mass loss at different time intervals. The slope of this curve is then used to determine an average wear rate. Examples of wear curves obtained in this test are shown in Fig. 9.14. The mass loss rate is converted to a volume loss rate by dividing by the density of the specimen. This volume wear rate is then normalized to the abrasive flow rate to provide the erosion value, which is defined as wear volume per gram of abrasive. The smaller the erosion value, the more wear resistant the material. Guidelines for the test duration are provided with the intention that the measurements be made in a period of stable wear behavior. Since 2 min or less is typically required for stabilization, it is specified that the first measurement be taken after 2 min. The test should be carried out for at least a total of 10 min but should not go beyond the point where the scar depth exceeds 1 mm. The reason for this limit is that beyond that depth the shape of the scar becomes significant in determining impact angle. Figure 9.15 shows a typical scar and the effect of wear geometry on angle of impact.

The data and information that are to be recorded and reported in conjunction with the erosion value are indicated in Fig. 9.16. When nonstandard test conditions are used,

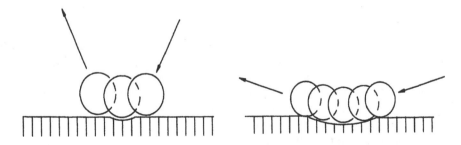

Figure 9.12 Comparison of near-normal and grazing-normal particle impact.

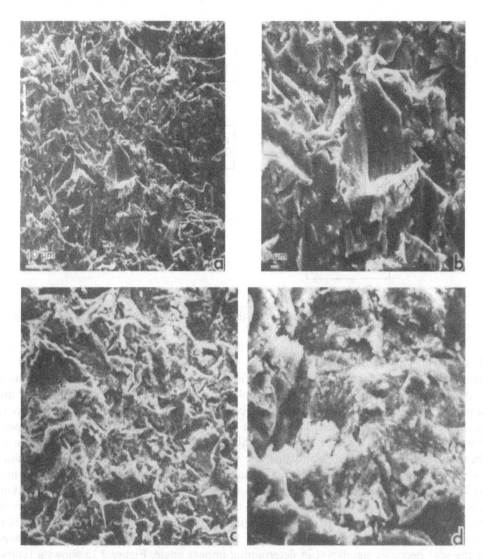

Figure 9.13 Morphology of particle erosion wear scars for grazing- ("A" and "B") and near-normal ("C" and "D") impact conditions. (From Ref. 115, reprinted with permission from ASTM.)

the test procedure calls for the testing of the reference material under those conditions that are performed and reported.

Studies have been done to understand the effect of various parameters on the test and to determine the level of control required. These studies also identified some of the factors, which can influence this mode of wear (9). Among these are particle velocity, abrasive or particle characteristics, particle flux, and temperature. Differences between the test and application conditions with respect to these parameters cause the test to provide a relative ranking rather than absolute performance. However, the erosion value generated in this test does provide some means of directly relating test and application. This is in terms of amount of abrasive. The erosion value is defined as wear per amount of abrasive. Therefore, knowledge of the amount of abrasive experienced or the amount

Table 9.1 Standard Test Conditions of ASTM G76 Test Method for Solid Particle Erosion

Nozzle tube	Dimensions	ID 1.5 mm±0.075 mm; minimum length 50 mm
	Orientation	Axis 90 ± 2° with specimen surface
	Position	10 ± 1 mm from specimen surface
Test gas		Dry air
Particle	Composition	Al_2O_3
	Size	50 μm
	Shape	Angular
	Feed rate	2.0 ± 0.5 g/min
	Velocity	30 ± 2 m/sec
Flow	Rate	8 L/min
	Presssure	140 kPa (may be different)
Duration		Minimum, 10 min; maximum, any acceptable provided crater depth does not exceed 1 mm
Temperature		18–28°

of abrasive per unit time in an application can provide an estimate of the wear or wear rate in the application. This information can also aid in assessing the relative severity of the test to the application and possible acceleration. Although additional aspects of the particle streams in the two situations are also needed to address these elements completely, such as type and velocity of the particles, and impingement angle.

In the standard tests methods for dry and wet sand and slurry abrasivity, the design of the test apparatus is treated in more detail than in the standard test method for this

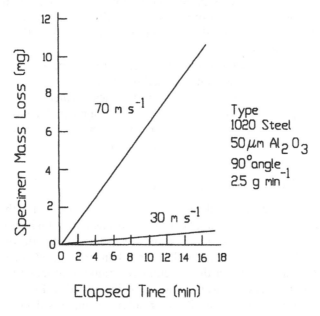

Figure 9.14 Wear curves obtained in the solid particle erosion test for 1020 steel at two different velocities. (From Ref. 115.)

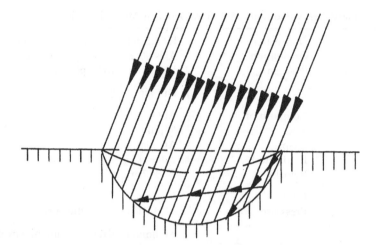

Figure 9.15 An illustration of the effect of erosion wear scar geometry on particle impact conditions. As a result of wear, the angles of impact are no longer constant across the surface.

erosion test. This is because of the nature of the wear situation involved. In the former two bodies pressed against one another under conditions of relative motion generate the wear. In these types of situations the alignment, rigidity, and consistency of speed provided by structure have significant influence on wear behavior. Hence, there is a need to control the structure to a high degree in those cases. In the present erosion test, however, the situation is obviously quite different. The pressing together of two bodies does not provide the wearing action but rather by the impingement of gas stream. As a consequence, alignment and characteristics of this stream are the controlling factors. Such items as the static alignment of the nozzle with respect to the wear specimen, the nozzle shape, particle flux, particle velocity, and other aspects of the stream are of importance. Since these are recognized in the method and tolerances are specified, the design of the rest of the apparatus is not critical.

This test also illustrates another aspect discussed in the general section on wear testing, which is the need to develop unique methods to measure and control certain parameters. Examples of this in the gas jet test are the measurement techniques for abrasive flux and abrasive particle velocity. These are both significant factors in the test and required development. A method for the former is presented and references to techniques developed for the latter are given in the ASTM standard, G76 (10–12).

9.2.5. Vibratory Cavitation Erosion Test

This test was developed to simulate the erosive wear caused by the formation and collapse of cavitation bubbles in applications associated with high-speed hydrodynamic systems. Surfaces of hydraulic turbines, pumps, propellers, and hydrofoils are exposed to this type of wear. This test has been used as a mechanism for studying cavitation erosion and has been used successfully in the selection and ranking of materials for applications where this type of wear is of concern. The test is illustrated in Fig. 9.17. The cavitation field generated at the surface of a vibrating specimen immersed in a liquid is used to generate wear. The wear surface of the specimen, which is submerged, is located close to the surface of the liquid, and an ultrasonic horn is used to vibrate the specimen.

A Test Report for G76

Material	Type	_____
	Composition	_____
	Hardness	_____
	Density	_____
	Heat/processing treatment	_____
Specimen Number		_____
	Initial roughness	_____
	Preparation/cleaning procedures	_____
Particles	Type	_____
	Size distribution	_____
	Shape	_____
	Purity	_____
	Source/method of manufacturing	_____
Test conditions	Carrier gas	_____
	Particle velocity and method of determination	_____
	Beam orientation	_____
	Particle flow	_____
	Particle flux	_____
	Temperature (particle/gas/specimen)	____ / ____ / ____
	Size/shape of eroded area	_____
	Test duration	_____
	Standard or nonstandard conditions	_____
Test apparatus	Description	_____
	Calibration frequency	_____
	Calibration material	_____
Reference material	Type	_____
	Description	_____
Erosion values	Average	_____ (mm^3/g of abrasive)
	Standard deviation	_____ (mm^3/g of abrasive)
Additional observations		_____

Figure 9.16 Data and information that are to be reported with the ASTM solid particle erosion test.

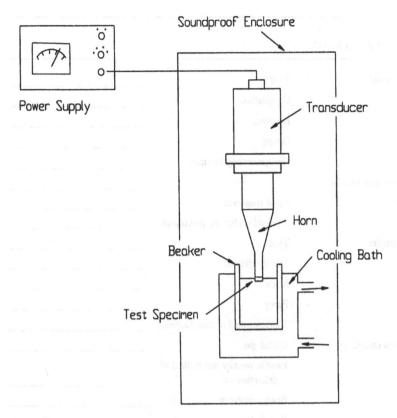

Figure 9.17 Schematic of the vibratory cavitation test apparatus.

A standard method for conducting this test is described in ASTM G32. The results of tests conducted with this procedure typically repeat within 20%.

Like the solid particle erosion test, the standard cavitation erosion test method focuses on the control in the immediate wear situation and not the overall apparatus. In addition to specimen geometry, other influences on cavitation include the temperature and properties of the fluid, the frequency and amplitude of the vibration, and the pressure above the liquid (13–21). Also, the method of specimen attachment can influence the coupling of the ultrasonic vibration between the specimen and the horn and therefore effects the cavitation. The standard method requires control of the frequency and amplitude of the vibration, control of the specimen and the liquid, including its temperature, attachment of the specimen to the vibrator, and pressure over the fluid. Procedures and techniques for implementing the test are given, as well as cautions regarding some of the common problems encountered with this type of test. Along with wear data, the method also requires the recording and reporting of additional information, including specimen characterization, test parameters (if the recommended standard ones are not used), identification of the liquid used, and observations of singular or unusual nature. The standard also specifies that tests on a reference material or materials should be done in any test program as a means of control. If nonstandard conditions are used, it provides a means of relating test conditions. Several reference materials are identified in the standard for this purpose.

The wear produced in this test is directly measured as mass loss. By using the density of the wear surface and the dimensions of the standard wear specimen, mass loss is converted to a wear depth, which is used as the measure of wear performance. The standard method requires the reporting of the mass loss as well as the wear depth. The test method involves the generation of a wear curve (i.e., wear depth vs. time) by interrupting the test at appropriate intervals. Typical curves for wear and wear rate from this test are shown in Fig. 9.18. While these curves are generally nonlinear, their shapes tend to vary. Both the intervals and overall duration of the test are dependent on the materials being evaluated and the test conditions. If possible, the duration should be long enough so that a maximum wear rate is produced. If that is not possible, different test times should be used so that the materials compared reach the same depth of wear. The intervals also should be selected so that a well-defined curve can be established. In practice, this means that the test can range from a few to 10 or more hours.

In several of the preceding tests, a wear curve was developed and fitted to some functional relationship. The best fit was then used to develop a single value that was used to compare wear resistance or quantify wear behavior. In the slurry abrasion test, for example, this was the wear rate after 2 hr. While it would be desirable to do something

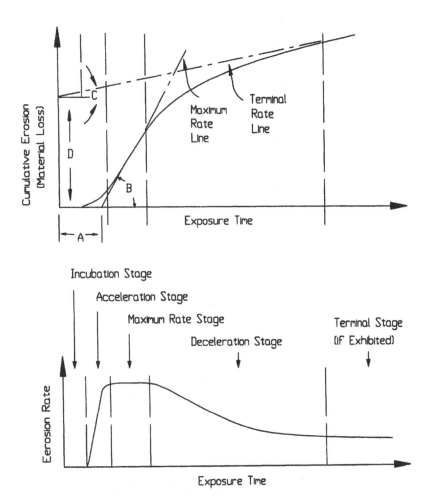

Figure 9.18 The general forms of the wear curve and the curve for erosion rate as a function of test time obtained in the vibratory cavitation erosion test. (From Ref. 117.)

similar in the case of cavitation, there is no consistency of behavior between materials to make such an approach generally feasible. While some materials exhibit a maximum erosion or wear rate in the test, others do not. In addition, the shape of the wear curve appears to be sensitive to the manner in which the test is conducted. This range of behavior can be seen in Fig. 9.19. Thus, the standard requires the reporting of the wear curve itself and that material performance be compared in terms of the relationship of one curve to the other. In any individual study, the wear curves or wear behavior might be similar enough that a single parameter or value can effectively be used to rank materials. In this case, the maximum erosion rate in the test or the terminal erosion rate of the test might be useful indicators of relative wear behavior.

This graphic comparison of wear behavior used in with this test is not as precise or desirable as the use of a single value to rank materials. However, it does illustrate the type of comparison that can be used and may be necessary to use in other wear studies or evaluations, where there is a wide range of behavior. In doing this type of comparison, there are certain conditions that tend to lead to confusion and errors in evaluations.

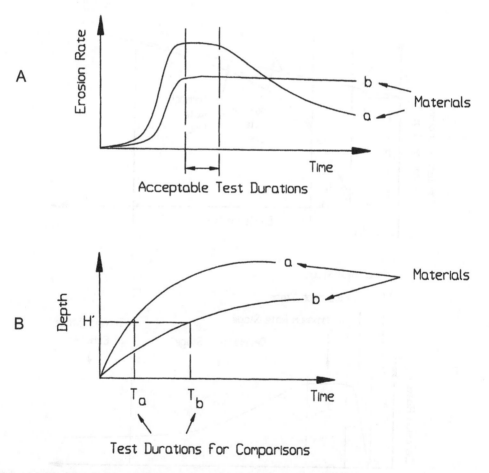

Figure 9.19 Illustration of the two methods of comparison used with the vibratory cavitation erosion test. When the materials exhibit a maximum erosion rate in the test, the maximum erosion rate is used, as shown in "A". For the case when there is not a maximum erosion rate in the test, the time used to develop the same wear depth is used, as shown in "B". (From Ref. 117.)

For example, curves can cross. In such a case, the relative rankings of the materials change, depending on the region the ranking is done (e.g., before or after the intersection). Another example is when one material shows a decreasing wear rate, while another shows an increasing wear rate. This implies that an intersection is likely at some point, even if it did not occur in the test. In cases like these, direct application of the results to a situation is risky. In general, such results indicate that further study or additional information is needed before the results can be applied. Frequently, this can be done by extending the test long enough so that the overall behavior can be broken up into a short- and long-term region, each being characterized by a particular level or amount of wear and morphology. The concept is that these two regions can be related to initial and long range wear in an application.

In effect, the ASTM standard for the cavitation test does that in terms of the guidelines for test duration, that is, test until the maximum erosion rate is achieved (long-term behavior) or test to common wear depth. The magnitude of the wear depth that is used would depend on the application that is being considered. These concepts are illustrated in Fig. 9.19. In general, when the use of a graphic comparison is needed because of the varied nature of the materials wear behavior, more extensive testing and careful consideration of the application are required. An alternate approach is to use another test to simulate the application, one in which the ambiguity does not occur.

9.2.6. Block-on-Ring Wear Test Using Wear Volume*

The basic configuration of this test is shown in Fig. 9.20. It is one of the more commonly used test configurations to study sliding wear and to rank materials in terms of resistance

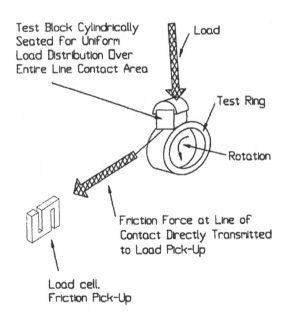

Figure 9.20 Schematic of a block-on-ring sliding wear test. Transducer to measure friction is not always part of apparatus.

*Another test method using this general configuration and wear rate is discussed in Sec. 9.2.16.

to sliding wear. While both the block and ring can wear in this test, the test is primarily used to evaluate the wear of the block material. This same test configuration has been used to evaluate lubricants (22). The methods of conducting the test, the data obtained, and the methods of analysis used are different for lubricant evaluations and wear studies. However, many of the aspects associated with control are the same. The test itself can be conducted under a variety of conditions of load, speed, lubrication, and even environments. When this test is used to rank materials, the ring material is typically fixed and the block material is varied. However, the wear of the block, which tends to experience the most pronounced wear in the test, can be influenced by the material of the ring. As a consequence, when used to rank individual materials for an intended application, the ring material should be one of the materials used in the application. If not, the correlation between test rankings and field performance is likely to be poor. Also, in relating wear behavior in the test to wear behavior in an application, it is necessary to consider the wear on the block and the ring, not just the wear on the block. When this is done and the test conditions provide good simulation of an application, material rankings obtained with this test have been found to correlate with field experience (23).

An ASTM standard for wear testing using this type of test has been developed (ASTM G77). These provide guidelines for conducting the test and analyzing and reporting data. Interlaboratory test programs using the procedures of ASTM G77 have indicated that the intralaboratory coefficient of variation for the block wear volumes is typically 20% for metals. The interlaboratory variations are larger, 30%. For some materials and test conditions, coefficients in the range of 10% have been obtained. The coefficient for ring volume tends to be significantly higher than those obtained for the block (e.g., two times higher). The coefficients of variation can vary with materials and test parameters. For example, with some plastics and short test times intra and interlaboratory coefficients of variations in the range between 30% and 60% have been found. For 10× longer tests, these coefficients reduce to the order of 10%. The variation associated with this test is partially the result of the sensitivity of this type of wear to a large number of parameters. It is also the result of measurement accuracy. The coefficient of variation for the width of the wear scar on the block, which is directly measured in the test and used to compute the volume, is significantly less (e.g., they are in the range 5–20%). However, for the geometries of the test, wear volume is related to the square of the width, which results in larger coefficients for this measure. For the ring, wear volume is determined by measuring a small change in a large mass. Because of the large variation associated with wear volumes in this test, it is generally recommended that several replicates (e.g., three or four tests) be done when using it to rank material pairs.

The basic test method is to press the block against the rotating ring and the wear on both the block and ring is measured after a specified number of revolutions. On the block, a cylindrical groove is generated as a result of the wear. The volume of the wear is determined by first measuring the width of the groove and to use this to calculate the volume. The geometrical relationship and the equation are shown in Fig. 9.21. The volume of wear for the ring is determined by mass loss and converted to volume loss by means of the density of the ring. While the standard test method does not specify a load or test duration, it does require a single load and number of revolutions be used when evaluating

ASTM G176 is a specific version of this method for plastic.

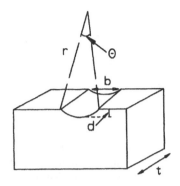

$$D = 2r$$

Scar Width $\quad = b = D \sin \dfrac{\Theta}{2}$

Scar Volume $\quad = \dfrac{D^2 t}{8} (\Theta - \sin \Theta)$

where $\quad \Theta = 2 \sin^{-1} \dfrac{b}{D}$

Scar Volume $\quad = \dfrac{D^2 t}{8} \left[2 \sin^{-1} \dfrac{b}{D} - \sin\left(2 \sin^{-1} \dfrac{b}{D}\right) \right]$

Figure 9.21 Method for determining the block wear volume from a measurement of the width of the wear scar. (From Ref. 118.)

materials. The number of revolutions and load used is to be reported with the wear volume measurements. The wear volumes are then used to rank materials in terms of their wear resistance. In addition, the standard test method requires friction measurements at the beginning, during and at the end of the test.

Elements of control are a significant portion of the standard test method. Principal areas of concern are apparatus construction and design and specimen geometry and preparation. Like the dry and wet sand tests, the overall construction of the apparatus can influence the results, through such aspects as dynamic characteristics, stiffness, ability to maintain alignment under dynamic conditions, to name a few. To control these factors, specific dimensions and tolerances are specified for critical elements, such as concentricity of the ring, specimen geometry, and bearings. Beyond this, the standard also specifies that a standard test be run to qualify a particular apparatus. This standard qualification test is found in ASTM's D2714.

Since the test can be performed with a wide variety of materials with different surface conditions, specific details regarding specimen preparation and cleaning are not provided. It is pointed out that this mode of wear can be very sensitive to the presence of surface contamination, surface composition, roughness, and oxide layers, and therefore these elements need to be controlled. It is pointed out that characterization of surfaces by such techniques as scanning electron microscopy (SEM) and electron dispersive x-ray (EDX) might be appropriate. If a lubricant is being used in the evaluations, appropriate control of the lubricant and method of lubrication is also needed.

The standard also prescribes specific procedures to be followed in the test as an additional means of providing control. For example, instructions as to how to handle

Figure 9.22 Black wear scars—conditions found in the block-on-ring test. "a" shows an ideal scar. "b" shows a scar with nonuniform edges as a result of galling and deformation. "c" shows the condition associated with crowned specimens and "d", misaligned specimens. (From Ref. 119, reprinted with permission from ASTM.)

the specimens once they are prepared are given, as are procedures for applying the load and bringing the ring up to speed. The user is also instructed to examine the shape and morphology of the wear scar on the block, pointing out features that should be observed and what actions should be taken if they are found. For example, micrographs in Fig. 9.22 show the effect of misalignment between block and ring, as well as an out-of-flatness condition on the block. When these conditions occur, the test should be rerun.

Also, evident in one of the micrographs (Fig. 9.22b) is an example of the effect that galling and plastic deformation may have on the appearance of the wear scar. This type of phenomena is a frequent characteristic of this test, particularly when a lubricant is not used and for certain types of materials and material pairs. These phenomena can distort the definition of the edges of the wear groove. When this distortion is small or moderate, as is shown in the micrograph, it is possible to estimate the true edge with sufficient accuracy for a valid test. Sometimes the distortion is so severe that this cannot be done, and a milder form of the test should be considered for evaluating the materials. This might involve using a lighter load, slower speed, and shorter duration. Even if this type of phenomena does not prevent accurate measurement of the edges, their significance in terms of simulation should be considered. If such phenomena are not relevant to the application, the test conditions should be modified to achieve better simulation. As a rule of thumb, the more "wear resistant" the category of the materials tested, the higher the loads and speeds that can be used without experiencing these types of phenomena. For example, higher loads and speeds can be used when ceramics or hardened steels are being evaluated, than when plastics or softer alloys are tested.

An extensive amount of information and data should be recorded and measured with respect to the test. The list of data that the standard recommends be reported with the test is shown in Fig. 9.23. This includes friction data. Comparison of friction behavior in repeated tests can be an indicator of the amount control in the test, as has been discussed.

While the test method allows the investigator to select parameters to best simulate the application, there are some features that potentially limit the degree of simulation that can be achieved. Since the block experiences significantly more rubbing than the ring, the test is limited in terms of its ability to simulate the relative amount of wear between the pair of materials. It is also a unidirectional test. The test also is one in which the contact pressure varies through the test. Initially, the contact can be described as a Hertzian line contact. Small contact area, high stress and significant subsurface stresses characterize this type of contact. Once a wear groove is formed in the block, it becomes a conforming

A Format for Reporting G77 Data

Test parameters

    Block material and hardness   _____

    Ring material and hardness   _____

    Initial and final roughness: Block   _____

                                Ring   _____

    Ring rpm   _____

    Lubricant   _____

    Test load   _____

    Test distance   _____

    Number of duplicates   _____

Results

    Block scar width   _____

    Block scar volume   _____

    Ring weight loss   _____

    Ring scar volume   _____

    Final dynamic coefficient of friction   _____

Optional information

    Block weight loss   _____

    Ring heat treatment   _____

    Block heat treatment   _____

    Lubricant composition   _____

    Initial static and dynamic coefficients of friction   _____

Unusual conditions   _____

Figure 9.23 A format for reporting block-on-ring test data.

contact. With this change, the stress system changes and the overall stress level decreases. As the wear increases, the area of contact continues to increase and the stress level continues to reduce. Since stress can influence wear, this characteristic can limit the degree of simulation that can be obtained with this test. For example, this aspect of the test could be a factor in the degree of correlation that can be obtained with conforming contact applications, such as with thrust bearings.

    Because of this aspect, wear rate tends to change its wear progresses with the result that a wear curve for this test configuration tend to be generally nonlinear. This results in the requirement that a single test duration be used for comparing materials. While this approach to the problem of nonlinear behavior has been found to be effective in material rankings with this test, it is not the only approach or the most complete approach that can be taken. An exposure that is associated with this approach is that if the nonlinear behavior or relationship varies with material pairs (24), it is possible to obtain different rankings with different test duration. This is because the wear curves can cross (Fig. 8.7). An alternate approach would be to develop a wear curve and base comparison on this, such as done in the standard liquid cavitation test. Since wear rates tend to become stable or quasistable (slowly decreasing) in this test, another approach is to use long-term wear rate, such as done described in Sec. 9.2.16.

To address this concern and other limitations, variations in the test procedure and parameters may be necessary. In such cases, the practices of the standard test can be used as guides as to what to control, how to insure a valid test, etc. For the nonlinear relationship problem, tests for two different durations can be performed to insure that the same type of relationship occurs for the different pairs. To better characterize the wear behavior of a pair of materials, tests interchanging block-and-ring materials can be performed. The tester can also be modified to provide oscillatory motion of the ring and thereby to provide better simulation to applications which are not unidirectional or which have different relative amounts of rubbing than that provided by the standard test.

As with several of the other tests the block-on-ring test may provide a ranking of materials for an application, but it does not provide an absolute measure of the wear performance in that application. Also, with the block-on-ring test, the relative separation in the rankings obtained in the test and in the field may not be the same. For example, a relatively large difference in performance in the test, perhaps a factor of two times, could be much less in the application (e.g., only 10% or 20%). Conversely, a relatively small difference in the test could correlate to a very large change in performance in the application. This lack of correlation is because of the nonlinear nature of the wear behavior that tends to be characteristic of this test. Finally, changing the test parameters can influence this type of correlation between the test and application.

9.2.7. Crossed-Cylinder Wear Test

This is a test that has been used to rank material pairs in terms of their resistance to sliding wear. It has been used for a number of years in industry, principally to evaluate tool steels and hard-surfacing alloys. Procedures and parameters used by the different laboratories tend to vary although a standard practice has been developed and issued as ASTM G83 for this type of test. The basic configuration of the test is shown in Fig. 9.24. One cylinder is held stationary and the other is pressed against it and rotated. The basic concept is to rank materials in terms of the wear produced after a fixed number of revolutions. Wear is directly measured by a mass loss technique but converted to volume loss for comparison. Studies, which used the standard practice, have shown that the coefficient of variation for intralaboratory test is within 15%, and for interlaboratory tests, 30%.

Unlike the block-on-ring procedure, which allows considerable flexibility in terms of test parameters and materials evaluated, the standard test method for the crossed-cylinder test is specific in terms of test parameters and limited in terms of the materials for which it can be used. The test is designed for unlubricated evaluations of metals but other materials are allowed, if they are sufficiently strong and stiff so that the specimens do not deform, fracture, or significantly bend under the load conditions specified. In general, this would exclude polymers. The standard test method has three procedures, which differ in terms of speed and duration to address different levels of wear behavior. Procedure A is the most severe test and is recommended for the most wear resistant materials. Procedure B is a

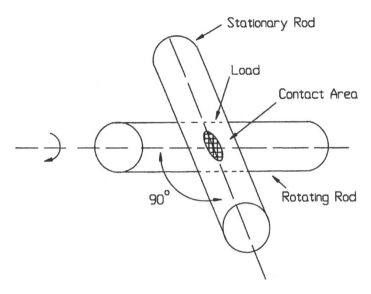

Figure 9.24 Diagram of the crossed-cylinder test.

shorter version of A, which can be used for less wear resistant materials that exhibit sufficient wear in the shorter period of time. Procedure C is a milder test (i.e., lower speed), that is run for a fewer number of revolutions. This was developed for the evaluation of materials that exhibited such severe wear under the conditions of A or B that valid or useful comparisons could not be made. This could be because of excessive heating under the more severe conditions, extensive galling or adhesion, which would influence the accuracy of the measurement technique, or complete wear-through of surface treatment layers. The selection of which procedure to use therefore depends considerably on the nature of the materials and associated wear behavior. It is possible that more than one procedure might be acceptable; however, the ranking and comparison of materials should be confined to within one test procedure. Inferring the relative behavior of materials in one test procedure, based on relative behavior in another, should not be done for several reasons. For example, using results from procedure C with results from either of the other procedures to establish a ranking should not be done, since the test conditions are different and their effect on material behavior is generally not known. Also, since the nature of the wear curves in this type of test tends to be of a variable, nonlinear nature, cross-comparison between procedures A and B may not be valid, even though load and speed conditions are the same.

This test method does not provide much flexibility in terms of providing simulation since there are only two speeds allowed. Basically the test simulates high-speed, dry sliding wear. The user of the test has to decide first of all whether or not the application can be described in those same general terms. If the answer is yes, the test provides first-order simulation. Second-order simulation would depend on the similarity of the standard test parameters and those of the application and the sensitivity of the materials to any differences in those parameters. Because of the unique geometry of the test, which is complicated by the wear of both cylinders, it is generally not possible to make an a priori judgement regarding second-order simulation. Therefore, comparison of wear scar morphology from the test with that from the application is one way of deciding on the degree of simulation and the likelihood of good correlation between the test and application. Similarity in the appearances generally implies that there should be correlation, even though there might be differences in specific values of the parameters.

The primary intention of the test is to characterize the wear resistance of self-mated pairs. In this case, the total volume of wear (i.e., the sum of the wear volumes obtained from the stationary and rotating member) is used as the measure of wear resistance. The test method allows testing with dissimilar metals as well, in this case the wear volume for both specimens should be reported separately. In addition, the method requires two tests with the dissimilar couple, with the position of the materials interchanged in each test. The sum of the wear volume obtained as the stationary and rotating specimen is reported and can be used to compare with self-mated behavior.

In addition to the quantitative measurement of wear, the test procedure requires that the worn specimens be examined for features that might make the test invalid, such as evidence of transfer, deformation, or distortion. If any of these occur to a significant level, the test is to be considered invalid and the test should not be used to evaluate the wear resistance of those materials.

Specimen preparation, cleaning, measurement procedure, as well as dimensions and tolerances of critical elements of the apparatus, are covered in the ASTM standard. In the block-on-ring test, a key element is the alignment of the block in the axial direction of the ring and considerable details as to how to insure proper alignment are provided in that standard. In the crossed cylinder wear test, however, the alignment criteria is not as critical, since it is a point contact but concentricity and run-out are major factors. Consequently, the test method provides considerable detail and comments regarding the needed tolerances and recommends specific chuck designs. It also identifies a qualification test and acceptance criterion to insure adequate performance and procedures.

There are several elements in this test that are similar to those in the block-on-ring test. One feature is that in both tests, the stress level decreases with duration and wear. A second feature is that the wear curves associated with both of these tests tend to be nonlinear and of a varying nature. A third similarity is that both involve the wear of two surfaces or bodies. To address the first two elements, both test methods employ the same general approach (i.e., using fixed numbers of revolutions to rank materials). Consequently, the general problems discussed regarding the extrapolation of block-on-ring test results to absolute performance in an application are the same for the crossed-cylinder test. The limited range of test parameters and the use of total wear to characterize the material couple further complicate the situation with the crossed-cylinder test. The more dissimilar the test and application conditions are, the less likely that relative rankings will be applicable. When there is only first-order simulation, only large differences in test results should be considered significant.

Another interesting observation with both of these tests is that the scatter associated with interlaboratory results are noticeably higher that with intralaboratory tests. This suggests a bias between laboratories with respect to the test but not lack of control. Such a bias could be attributed to slight but consistent variations in procedures or ambient environments. Slight variations in the testing apparatus (e.g., in design and construction, alignment, load control, and vibration) also can be factors. Assuming this is the major factor, the studies done in terms of these two tests suggest that these types of machine variations can cause 10–15% scatter in sliding wear behavior. Of course much larger variation can result with poorly designed and built apparatuses. This in turn emphasizes the need for proper design and construction of wear apparatuses if minimal scatter is to be achieved.

9.2.8. Pin-on-Disk Wear Test

This is another configuration that has been used extensively to study wear and to rank materials. It is viewed as a general test that can be used to evaluate the sliding wear behavior of material pairs. Its correlation with an application depends on the degree of simulation that the test parameters have with those of the application. The basic configuration is shown in Fig. 9.25. A radius-tipped or flat-ended pin is pressed against a flat disk. The relative motion between the two is such that a circumferential wear path on the disk surface is generated. Either the pin or the disk can be moving. The test parameters that have been used with this test vary. The ASTM standard for this test, ASTM G99, which does specify the use of a rounded pin, does not specify specific values for the parameters, but allows those to be selected by the user to provide simulation of an application. The parameters that can vary include size and shape of the pin, load, speed, and material pairs. The test can also be done in a controlled atmosphere and with lubrication. Like the block-on-ring and crossed-cylinder tests, stress levels in the rounded pin version of this type of test change during the course of the test as a result of the wear. Consequently, the relationship between wear and duration or amount of sliding tends to be nonlinear. For material ranking and comparison, the ASTM standard recommends measuring the wear on both members after a fixed number of revolutions. It is also recommended that with dissimilar pairs of materials that two tests be performed with the materials changing positions in the test. The standard allows the use of wear curves for comparison. This is particularly useful if nonlinear behavior is to be taken into account. When this approach is used it specifies that new specimens are to be used for each data point on the curve. The test should not be stopped for intermediate wear measurements and restarted because of potential problems with alignment, disturbance of debris and surface films, and introduction of contamination.

With flat-ended pins, an additional concern is initial alignment. In such a case, it is necessary to allow the specimens to wear-in (so that uniform contact is achieved) before useful data can be obtained. This is illustrated in Fig. 9.26. The block-on-ring test has a similar problem associated with alignment in the axial direction of the disk. With this type of pin, the duration has to be long enough so that the wear-in wear is negligible

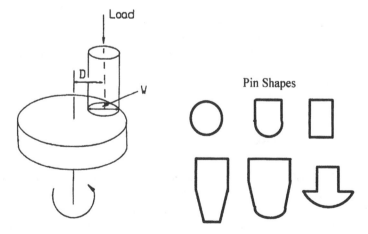

Figure 9.25 Diagram of pin-on-disk test and cross-sections of pin shapes used with this type of test. With curved surfaces wear is generally confined to the curved region.

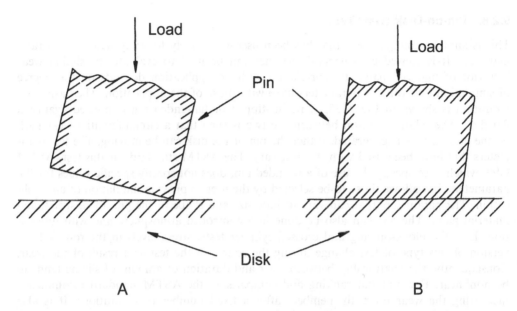

Figure 9.26 Misalignment in the pin-on-disk test when a flat pin is used. "A", initial; "B", after wear-in.

in comparison to the final wear, if final wear volume is to be used as the measure of wear resistance. An alternative approach is to use wear rate in which case the test has to be long enough so that a stable wear rate is obtained. Because it eliminates the alignment problem, a rounded pin is the preferred configuration. With a rounded pin useful wear data can be obtained after small amounts of sliding and thereby provide a continuous curve. If the pin is flat-ended, this would not be possible since the initial portion of the wear curve would be strongly influenced by the misalignment between the pin and the disk.

The ASTM test method allows both geometrical and mass loss methods for determining wear but in either case the measurement should be converted to volume loss for reporting. With mass loss, this is to be done by dividing the mass loss by the density. With the geometrical approach, this is done by converting a measured linear dimension, to a volume using the appropriate relationship for the geometry of the wear scar. For example, in the case of negligible wear on one member and a spherical-ended pin, the width of the wear scar can be used to compute the volume by means of the following equations:

$$V = \frac{\pi}{64} \times \frac{W^4}{R} \quad \text{pin wear} \tag{9.3}$$

$$V = \frac{\pi}{6} \times D \times \frac{W^3}{R} \quad \text{disk wear} \tag{9.4}$$

where V is the volume of wear; W is the width of the flat on the pin or the width of the wear track on the disk; D is the radius of the wear track; R is the spherical radius of the pin. In both cases, the wear scar is either the volume of a spherical cap of cord W (pin wear) or a groove whose profile is a circular section of cord W (disk wear). This is shown in Fig. 9.27. For situations in which there is wear on both surfaces or in which the radius end of the pin is not spherical, similar relationships have to be identified or

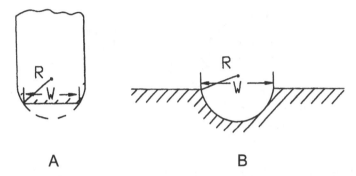

Figure 9.27 Wear scar cross-sections in the pin-on-disk test with a spherical-ended pin. "A" shows the situation when there is a negligible wear on the disk; "B", negligible wear on the pin.

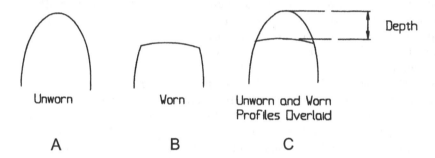

Figure 9.28 Overlay technique to determine pin wear depth when there is significant wear on the disk. "A" is the profilometer trace over the center of the spherical surface before the test; "B", the trace over the same location after the test. "C" shows "A" and "B" overlaid to determine the depth.

developed. In these cases, wear depth rather than width might be the best direct measurement of wear. For the pin, profilometer traces before and after wear can be used to determine depth by means of an overlay technique illustrated in Fig. 9.28. For the disk, a profilometer trace across the wear scar can be used. This is shown in Fig. 9.29. Once this is done, appropriate geometrical relationships or numerical methods can be used to determine wear volumes (37,112). More examples of this geometrical approach can be found in the design and problem solving examples in EDW2E.

While this geometrical technique is more complex than the mass loss technique, it is often a more sensitive technique when small amounts of wear are involved, as may occur

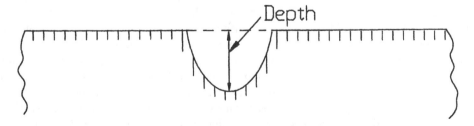

Figure 9.29 Determination of disk wear scar depth from a profilometer trace.

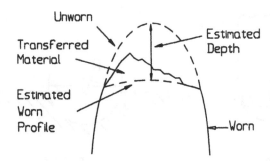

Figure 9.30 Illustration of a technique to estimate wear depth when transferred material is present.

in this test. Also in this test, as well as in the two previous ones for sliding wear, transfer can occur to such a degree that ability to accurately measure the wear is greatly decreased. Sometimes transfer is so great that the test is an invalid method for ranking materials. With proper interpretation of profile measurements, however, it is sometimes possible to distinguish build-up from wear so that a valid result can be obtained. An example of such an interpretation is shown in Fig. 9.30.

The ASTM standard method addresses specimen preparation, cleaning, tolerances, loading, and measurement techniques. It also requires that the wear volumes of both members, test conditions, a complete description of the materials involved, and the preparation and cleaning procedures used should be reported. Test parameters should include load, speed, temperature, roughness, dimensions, and shapes. It also recommends that the coefficient of friction be measured in these tests and that these values be reported. Initial and final values for the coefficient of friction should be given and any significant changes during the test should be noted.

The test method also provides specific cautions. For example, it is pointed out that while the test can be performed with the disk either horizontal or vertical, the results can be different. The reason for this is that in the vertical position wear debris can be removed from the system by gravity, while in the horizontal position it cannot. It also advises against the use of wear measurements made with continuously monitoring transducers. Transfer films, wear debris, and thermal effects can influence such measurements, and, as a result, be erroneous. The method, per se, does not specify either the magnitude or the method of loading, but it is pointed out that the method of loading can be a factor in the wear. For example, differences have been observed between tests that have used a dead weight loading method and those which have used a pneumatic method. This is attributed to differences in the stiffness associated with these approaches and the presence or absence of significant vibrations.

Interlaboratory test programs have been conducted for this test method. These studies have shown that when the procedures and techniques of the standard are followed intralaboratory tests should be repeatable within 20%. If tests do not repeat within that, it is recommended that some investigation should be done to determine the reason for the increased scatter. Possibilities are apparatus problems, poor discipline in performing the test, or variations associated with the materials used. Interlaboratory variation is of the order of 40%. Both of these variations are similar to those associated with the previously discussed tests for sliding wear. This difference in the coefficient of variation between intra- and interlaboratory implies that machine-to-machine differences can contribute significantly to wear behavior.

Correlation with an application is dependent on the simulation associated with the test parameters. The usual techniques for addressing simulation need to be pursued (e.g., comparison of wear scar morphology, selection of loads and speeds, etc.). Acceleration can be associated with this type of test as well and the actual degree depends on the relative values of the parameters. However, very little acceleration may be associated with test conditions that provide good simulation. Any attempts to provide acceleration should be investigated carefully.

9.2.9. Test for Galling Resistance

Galling is a severe form of wear characterized by macroscopic material transfer and removal or formation of surface protrusions. It generally occurs in sliding systems, which move slowly or intermittently and are either poorly lubricated or not lubricated. Seals, valves, and threaded components are examples of applications, which often exhibit this type of wear. However, it can also occur in gears and bearings under heavy loads. ASTM G98 is a standard test method for ranking material pairs in terms of their resistance to galling. It is an accelerated test in that the conditions of the test were selected to promote galling and not to provide a simulation of an application. Generally, the rankings that have been obtained with this test have been found to be useful guides in selecting material pair for applications, which are prone to galling (25,26).

The basic test configuration is shown in Fig. 9.31. It consists of a flat-ended cylinder pressed against a flat surface. The test is performed by slowing rotating the cylinder through 360° and then separating the two members. The surfaces are visually examined for evidence of galling. If there is none, the test is repeated with new surfaces at a higher load. If there is evidence of galling, the test is repeated at a lower load. This procedure is followed until two load levels bracket the occurrence of galling and the load midway between the two is used to calculate what is called the threshold galling stress. This is used to rank the material pairs. The higher the stress, the more galling resistant the pair.

The size, flatness, and roughness of the specimens and maximum load intervals are specified. The speed of rotation in the test is controlled by specifying the duration of the rotation. The method allows the use of any apparatus that can maintain a constant load to

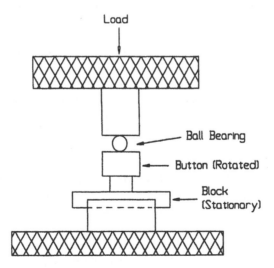

Figure 9.31 Configuration of the test for galling.

A Format for Reporting G98 Data

Materials	Button	Composition	_____
		Hardness	_____
		Heat treatment	_____
	Block	Composition	_____
		Hardness	_____
		Heat treatment	_____
Specimens	Preparations and cleaning		_____
	Procedures	Button	_____
		Block	_____
	Initial roughness	Button	_____
		Block	_____
Environment	Atmosphere		_____
	Temperature		_____
	Humidity		_____
Test system description			_____
Test conditions	Interval		_____
	Rotation time		_____
Threshold galling stress for couple			_____
Unusual conditions			_____

Figure 9.32 Data to be reported in the ASTM test for galling.

the accuracy required, provide proper alignment, and rotational control. Since galling is an adhesive phenomena, proper cleaning of the specimens is a critical factor in this test and this is emphasized in the test procedure. The information that should be reported with results is shown in Fig. 9.32. The method also provides guidance as to how to examine the surfaces for evidence of galling. Examples from this test are shown in Fig. 9.33.

This test is an example of one in which the appearance of the worn surface is used to evaluate the wear resistance of materials. When such an approach is used, it is necessary to define the type of technique to be used in evaluating the appearance. One aspect of this is magnification. Different techniques could enhance or degrade the identification of certain features and variations in precision would result. In addition, different techniques might be sensitive to different aspects of the phenomena and in effect would result in varying criteria. To address these concerns, the test method specifies unaided visual inspection for determining whether or not galling has occurred. While qualitative approaches, like the one used in this test, are more subjective than quantitative approaches and as a result tend to be less precise, they can be effective and consistent. Good inter and intralaboratory

Figure 9.33 Examples of wear scars produced in the test for galling. "A" shows examples of surfaces exhibiting galling in the test. "B" shows the results of a test sequence used to determine the galling threshold. (From ASTM G98, reprinted with permission from ASTM.)

reproducibility has been found in interlaboratory testing done for the development of the ASTM standard. Experience has shown that a 5–10 kpsi difference in the galling threshold stress is needed before a difference in field performance is observed. Interlaboratory testing programs, utilizing this standard method, generally have shown repeatability within that range.

9.2.10. Rolling Wear Test

A configuration that has been successfully used to address rolling wear is illustrated in Fig. 9.34. Basically, it consists of a pair of driven rollers pressed against one another. The typical procedure is to visually monitor the condition of the roller surfaces and determine the number of cycles for a selected level of surface damage to occur. This could be the appearance of cracks, surface texture change, or spalls. Figure 9.35 illustrates examples of this type of damage and wear. These tests are usually quite long, extending for days or weeks and inspections are done on a periodic or scheduled basis. This is another example of the use of appearance criteria in a wear test. The longer the number of cycles, the more resistance the pair is to rolling wear.

Critical elements of this test are control of the surface velocities of the two rollers, alignment of the rollers, and geometrical tolerance of the rollers. With respect to the last,

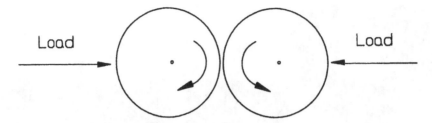

Figure 9.34 Basic configuration used in rolling wear tests.

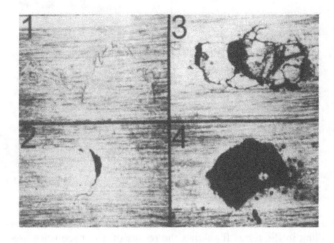

A

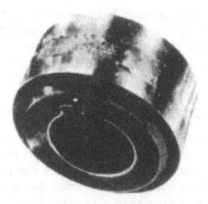

B

Figure 9.35 Examples of wear produced in a rolling test. "A" shows various stages in the formation of a single spall. "B" shows the appearance of the surface of a roller with extensive spalling. ("A" from Ref. 18, reprinted with permission from NLGI; "B" from Ref. 27, reprinted with permission from Penton Publishing Co.)

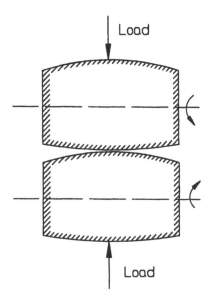

Figure 9.36 Configuration of a rolling wear test in which curved rollers are used.

particular attention has to be paid to the edge conditions of the rollers so that a significant stress concentration condition does not occur. This means that the edges of rollers should be well rounded. Use of rollers of the same length can help to minimize this exposure, as well. Another approach that has been used is to use slightly curved or crowned rollers, as is shown in Fig. 9.36. This type of test has been used to address conditions of pure rolling, in which case the surface velocities of the two rollers must be identical. In addition, the test has also been used to address conditions of mixed rolling and sliding. In this case, the relative velocities must be controlled so that the proper ratio of sliding to rolling is achieved and maintained. There are two elements to controlling the velocity; one element is the rotational speeds of the two rollers. The other is the radius of the rollers. In addition, control of the preparation and cleaning of the rollers are important, as well as the consistency and uniformity of the materials and lubrication, if used.

This test has been used for the evaluation of material pairs for rolling applications such as gears, cams, roller bearings, and ball bearings. In these types of applications, additional forms of wear might also be present. For example, different regions of gear teeth are exposed to different modes of wear. Rolling predominates along the pitch line, while sliding predominates at other locations along the tooth profile. Nonetheless, generally good correlation has been found between this test and actual performance for those regions where rolling is the major characteristic. One way in which this type of test has been used is to develop data in conjunction with a model for rolling wear (27,28). The basic concept of that model is that there is a power law relationship between the number of cycles to failure and the stress level under which rolling takes place, as per the following equation:

$$N = N_o \times \left(\frac{S}{S_o}\right)^n \tag{9.5}$$

N is the number of cycles to failure at stress level S and N_o is the number of cycles to failure at stress level S_o. The parameters of the model that need to be determined experimentally are N_o and n. This can done by performing a series of tests at different stress

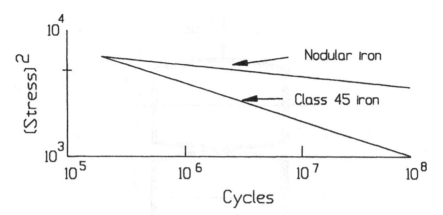

Figure 9.37 Examples of the relationships between stress and number of revolutions obtained in rolling wear tests with different materials. (From Ref. 27.)

levels and determining the number of cycles to failure for each of the stress level. By fitting these data to Eq. (9.5), the values can be determined or a curve defined that represents that type of equation. Examples of several curves characterizing pairs of materials are shown in Fig. 9.37. This is a log–log plot and Eq. 9.5 is a straight line on such a plot. Consequently, a minimum of two tests at different stress levels is required to define the parameters. It is generally desirable to do repeated tests and ones at additional stress levels. The differences in stress levels and number of repetitions should be large enough that a good estimate of n can be obtained. Changing geometry, size, or load can vary stress level. However, it is usually more convenient and preferred to perform tests at different loads, since specimens can then be made to a single size and shape. Changing size and shape may introduce or change other factors, particularly the amount of slip and alignment, which can result in increased variability or scatter.

The same basic test and procedure is used with different amounts of sliding introduced. Examples of curves obtained for these mixed conditions are shown in Fig. 9.38. Again, the data are fitted to Eq. (9.5). In general, different values of N_o and n are obtained for different percentages of sliding. The percentage of sliding is defined as

$$\% \text{ Sliding} = 2 \times \frac{|V_1 - V_2|}{|V_1 + V_2|} \times 100\% \tag{9.6}$$

where V_1 and V_2 are the surface velocities of bodies 1 and 2, respectively. At some percentage of sliding, the morphology of the wear changes to one more characteristic of sliding wear (e.g., scratches, scuffed appearance, adhesion, etc.). This would represent a limiting value for the rolling wear model. This transition point can be established by performing tests at different percentages of sliding and looking for this type of change in morphology. Since the value of this transition point can vary significantly with materials pairs (e.g., 9–300%), it is often of engineering significance.

Testing in this fashion (i.e., to determine values of parameters of a model) provides the test with the element of being able to determine absolute performance, as well as acceleration. Eq. (9.5) can be used to predict or project actual performance if the stress level in the application is known. Acceleration is provided by being able to define N_o and n in shorter periods of time by performing tests at stress levels above those in the application. For the simulation of application, it is important to insure that the stress levels in the

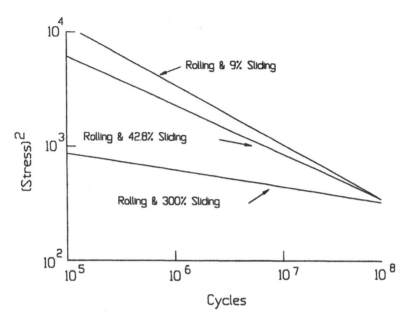

Figure 9.38 Examples of the relationships between stress and number of revolutions obtained in rolling tests involving slip. (From Ref. 27.)

test remain in the same range as that of the application. Generally, this means that the loads used in the test should result in stresses in the elastic range. If the test loading is such that plastic deformation occurs, the test method is not valid. This can be a practical problem for some materials, like thermoplastics. For these materials, stress levels required to avoid creep or deformation might result in very long test times and it may not be a practical way to evaluate these materials.

9.2.11. Reciprocating Pin-on-Flat Test (Oscillating Ball-Plane Test)

Generically this test is very similar to the pin-on-disk test and used for the same purposes. It differs only in the type of motion. The generic features of a reciprocating pin-on-flat or ball–plane test are shown in Fig. 9.39. One difference between the flat and disk tests is the shape of the flat member of the contact. In the pin-on-disk, it is a disk to accommodate rotation, while with the flat and plan test configurations, it is normally a rectangular block or flat specimen. However, the fundamental difference is with the type of motion that each provides. The motion is generally unidirectional at a constant speed in the pin-on-disk test. In the pin-on-flat test and ball-on-plane test, there is a reversing of the direction of sliding and the speed may vary throughout the cycle. One of the consequences of the change in directions is that each cycle contains acceleration and deceleration portions, an element that is not present in the pin-on-disk test. The velocity profile tends to vary with different apparatuses and depends on the nature of the drive mechanism used. For example, the profile is sinusoidal if a rotating eccentric is used. If a linear stepper motor is used, it would have a square wave profile. These differences in the motion can influence wear behavior for a variety of reasons including the influence of debris, build-up of transfer and third-body films, lubrication, and fatigue wear mechanisms (which can be influenced by stress reversals). Consequently, the motion in one type of test could provide

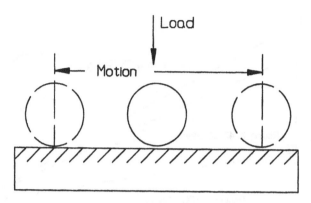

Figure 9.39 Configuration of reciprocating pin-on-flat test, using a spherical specimen as the "pin". The shapes used for the pin are the same as with the pin-on-disk test. (See Fig. 9.25.)

better simulation for an application than the motion in the other type. While this potential exists and must be recognized, it generally does not appear to be a major factor. Both tests have been used effectively to address wear concerns in both unidirectional and oscillatory applications.

It is also desirable to measure friction in conjunction with a pin-on-flat or ball-on-plane wear test, as it is with the block-on-ring and pin-on-disk tests. Therefore it is a good practice to incorporate this capability into these types of apparatuses. Many of the ones described in the literature (29) have this capability. One advantage or use of these oscillating tests is that its oscillatory nature makes it suitable for the simulation of fretting or fretting corrosion situations. This is done by reducing the amplitude of the motion to the range of associated with fretting (30,31).

Like with the pin-on-disk test, these reciprocating tests can be used to rank material pairs in terms of wear resistance. ASTM G133 describes a standard test method for such a purpose, which utilizes a sphere or spherical-ended pin. Like the ASTM test method for pin-on-disk, ASTM G99, wear volume after a fixed amount of exposure (number of cycle) is used. The coefficients of variation from interlaboratory test programs are between 20% and 30% within a laboratory and about 50% between laboratories. This test method and ASTM G99 (pin-on-disk test) are useful guides for conducting these types of tests. The ball–plane test has also been used in a different manner to address engineering wear situations (30–35). With these uses the same methods for providing control apply but with some modifications that are associated with the taking and analysis of the data. However, the measurements made and the analysis techniques tend to be different for these two uses. If the standard method is used, the volume of wear generated after a specified amount of sliding is used to rank the material pairs (with the caution that the wear curves in these tests are frequently nonlinear). In the engineering use, a linear wear dimension (e.g., such as scar width or depth) is often used and a wear curve developed, rather than a single measurement.

The following is a description of a method that can be used for engineering evaluations with a ball–plane test and extended to the pin-on-disk test as well. In addition, some elements of this approach can also be applied to the block-on-ring test. In many of the engineering applications of this test, the primary wear measure is the depth of the wear scar. This is usually determined by means of a profilometer measurement. Fig. 9.40 shows typical traces for the ball and the plane. With the plane, the trace automatically

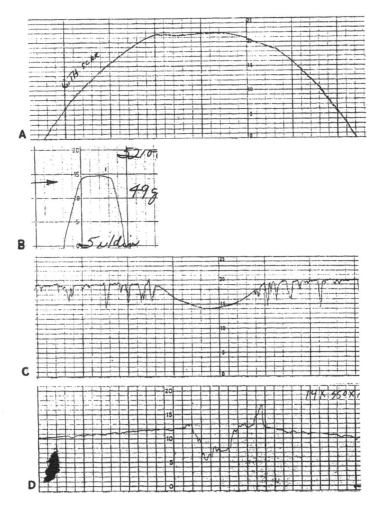

Figure 9.40 Profilometer traces across wear scars occurring in ball–plane wear tests. "A" and "B" are for the ball specimen; "C" and "D", for the plane surface. The radii of the spheres used and the magnifications of the traces are different.

provides a reference level to determine wear depth. For the ball, some technique comparing the unworn profile with the worn profile is often used. A graphical over-lay technique is shown in Fig. 9.41 for a flat wear spot on the ball and for a more general condition in Figs. 9.28 and 9.30. Analytical techniques, which are based on the measurement of the width of the spot, can also be used (36). To improve accuracy with an overlay technique, it is desirable to use unworn profiles of the actual specimen rather than a theoretical or nominal shape.

The method involves the use and analysis of a wear curve, generally plotted on a log–log scale. Conducting tests of different duration generates the wear curve. The intervals and overall duration of the test vary with the situation or application being addressed. The concept is to generate a well-defined wear curve which is representative of the wear in the application. As a rule, this means adjusting the total measurement interval to extend from the smallest amount of sliding required to produce a measurable

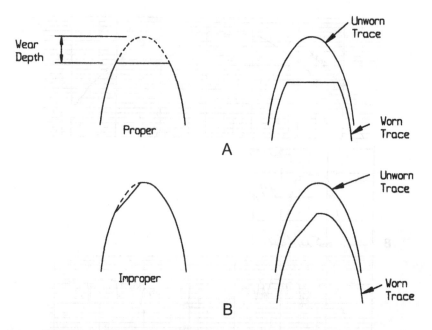

Figure 9.41 Illustration of the overlay technique used to determine wear depth on a curved surface. "A" shows the proper technique; "B", an improper technique. In using the technique, the wear scar should be at the apex of the trace.

wear depth to that required to produce a significant increase in the wear depth (e.g., for example an order of magnitude increase or more). It is generally appropriate to select logarithmic or half-logarithmic intervals since the wear depth typical increases in a less than linear fashion. To characterize the wear behavior the data are then fitted to a power relationship

$$h = C \times S^n \tag{9.7}$$

where h is the depth of wear and S is the amount of sliding, number of cycles, or time. This is usually best done by plotting data on log–log paper and fitting the log-form of Eq. (9.7) to the data. Best-fit values for C and n are used to characterize the wear behavior of the system. Typical wear curves obtained in such a fashion are shown in Fig. 9.42. In effect, the standard test method only uses C for this characterization, which presumes that n is a constant.

There can be poor correlation between the wear data and Eq. (9.7) in some instances, as in one of the cases shown in Fig. 9.40, curve 2. In such cases, the wear data are fitted in a piece-wise fashion. Usually, it is sufficient to separate the data into two regions, such as initial and long-term behavior. The data in each region are fitted by an expression of the form of Eq. (9.7) (Fig. 9.43). In such cases, there is a significant difference in n for the two regions. Generally, this piece-wise fit is necessary as a result of transitions in wear behavior, which can be associated with observed changes in friction and wear scar appearances. In some cases, additional regions may need to be considered in the same manner (34).

Once the wear curve is defined in terms of one or more pairs of C and n values, the results can be interpreted in terms of various models or theories. These models can then be

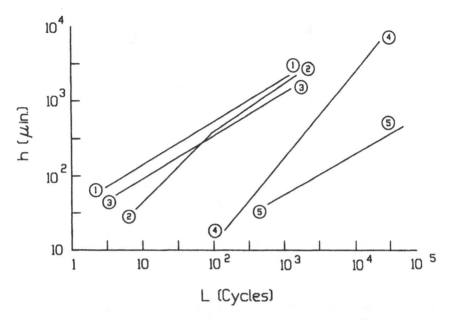

Figure 9.42 Example of wear curves obtained in ball–plane tests. Curves 1, 2, and 3 are for the depth of wear on a steel sphere sliding against ceramic flats of different roughnesses and with different lubrication conditions. Curves 4 and 5 are for the depth of wear on two different steel flats being worn by ceramic spheres of different radii and with different loads. (From Ref. 136.)

extended to the application. The following example for the case of a flat spot being worn on a ball illustrates this type of consideration. Because of the geometries involved, a linear relationship between S and wear volume results in a value of n close to 0.5. If the value obtained for n is close to this, it would imply stable wear behavior, as well as the wear being consistent with models for abrasive and adhesive wear. However, if a value of n significantly greater than 0.5 were obtained, this would imply that there is some transition in wear occurring, as is illustrated in Figs. 9.43 and 4.46. Likewise if a value of n lower than 0.5 is obtained, this suggests a wear mode, referred to as a variable energy wear mode, which is associated with fatigue wear and stress-dependent, is dominant (24).

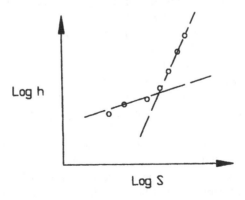

Figure 9.43 Example of the piece-wise fit of wear data to relationships of the form, Wear $= K \times$ Usage^n.

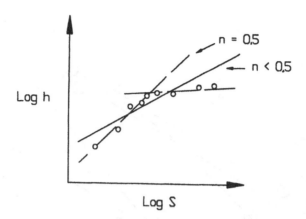

Figure 9.44 The effect of a transition to a milder mode of wear on the apparent exponent. The solid curve represents the best fit of all the data. The dashed curves show a piece-wise fit of the same data, assuming a transition in wear behavior.

Such modes are possible, since stress levels decrease with wear with a ball–plane geometry. Values less than 0.5 can also imply that a transition to a milder mode of wear is occurring as shown in Fig. 9.44. Interpretation of the wear curves in this fashion can then be used as a basis for predicting absolute wear behavior in an application and providing some element of acceleration to the wear test. For such uses it might be necessary to convert Eq. (9.7) to the equivalent one for wear volume by using the appropriate geometrical relationship between depth and volume.

Obviously the same type of approach can be applied to other wear tests, like the pin-on-disk or the block-on-ring. Wear on the counterface has to be taken into account with this approach as well. This might require the use of two wear curves to characterize the system, which can complicate the evaluation and extension to an application. However, in many cases, the wear on one member is negligible or it might be possible to adjust the test situation so that this is the case. In both the pin-on-disk and the ball-on-plane tests, the relative wear can be changed by adjusting such factors as load, amplitude of motion, and location of the materials in the wear test. The occurrence of significant wear on the counterface at some point during the test might result in transition in the wear behavior of the primary surface. The use of this type of approach is illustrated in the discussions of other tests and case studies in EDW2E.

In the discussions of the three sliding tests, which utilize initial nonconforming contacts, the comment has been made that the wear curves are typically nonlinear. The general reason proposed for this nonlinear behavior is the modification of the contact geometry with wear and, more specifically, the variation in stress level with wear. On the one hand, this is a complicating feature of the test, as has been mentioned in the discussions of these tests. On the other hand, these tests provide a means of investigating the sensitivity of wear to these parameters. A test geometry, in which the geometry of the contact or the stress level does not change, such as a thrust washer configuration or a rectangular block on a flat, would miss such factors. For example, the variable and constant energy wear modes that have been observed with a ball–plane test would not have been evident in the conforming tests (24,37). While simulation might require the use of conforming and constant area configurations, the nonconforming tests tend to provide a much more complete valuation of wear behavior because of these sensitivities.

9.2.12. Drum Wear Test

The test apparatus for this test is shown in Fig. 9.45. This test was developed to address wear problems associated with such materials as papers, printer ribbons, and tapes (38–40). These materials tend to be abrasive and can wear hard, wear-resistant materials (e.g., hardened steel, tungsten carbide, and diamond). At the same time, the wear resistance of these materials is very low in comparison to that of the counterface materials used in most applications. The use of more conventional test configurations (such as pin-on-disk or block-on-ring, in which one of the members could accommodate the mounting of paper, tape, or ribbon samples) generally results in little wear of the wear specimen but significant wear of the tape, ribbon, or paper specimen. In addition to this, the abrasivity of these materials tends to decrease with wear. As a result, it is generally not possible with these types of tests to either determine the wear resistance of the counterface or to get an accurate measure of the abrasivity of the paper, tape, or ribbon. Furthermore, in many of the applications, it is the counterface which experiences significant wear and the paper, ribbon or tape experiences minor wear. Consequently, common and conventional test configurations do not provide good simulation. The drum test apparatus was designed and the test method developed to provide a large amount of surface area of the paper, ribbon, or tape, against a relatively small amount of wear area for the wear specimen and to provide simulation in terms of loads, speeds, and relative wear.

While this apparatus was developed to specifically address wear between magnetic heads and paper imprinted with magnetic characters and bar code, it can be used with any web-like materials. This apparatus, like the slurry abrasivity apparatus, can be used either to determine the abrasivity of materials or to determine the wear resistance of materials to this type of wear. Several examples of its use are discussed in the literature and test results have been found to correlate with a variety of applications (e.g., wear of magnetic heads, type surfaces in printers, punches, and guiding surfaces for papers, ribbons, and tapes) (39–43). A standard test procedure (ASTM G56) has also been established with this apparatus to characterize the abrasivity of printer ribbons. While details of the test procedures associated with these applications do vary, there are some common features and elements. A review of the procedures in ASTM G56 serves to identify most of these.

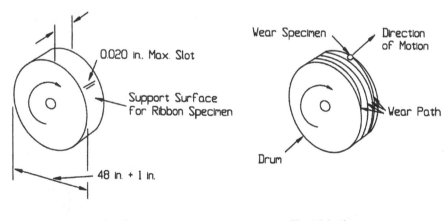

Figure 9.45 Basic configuration of the drum wear test for web material.

In this type of wear test, the ribbon or other web material is wrapped around the periphery of the drum and the wear specimen is loaded against the wrapped surface of the drum. As the drum rotates, the wear specimen moves across the surface of the drum in an axial direction. The resultant wear path on the surface of the drum is a helix. The values of the load, rotational speed, and cross-feed speed of the specimen, as well as the shape of the wear specimen, can be varied to provide simulation. These parameters also influence the wear behavior in the test. For the standard test to determine ribbon abrasivity studies were performed to investigate the influence of these parameters on the wear and specific values were selected for the standard.

To determine ribbon abrasivity, a 52100 hardened steel sphere is used as the wear specimen. The abrasivity of the ribbon is quantified in terms of a wear coefficient, K, which is given by the following equation:

$$K = \frac{V \times p}{P \times S} \tag{9.8}$$

where V is the volume of the wear produced on the sphere after a sliding amount, S, under a normal load of P. p is the penetration hardness of the 52100 steel. The higher the value of K, the more abrasive the ribbon is. A specific duration for the test is not specified and can be varied, provided the amount of sliding is sufficient to produce a sufficiently large flat spot on the wear specimen. Examples of typical wear scars generated in this test are shown in Fig. 9.46. The wear scars are not always perfectly flat and round. There is typically some rounding of the edges of the wear scar and the scar tends to be elongated in the direction of sliding, particularly for small amounts of wear. The volume of wear can be determined by any method but the standard method uses a profilometer technique that involves comparing initial and worn traces in two orthogonal directions. One set of measurements is taken in the direction parallel to the sliding; the other, perpendicular to the sliding. Using data obtained from these comparisons, the wear volume is calculated by means of the equations provided, which are based on the geometry of the specimens and wear scars. One technique for comparing the unworn and worn traces and the equations used to determine volume are shown in Fig. 9.47.

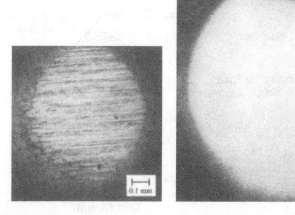

Figure 9.46 Examples of wear scars produced on steel surfaces by printer ribbons, using the drum test. (From Ref. 120.)

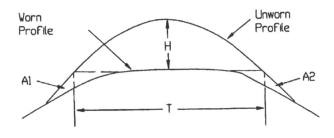

$$V' = 1.0472\,H^2\,(0.3750 - H) + 1.5708\,T\,(A_1 + A_2)$$

$$V = (V'_\parallel + V'_\perp)/2$$

Figure 9.47 Method for determining wear volume on spheres using a profilometer overlay technique. V' is the volume obtained from a single trace. V is the average volume used for determining the wear coefficient. (From Ref. 112.)

It can be seen that this method allows for wear scars that are not perfectly flat. The wear volumes determined from the two orthogonal measurements are averaged to account for noncircular wear scars.

The coefficient of variation for K that has been obtained with this method is in the range 5–25%.

While K has the form of an abrasive wear coefficient that is independent of the abraded material (e.g., Eq. (3.95)), it is not completely independent. As discussed in the section on abrasive wear, 3.8, the wear of a material can be influenced by its relative hardness to that of the abrasive. The wear is much less if the abraded surface is harder than the abrasive. Ribbons and inks can contain particles of different hardness. Consequently, while the abrasion by softer abrasives may be negligible with a hard specimen, it may be significant with a softer specimen. In addition since there is the potential for corrosion with ink ribbon, chemical aspect of the wear processes can be different for different materials, making K material sensitive. Consequently, the standard test does not provide an absolute measure of the effective abrasivity of the ribbon but a general ranking of the ribbons. Slight variations in ranking would be anticipated with the use of a different wear specimen. Hardened 52100 steel spheres were chosen for the standard specimen since they are easily obtainable with good control of size, composition, and hardness. Furthermore, the hardness is representative of many of the materials used in the applications and chemically it is not unique or particularly unusual.

To account for this potential variability and to determine an abrasive coefficient for different ribbon and material combinations, tests with other materials are performed. When this is done, another wear coefficient, K', is used. K' is given by

$$K' = \frac{V}{P \times S} \tag{9.9}$$

Values of K' obtained 52100 and stainless steel specimens are shown in Table 9.2a, along with K values for these ribbons. These show some variation in the relative rankings. Since stainless steels tend to be more corrosive resistant, the primary difference in K' values and rankings obtained with these two material is attributed to oxidative effects. This is

Table 9.2a K and K' Values for Different Ribbons

Ribbon	K (10^{-6})	K' $(10^{-12}$ in.^{2}/lb)	
		52100	420 Stainless steel
A	11.0	10.2	1.6
B	8.0	7.5	1.2
C	7.5	7.1	0.8
D	7.1	6.7	0.4
E	5.6	5.5	1.6
F	5.0	4.7	1.6
G	4.2	3.9	1.2
H	3.2	3.1	0.8
I	2.9	2.8	0.4
J	2.8	2.8	0.4
K	2.6	2.4	0.8
L	1.4	1.2	0.3
M	0.6	0.6	0.4
N	0.6	0.6	0.3
O	0.2	0.2	0.4
P	0.03	0.03	0.16

supported by the similar values obtained for K' when they are tested in a noncorrosive abrasive situation (Table 9.2b). K' can be viewed as a coefficient in a linear equation for abrasive wear,

$$V = K' \times P \times S \tag{9.10}$$

K' is the abrasive coefficient for the particular ribbon/material pair tested and has the dimensions of volume/load-distance. In this context, the test provides an absolute measure of abrasion resistance or abrasivity and can be used to predict wear in applications utilizing that pair.

In selecting the parameters for this type of test, a major factor is the amount of damage or wear that is produced on the web material. Obviously the load, speed, and shape of the slider should be such that the abrasive media is not torn or otherwise damaged in a macroscopic way. Beyond this there is the additional concern with microscopic damage or changes that can occur and influence the abrasivity of the ribbon, tape, or paper being tested. A general feature of these materials is that their abrasivity changes with use (39). For example repeated tests on the same ribbon specimen, utilizing the parameters of ASTM G56, show a significant decrease in K values. A factor of 1/2 is usually

Table 9.2b Comparative Data for 52100 and 420 Stainless Steel Under Dry Abrasive Conditions with the Same Apparatus

	Volume $(10^{-6}$ in.^{3})	
	52100	420 Stainless steel
Sand	2	2.4
Paper	0.003	0.008

found with a single repeat. Paper surfaces show even larger changes (e.g., an order of magnitude or greater has been observed under some conditions). A major element in controlling this effect is selecting the cross-feed speed of the specimen. Depending on the size of the contact zone, which increases with wear, and the advance per revolution of the drum, different ratios of virgin to used web surface can occur within the contact zone. This is illustrated in Fig. 9.48. In practice, it is not necessary to completely eliminate the overlap condition. It is sufficient to determine a cross-feed speed that is large enough so that the test is insensitive to this effect and to optimize the use of the surface area of the web material (e.g., little or no gaps between the wear tracks). The choice is influenced by the sensitivity of the particular material to potential damage, the size of the wear scar typically produced, and the load. Therefore, value that is used can vary with the type of evaluation being done. The particular value for the ASTM standard was selected empirically and based on such considerations (39).

In general, the selection of the test parameters that can be used in this type of test will vary with the type of web material being involved. For example, while the values of the test parameters used in ASTM G56 are appropriate for ribbon, tests with papers have

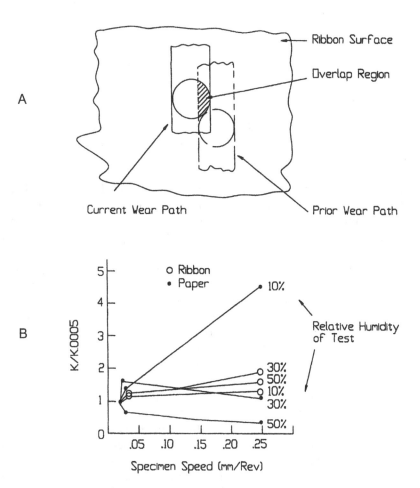

Figure 9.48 The diagram illustrates the overlap that can occur in the drum test. The graph shows the effect that this overlap can have on wear. (From Ref. 39.)

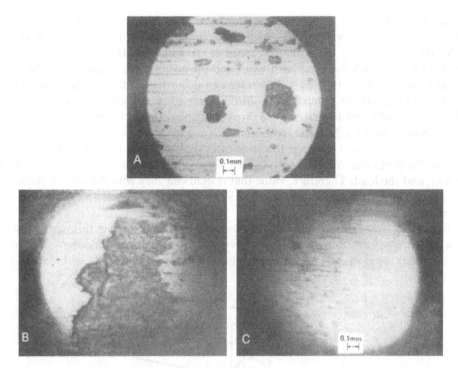

Figure 9.49 Micrographs of worn surfaces produced in drum tests against papers, showing the build-up of paper wear debris on the surface ("A" and "B") and the distortion of the wear scar as a result of this build-up (C). (From Ref. 39, reprinted with permission from Elsevier Sequoia S.A.)

found them to be inappropriate since paper tends to be damaged more easily. In addition, the accumulation of paper debris in the contact region can alter the wear behavior significantly. Examples of the distortion in wear behavior that can occur with paper tested under the wrong conditions with this type of apparatus are shown in Fig. 9.49. As a result, tests at lower speeds and loads but with higher cross-feed speeds are recommended for papers. In addition, humidity and moisture content have been found to be significant in tests with paper and these elements have to be controlled as well (44).

In addition to reporting the test parameters with the wear data, any sample preparation procedures and conditioning should be reported. The temperature and humidity conditions under which the tests were performed should also be given.

This type of test provides an interesting aspect regarding simulation and illustrates the latitude that can be associated with simulation when there is sufficient understanding of the wear situations. The test configuration suggests simulation of applications in which paper or ribbon slides over a surface or where a component slides over a paper or ribbon surface, such as is illustrated in Fig. 9.50. Correlations applications of this type have been demonstrated for this and similar tests (41,43,45,137). In addition to these applications correlation has also been found with the wear of typefaces in printers, where impact is the predominate characteristic of the contact (46). Studies have shown that the basic wear mechanism in such cases is abrasion as a result of micro-slip between the paper or ribbon surface and the typeface that occurs during impact. The impact nature of the contact determines the amount of sliding and the load during the period of contact but does not directly cause

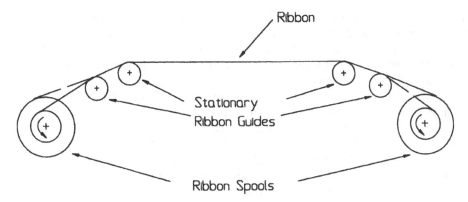

Figure 9.50 Schematic of a system used to spool ribbon through an impact printer. Wear of the ribbon guides are a concern.

any wear. This wear test is used for these applications to determine an abrasive wear coefficient for the paper/ribbon-type pair, which is then used in conjunction with an abrasive impact wear model to predict wear in the application. When the test is used in this fashion, it provides an acceleration factor that reduces evaluation time by a factor of 10^{-2}–10^{-3}. This illustrates that with increased understanding of the wear situations, it is possible to develop and utilize tests that focus on intrinsic behavior and are more laboratory-like than robot-like. Large acceleration factors often can be obtained in these types of approaches.

9.2.13. Thrust Washer Test

The thrust washer test is one that has been used to characterize the stable wear behavior of plastics sliding against metals surfaces. The basic contact condition of the test is shown in Fig. 9.51 and is similar to a thrust washer configuration. Generically this type

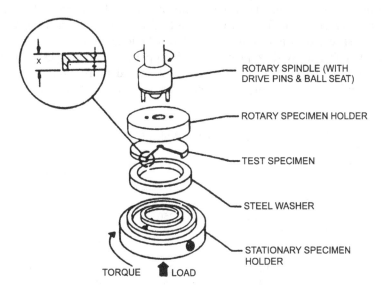

Figure 9.51 Configuration of the thrust washer test for sliding wear. (From ASTM D3702, reprinted with permission from ASTM.)

of test is different from the other sliding wear tests described in that after a break-in period the pressure and area of contact remains constant with wear, both surfaces experience the same amount of rubbing, and the area of contact is large. These features tend to better simulate those associated with many bearing applications for self-lubricating polymers than the spherical or cylindrical type of contacts used in the other tests. Since it is not possible to insure perfect alignment of the ring and the plate used in the test, a break-in period is required to achieve a conforming contact. During this break-in period wear data are usually not taken. While this type of test can be performed in a variety of ways and with a variety of conditions and materials, the test has been used mainly to characterize the wear behavior of self-lubricating materials (i.e., plastics), against metal surfaces. An ASTM test method, D3702, has been developed around this test for that purpose.

ASTM D3702 specifies the stationary ring to be made of 1018 steel, hardened to R_c20 and have a 16 μin. R_a surface roughness. The plate or rotating wafer, as it is called in the test method, is the polymer specimen. Dimensions and tolerances are specified for both. The test method allows the use of several combinations of speed and load, which are representative of the range of PV [pressure (P) times velocity (V)] values that are found in many applications where plastics are used. A list of those conditions is given in Table 9.3. Procedures and guidance for cleaning and handling of the specimens are also provided. In addition, a 40-hr break-in period is specified prior to obtaining wear data. The break-in load and speed are the same as for the rest of the test. After the break-in period, the polymer wear specimen is removed and cleaned with a lint-free cloth and initial thickness measurements made. Then the specimen is remounted, loaded, and the test run for a predetermined amount of time. At the end of the test, the plastic specimen is again removed, cleaned with a lint-free cloth, and remeasured. The change in thickness is then converted to a linear wear rate by dividing the change in thickness by the duration of the measurement period. This depth wear rate is what is used to characterize the wear of the self-lubricating material.

The procedure requires the specimen to equilibrate for 1 hr at room conditions prior to the measurements. The thickness measurements are to be taken at four points, 90° apart, and the average of the four is to be used to determine the wear rate. While a fixed duration is not specified, it is recommended that it be long enough so that the thickness change exceeds 0.004 in. With a properly built apparatus and proper implementation, interlaboratory testing has indicated that this test should be repeatable to within 20%. If scatter beyond this is encountered, some further investigation should be done to

Table 9.3 Combinations of Loads and Velocities Which May Be Used in ASTM D3702 Thrust Washer Test

Rotational speed (rpm)	Rubbing velocity (ft/min)	Loads (lb)			
		1250	2500	5000	10,000
		PV (psi-ft/min)			
36	10	25	50	100	200
180	50	5	10	20	40
900	250	1	2	4	8

determine the reason for it. This should include an examination of the apparatus and techniques. The interlaboratory testing programs have also shown that an interlaboratory scatter should not exceed 30% when the same test parameters are used.

While not a requirement with the standard method, it is desirable to include in the data reported, the particular load and speed used, the ambient conditions of the test, and adequate descriptions of the materials used, along with cleaning and preparation techniques. Comments regarding the appearances of the both the wear surface and the counter-face, both at the end of the break-in period and the end of the test, are also useful.

The ASTM test method is designed to determine a measure of the stable wear rate of plastics in the mild wear regime or under the P–V limit for the material. The test method itself does not provide a measure or determination of the P–V limit. However, by performing tests at different conditions of load and speed outside of the range specified, the P–V limit can be determined (47). While the test has been found to be of use in ranking plastics in terms of general characteristics, correlation with applications depends on the degree of simulation between the test and the application and need to be addressed on a case-by-case basis.

Since the test method limits the counter-face material to a single material and surface condition, the standard test does not provide an absolute measure of wear performance. Roughness, as well as composition and hardness of the counter-face are known to influence the wear rate of plastics (48). Such effects were discussed in the sections on tribofilm wear and tribosurfaces, 3.7 and 4.3, respectively. While the standard test method does not directly provide a measure of absolute wear performance, the format of the test does have the potential to do so. This is because the basic test method can be used to determine the coefficient of a wear model. The fundamental modification of the test method, required for this, would be the use of different counter-face materials so that actual material pairs and interface conditions could be evaluated. The following illustrates this use.

The underlying model or wear relationship for this type of use would be the following:

$$v = K \times L \times S \tag{9.11}$$

where v is the volume of wear; L, the load; S, the amount of sliding; K, a wear coefficient. Letting t be time, this equation can be reduced to the following for a conforming contact with constant surface area, A, and sliding at a velocity, V:

$$\frac{v}{A} = K \times \left(\frac{L}{A}\right) \times (V \times t) \tag{9.12}$$

$$h = K \times (P \times V) \times t \tag{9.13}$$

$$K = \frac{h}{t} \times (P \times V)^{-1} \tag{9.14}$$

P is the contact pressure. K in this model would be the wear rate determined in the test, WR, divided by the PV value used in the test,

$$K = \frac{\text{WR}_{\text{test}}}{PV_{\text{test}}} \tag{9.15}$$

Using Eq. (9.16), this wear test then provides a means of predicting wear in an application for which the model is valid (i.e., where there is adequate simulation)

$$v = \left(\frac{WR_{test}}{PV_{test}}\right) LS \qquad\qquad (9.16)$$

where L and S are the load and sliding distance in the application. Variation of K as a function of PV, velocity, and load can be addressed by testing at different velocities and loads. Similar approaches can be used for other test configurations. In general, the requirement for this to be done is that the test method associated with these either have to result in a wear rate or wear curve or be modified to provide one or the other.

While this test can be used with materials other than plastics, the specific values of the test parameters and dimensions of the specimens are likely to be different. One reason for this is to provide simulation since the loads and speeds in D3702 were selected to simulate typical conditions involving the use of plastics. Another reason is to insure proper break-in. Generally, metals and ceramics are stiffer than plastics and, as a result, initial alignment requirements are tighter. Also, different types of materials respond differently to different types of break-ins. For example, it might be necessary to break-in the surfaces with an increasing load, rather than at the test load, as is done in some versions of the block-on-ring test. A third reason is that with other material pairs the coefficient of friction can be much higher than with self-lubricating materials. This can influence heating effects and the design of the apparatus. Tests with dry, clean, metal surfaces will generally require a much more rugged design or the use of smaller specimens and lighter loads.

9.2.14. Hostile Environment Ceramic Tests

When pin-on-disk or reciprocating pin-on-flat tests are applied to the evaluation of ceramics, the test apparatus and procedure are generally more complex than with other materials. This is basically because the tests are frequently done in controlled and varied atmospheres and at elevated temperatures. For this, the apparatus must then include an atmospheric chamber, heating elements, and a means of controlling and monitoring atmospheric composition, and temperature. The apparatus also has to be designed so that sliding and loading can be provided and controlled with the specimens being inside a chamber. In addition, the apparatuses normally have friction measurement capability, since the monitoring of the friction behavior during the test is typically done with this type of material.

Testing in unique atmospheres and at elevated temperatures is done to simulate the conditions under which ceramics are frequently used. Such simulation is required since ceramics are reactive at elevated temperatures and it has been found that their wear and friction behavior are very strongly influenced by these reactions, as well as by surface layers of various types. Since the coefficient of friction associated with these materials is also affected the nature of the surface layers formed and changes in reactions, it is generally desirable to monitor friction in these tests as well (49–53). As has been pointed out in Chapter 5 on friction, such monitoring of friction behavior is generally useful in understanding wear behavior.

In addition to the simulation of temperature and atmosphere in these pin-on-disk tests, the values of the other parameters, such as load, stress, speed, and material preparation, are also selected to provide simulation. With this degree of simulation the pin-on-disk tests provide a convenient way of identifying and studying major wear phenomena associated with ceramics (54). While this is the case, the degree of simulation that is typically achieved in these tests is usually not sufficient for the results to be directly applied to an application. First-order simulation is typical of these tests because of the impossibility with a simple pin-on-disk test of completely simulating the complicated conditions of temperature, environment, stress, and geometry that occur in the many of the applications for ceramics (53). As a result, these tests are frequently used as the initial portion of a graduated testing program (50,55,56) in which pin-on-disk test is used to provide a coarse ranking of materials and to select materials for further evaluations in more simulative and complex tests. The results of these more simulative or robot-type tests are then correlated with an application (56,57). For example, the pin-on-disk test can be used to select material candidates for an engine test, which is then used to rank materials for such an application (50).

The duration of the tests used for ceramics varies from a few minutes to several hours. The volume of wear that is produced during this time period is often used directly as the measure of the wear performance. In some cases, this is converted to a wear rate by dividing by time or to a wear coefficient by dividing by the product of the load and distance of sliding. Either of these is then used to rank materials. In addition to these quantitative measures of wear, the surfaces are typically examined in a variety of ways to identify reaction products, films, cracks, and morphological features. In some cases, the results of these examinations are used to rank material performance, either in conjunction with the quantitative measurements or by themselves. Which approach is used, as well as the duration of the test, is usually determined by the information that is desired from the test, the properties of the materials, and the nature of the wear behavior found. For example, the formation of a particular compound during sliding might be the selection criteria for a higher level test and it is necessary to determine whether or not this occurs. In another case, it might be the comparison of wear rate that is of interest. The observations and measurements made in conjunction with these tests are not confined to the pin but are done for both surfaces.

The normal procedures, practices, and elements that are associated with the proper performance of a pin-on-disk or reciprocating pin-on-flat test are applicable to the testing of ceramics. Because of the more complex nature of these tests and the unique nature of ceramics, certain elements deserve additional focus. One aspect is that there are usually more elements to control and monitor in tests with ceramics. In addition to environmental factors, material processing and preparation steps are often major factors in these cases. For example, since moisture can be a significant factor in wear and friction behavior of ceramic surfaces, a bake or drying step is usually recommended as part of the cleaning process (49). Another example of this type of concern is with the machining and finishing processes used. Because of the brittle nature of ceramics, their wear behavior is significantly influenced by the presence of micro-cracks or residual strain in the surface region. As a result, processes, which tend to produce such damage, should be avoided or properly controlled. If they cannot be avoided, care should be taken to either reduce the amount of damage to an insignificant level or to control it sufficiently that consistent behavior is obtained. Because of these additional concerns, it is usually desirable to develop a reference test which utilizes a well-controlled ceramic and to use it to monitor overall test consistency. Furthermore, the information that is reported with the data should include information regarding these additional elements, such as the specific atmospheric and temperature conditions of the tests, preparation procedure, and initial condition of specimens. Also it is not only desirable to utilize friction measurements and additional forms of surface analysis in conjunction with these tests but also to report this information as well.

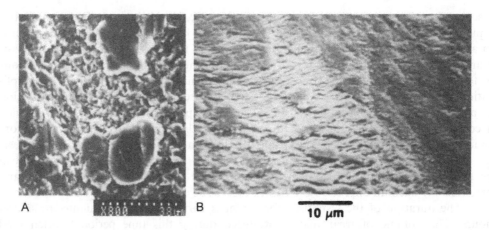

A   B   10 μm

Figure 9.52 Examples of initial and long-term ceramic wear scar morphology in pin-on-disk tests. "A" illustrates the relatively coarse features that can occur initially. "B" illustrates the morphology associated with larger amounts of sliding. ("A" from Ref. 121; "B" from Ref. 122; reprinted with permission from ASME.)

As with all tests which utilize a pin with a curved surface, the stresses change with wear in ceramic pin-on-disk tests and can result in nonlinear wear curves or behavior. There can be a unique aspect to this complication with ceramics. With the initial point contact geometry of these tests, significant subsurface stresses exist to a depth comparable to the diameter of the contact area. It is possible for cracks to develop in this region. With growth, these cracks will result in the formation of large wear particles. As the wear progresses, the contact geometry changes from a point to a conforming contact, with the result that subsurface stress is reduced. Cracks will now tend to form on and closer to

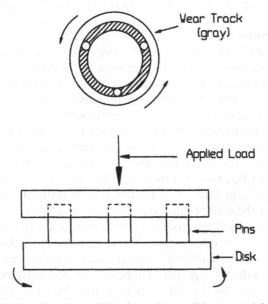

Wear Track (gray)

Applied Load

Pins

Disk

Figure 9.53 Configuration of the three pin-on-disk test used for ceramics.

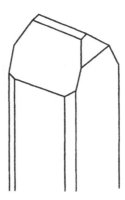

Figure 9.54 Pin geometry used to investigate the wear behavior of ceramics for machining applications.

the surface and the resultant wear particles will be smaller. The effect of this is on wear is illustrated in Fig. 9.52. The formation of large initial wear particles can influence wear behavior in two ways. One effect is an initial high wear rate as a result of wear taking place in the form of larger particles. This is a short-range effect and can easily be taken into account by dividing the wear behavior into initial and stable regions. A second effect influences long-term behavior and is more difficult to address. This can occur by these large wear fragments staying in the contact area and influencing subsequent wear behavior. Another possibility is that the surface morphology, which is produced in this initial period, to influence subsequent wear, such as through a roughness effect. Such effects on long-term behavior can have significant impact in terms of the degree of simulation that can be achieved. A test, where these effects occur, would not be expected to correlate well with an application, which does not experience such an initial period.

One way of approaching this type of concern is to use a combination of load and radius such that the initial subsurface stress is below those needed for crack formation; another way is to modify the contact geometry. Two approaches that have been used involve a modification of the pin-on-disk test to provide better simulation. One approach, which is used to simulate piston ring applications for ceramics, involves the use of three flat pins instead of one rounded pin (50) (Fig. 9.53). In this test, the pins simulate the ring and the disk simulates the cylinder. With this test, a break-in period is used, prior to taking data, which is typical of all tests which use a flat-on-flat contact. A second type of modification is for the use of ceramics as cutting tools (55). In this case, the end of the pin has a shape simulating a cutting tool (Fig. 9.54). Typically, with these modified tests, measurement and analysis techniques are the same as of those used with the more standard pin-on-disk tests.

9.2.15. Liquid Impingement Erosion Tests

This type of wear is produced when jets or droplets of liquid impact a solid surface and is similar to that produced by cavitation erosion. Wear on airplane windshields and airfoil surfaces as a result of rain are examples of this type of wear. In this type of wear, the liquid provides pressure pulses to the surface. A variety of wear apparatuses have been used to evaluate materials in terms of their resistance to this type of wear. Two examples are shown in Figs. 9.55 and 9.56. Generally the test configurations used involve a jet impacting

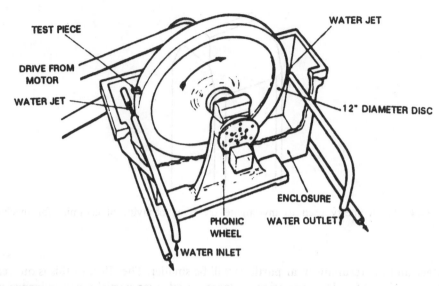

Figure 9.55 A small, low-speed apparatus used to investigate liquid impingement erosion. (From Ref. 122, reprinted with permission from ASTM.)

a specimen or a specimen moving through a droplet field. There are some significant differences between these two types of tests. With jets, the impact is generally focused on a specific region. With a droplet field, the impacts are distributed randomly over a large area. Repetitive impact tests tend to be more severe than distributed impact tests. With either type the recommended practice is to develop a wear or erosion curve relating damage to time or amount of liquid impingement. The measure of the damage varies, depending on the nature of the materials and the function of the materials. For bulk materials, where dimensional changes are the only concern, mass loss is frequently used. For optical applications, light transmission characteristics might be used.

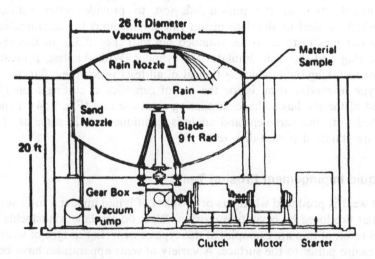

Figure 9.56 A large, high-speed apparatus used to investigate liquid impingement erosion. (From Ref. 122, reprinted with permission from ASTM.)

The curves that are generated in these tests are characteristically nonlinear but tend to have a typical shape, "A" in Fig. 9.57A. Other possible shapes from these tests are also illustrated in this figure. In the typical case, there is an initial incubation or low erosion rate period. This is followed by a second region in which the erosion rate increases to a maximum value. A third region occurs in which the erosion rate decreases to some lower, stable value. This is illustrated in Fig. 9.57B. The duration of the incubation period, the maximum erosion rate, or the stable erosion rate are normally used for ranking purposes. The significance and selection of each depends on the application. In optical applications, the duration of the incubation period is more important than the rates. For applications in which long-term behavior is important, the stable erosion rate is the most significant. With anomalous behavior, curves "B", "C", and "D" in Fig. 9.57A, these features may not be present. In this case, other aspects of the curve are used for comparison purposes. In these cases, maximum erosion rate or cumulative damage are the more

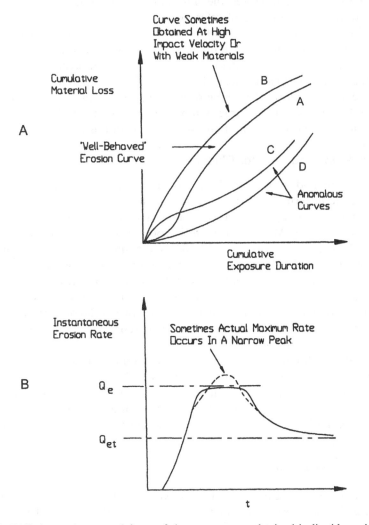

Figure 9.57 "A" shows the general from of the wear curves obtained in liquid erosion tests. "B" shows the typical behavior of erosion rate during the course of an erosion test. (From Ref. 123.)

commonly used feature for comparison. This method of analyzing liquid impingement data can either be done with erosion curves that relate cumulative damage to test duration or time or with curves that relate cumulative damage and amount of impinging liquid. The latter approach has an advantage. When damage is correlated with cumulative amount of impinging liquid, rather than accumulated test time, a basis for relating results from different tests and between tests and applications can be established.

When thin coatings are evaluated in these erosion tests, it is sometimes difficult to quantify intermediate damage and, as a result, erosion curves cannot be generated. In these cases, time to wear-through of the coating is used. This point is determined by monitoring the coating during the test.

A general method of conducting and analyzing these tests is given in the ASTM G73. It is pointed out in the ASTM standard that these tests should not be carried out beyond the point that the wear depth of the scar exceeds the width since significant changes in impact angles tend to occur beyond this point (see Fig. 9.15.) The standard also focuses on the need for control, discusses the information that should be reported with the results, and recommends the use of reference materials to normalize the erosion curve parameters. When the procedures recommended in this standard are followed it is generally found that similar rankings are obtained from different tests when the differences in material behavior are greater than 20%. For smaller differences rankings generally depend on the test and test parameters. General correlation is also found with applications as well, particularly in terms of ranking or screening materials. Absolute performance is less predictable, because of the large number of factors involved and the difficulty of describing field or application conditions associated with this type of wear.

9.2.16. Block-on-Ring Test for Plastics

A standard test method using a block-on-ring configuration has been developed for the ranking of plastic in terms of their resistance to sliding wear, ASTM G137. While test parameters and the configuration are different than those in thrust washer test used to evaluate plastics (ASTM D3702), there is good correlation with rankings obtained. The prime advantage of G137 over D3702 is the time it takes to complete a test. G137 is a significantly shorter test. The test method allows the use of different materials and roughness for the ring, different loads, different temperatures or environments, and different speeds, though a maximum speed is specified. The test procedure requires reporting of these parameters along with the results. In interlaboratory testing programs the coefficient of variation ranges between 45% and 106% with a laboratory and between 84% and 106% between laboratories.

The significant difference between this test method and the more general test method described in Sec. 9.2.6 is that wear rate is used rather than wear volume. The test consists of interrupting the test at intervals for measuring mass loss of the specimen and developing a wear rate curve. A specific wear rate for each interval is computed, using the following equation, and plotted as a function of time. Such a curve is illustrated in Fig. 9.58:

$$\mathrm{WR_s} = \left(\frac{1}{NV\rho}\right)\frac{\Delta m}{\Delta t} \tag{9.17}$$

where $\mathrm{WR_s}$ is the specific wear rate for the interval. N is the load. V is the velocity. ρ is the density. Δm is the mass loss for the interval and Δt is the duration of the interval. This is done until a steady state is reached. The steady state is defined as region in which the specific wear rate curve becomes flat with less than 30% variation in specific wear rate for the

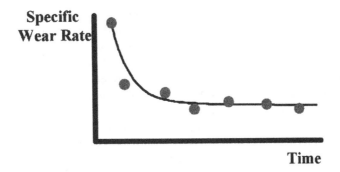

Figure 9.58 Illustration of a wear rate curve obtained with the polymer block-on-ring test, showing the decrease in wear rate usually observed in this test. The points represent values of the specific wear rate obtained for the individual test intervals.

intervals. This is done using a minimum of six intervals of which three must be in the stable region. The method also specifies that the total test time in the stable region be a minimum of 18 hr. The average of specific wear rates for the intervals in the stable period is then used as the measure for characterizing the plastics wear resistance.

In this test, variations in the wear rate in the stable period can be from two sources. One is simply from experimental variations and measurement accuracy. The other is a periodic fluctuation in wear rate that some plastics exhibit in an otherwise period of stable wear behavior. This type of behavior is called oscillating wear.

9.2.17. Impact Wear Tests

Since impact wear testing has not received the attention that sliding and rolling wear testing has received, there are no broadly used tests for impact wear. However, a number of different methods and apparatuses have been used to study normal and compound impact wear and to compare the resistance of materials to these types of wear (58–63). Apparatuses that have been used generally can be grouped into two generic categories, pivoting and ballistic. These are illustrated in Fig. 9.59. For simple or normal impact, that is, no sliding involved, the flat member is stationary. For compound impact, that is, combined impact and sliding, the flat member rotates or oscillates beneath the hammer or projectile. With impact wear tests the wear specimen is generally softer than the counterface and can be either the flat or the moving hammer or projectile. Testing methods and techniques for measuring wear and comparing materials are similar to those used for sliding wear.

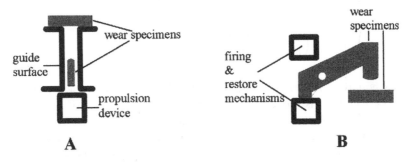

Figure 9.59 Schematics of ballistic ("A") and pivoting ("B") impact wear testers.

Materials are characterized using either wear after a fixed number of impacts, wear rates, or wear curves. Weight-loss and geometric methods are used for determining wear. Generally the same type of parameters need to be controlled and reported with the wear data. The number of impact usually required for these tests is large and high repetition rates are required or at least desirable for impact wear apparatuses. The number of impacts used in wear tests for evaluating materials for engineering applications is usually somewhere between 10^6 and 10^8 impacts (64).

By their very nature, designs of impact apparatuses tend to be more complex than those used for sliding. Because of the dynamic nature of the contact in impact wear testing, there is usually the need for greater concern with the design of the apparatus and shape of the impactor to insure adequate repeatability. Fundamentally, this is because impact stresses and loads are not sole determined by the momentum or energy of the hammer or projectile. They are also affected by a number of other factors, including the geometry of the contact, possible rotation of the projectile, and stiffness of the hammer. Fretting in impact situations can also affect wear and is generally undesirable in impact test. Consequently, the stiffness of the apparatus can also affect wear behavior and repeatability (65–67). While flat-surface hammers and projectiles have been used, curved surfaces are preferred because of alignment problems.

9.2.18. Tests for Paint Films

Examples of the need to use several wear tests to evaluate materials for a wear situations are a group of tests used to evaluate paints for automotive applications (68–70). These tests also illustrate the use of measures other than wear volume, wear rate, or duration for the evaluation of the wear resistance of materials. Studies of the damage found on painted surface of external surfaces of automotive applications lead to the identification of four modes of wear or damage. One is erosion, resulting from the impact of small particles. The other is abrasion, where hard particles are drawn across the surface. A third, called friction induced damage, is the pealing of the surface as a result of rubbing contact. The fourth is an impact wear resulting from the impact of large stones. Four different tests are used to simulate these conditions and in the evaluation of paints and paint systems for these applications.

To simulate the erosion mode a solid particle erosion test is used (69). This test is shown in Fig. 9.60. The metric that is used in this test for characterizing wear resistance or erosion resistance is the mass of particles required for the removal of the coating, Q_c. The larger this value the more resistant the paint.

Studies have shown that this test results in the removal of the coating in a circular spot of increasing size, as shown in Fig. 9.60B. These studies have also shown that Q_c and r, the radius of the spot, are related by Eq. 9.18.

$$r = \left(\frac{h}{\beta}\right) \ln\, m - \left(\frac{h}{\beta}\right) \ln\left(\frac{2\pi h^2 Q_c}{\beta^2}\right) \tag{9.18}$$

In this equation, h is the stand-off-distance between the nozzle and the surface, as shown in Fig. 9.60A. m is the mass of erodent resulting in a spot of radius r. β is called the focus coefficient, which defines the divergence of the particle stream and is dependent on the nozzle roughness and the nature of the particles, not on the coating. By monitoring the growth of the radius with r with the amount of erodent, both β and Q_c can be determined.

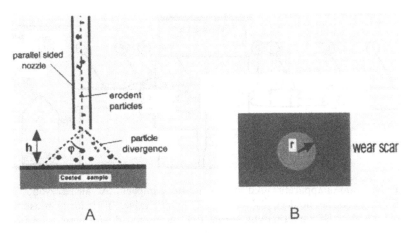

Figure 9.60 "A" shows a schematic of the erosion test used to evaluate painted surfaces. "B" illustrates the wear scar produced in this test. (From Ref. 69.)

A micro-abrasion test, shown in Fig. 9.61, is used to evaluate abrasive wear resistance (69). In this test, a ball is loaded against the paint surface and rotated. A slurry of abrasives is fed to the nip and abrasive particles are dragged across the painted surface. This test results in a wear scar, which has the shape of a spherical segment of a sphere the same size as the ball. The abrasion resistance of the surface is given by an abrasive wear coefficient, κ, defined as

$$\kappa = \left(\frac{\pi b^4}{64R}\right)\left(\frac{1}{SL}\right) \tag{9.19}$$

b is the diameter of the wear scar. R is the radius of the ball and S and L are the total amount of sliding and the load used in the test. The smaller value of wear coefficient the more abrasive resistant the paint.

To simulate stone impact a test apparatus was developed, which fires a projectile at a painted surface (68). This apparatus is shown in Fig. 9.62. While actual stones can be fired with this apparatus, tests used for evaluating material generally use a standard projectile, such as ceramic cylinders. Materials are evaluated in terms of the area of paint removal

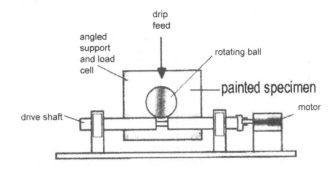

Figure 9.61 Diagram of micro-abrasion test used for the evaluation of painted surfaces. (From Ref. 69.)

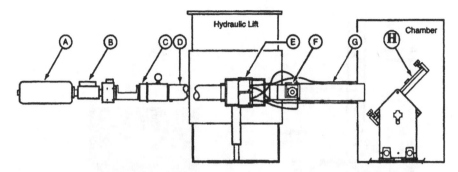

Figure 9.62 Diagram of apparatus used to simulate stone impact. "A" air reservoir; "B" fast acting valve; "C" breech (sabot holder); "D" barrel. "E" muzzle valve, "F" velocity measuring system.; "G" insulated muzzle; "H" target mount. (From Ref. 68, reprinted with permission from Elsevier Sequoia S.A.)

from a single impact. The smaller the area, the greater resistance the paint system has to this type of wear.

The test for friction-induced damage involves dragging a rounded slider across a painted surface, under condition that will cause peeling (70). A schematic of the test apparatus for this test is shown in Fig. 9.63. Figure 9.64 provides an illustration of the method. During the test the friction force is monitored and a curve is generated of friction force vs. sliding distance, as illustrated in Fig. 9.65. The area underneath this curve is the energy expended in the test. The wear resistance of the coated is rated by the ratio of the pealed area to the energy expended. The smaller this value, the more wear resistance the paint. Figure 9.66 shows a comparison of several coatings as a function of temperature using this parameter.

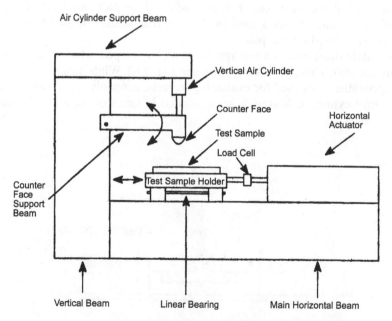

Figure 9.63 Diagram of apparatus used to simulate friction induced damage of paint surfaces. (From Ref. 70, reprinted with permission from Elsevier Sequoia S.A.)

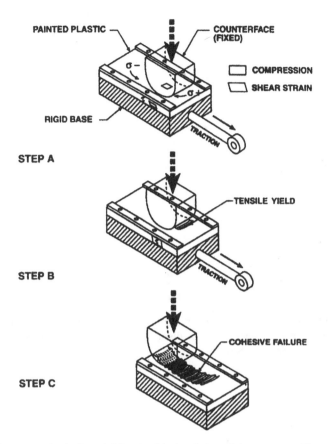

Figure 9.64 Phenomenological model for the friction induced damage test. The specimen moves to the left with the counterface in its stationary position. Step A: normal load is applied and motion initiated. Step B: onset of damage. Step C: conclusion of test, resulting in friction induced damage. $\sigma+$ denotes tension; $\sigma-$ denotes compression; τ denotes shear. (From Ref. 70, reprinted with permission from Elsevier Sequoia S.A.)

9.2.19. Scratch Test

Scratch tests are primarily used to investigate the effect of various material parameters with respect to single-cycle deformation processes (71). Typically, these tests involve pressing a hard, sharp stylus against a flat specimen of the material to be evaluated and either moving the stylus along the surface or moving the surface underneath the stylus. The groove is measured and used to characterize the wear resistance. Larger grooves correspond to lower resistance. Styluses with angular and rounded tips are used. In addition to these sliding scratch tests, pendulum apparatuses are also used. In these, a stylus is attached to a pendulum, which is dropped with a know amount of energy. The stylus is set to engage the surface and move across it, creating a groove. The pendulum is then captured at its peak and before it can swing back across the surface.

Wear behavior in these tests is often additionally characterized in terms of two other parameters. One is the ratio of the volume of material removed to the volume of material displaced. This ratio is sometimes called the removal coefficient, degree of wear, or abrasive fraction. The second is specific grooving energy, which is the energy dissipated in

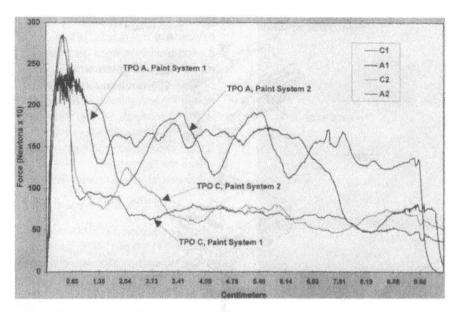

Figure 9.65 Examples of force–distance curves obtained in the test for friction induced damage. Areas under the curves are the energy lost or dissipated in the test. (From Ref. 70, reprinted with permission from Elsevier Sequoia S.A.)

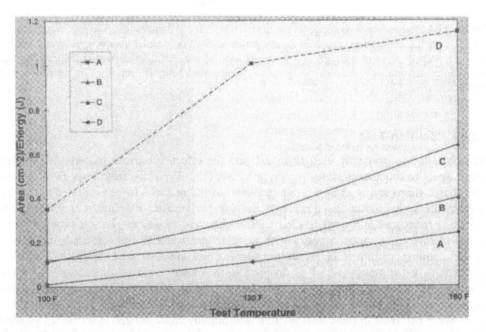

Figure 9.66 Example of the comparison of painted surfaces in terms of their resistance to friction induced damage. Resistance is inversely related to the ratio of the damaged area to the energy dissipated measured in the test. (From Ref. 70, reprinted with permission from Elsevier Sequoia S.A.)

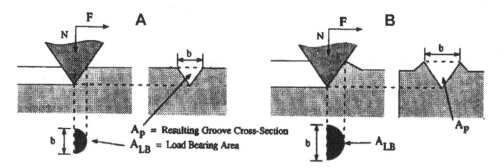

Figure 9.67 Illustration of a scratch hardness test using a conical stylus. "A" illustrates the test when pure cutting is involved; no buildup of ridges. "B" illustrates the test when plowing or deformation also occurs; there is the ridge formation. Note that part of the load, N, is supported by the ridge. (From Ref. 71, reprinted with permission from ASM International.)

forming per unit of mass removed. For sliding scratch test, this requires obtaining a force–displacement curve, such as the one used in the test for friction induced wear of painted surface (Sec. 9.2.18). With the pendulum type of test, the energy dissipated can be determined from the difference in the initial and final height of the pendulum.

These types of tests are also used for two other purposes related to wear. One is to measure adhesive strength of coatings. The other is to obtain a scratch hardness number for materials (72). For determining adhesive strength of coatings, the load at which debonding is observed to occur is determined by performing tests at increasing loads. This load is then directly used to compare materials or converted to a more fundamental measure of bond strength. In a scratch hardness test, the width of the groove is measured and used to compute a hardness value. As with an indentation hardness test, hardness in a scratch test is defined as the ratio of the load to the area supporting the load. This is illustrated for a scratch test and indentation test in Fig. 9.67. Values vary with the shape of the stylus. In scratch tests, there are two general shapes typically used for this purpose. Shapes with circular cross-sections, such as cones, spheres, and parabolas, are one type. Square-base pyramid shapes are the other type. The equation for scratch hardness using the former geometries is

$$H_s = \frac{8N}{\pi b^2} \tag{9.20}$$

For the latter,

$$H_s = \frac{4N}{b^2} \tag{9.21}$$

In these equations, N is the load and b is the width of the groove.

9.2.20. Wear Tests for Coatings

In general, the wear resistance of coatings is evaluated using the same tests that are used with bulk materials. While this is the case, test parameters are usually different when these tests are used to evaluate coatings. Generally, test parameters are milder so that wear in the test can be limited to the coating. Depending on thickness, typical geometrical methods for measuring wear can be used. Mass loss methods are also possible, if densities are known. For thin coatings, special techniques might be required. An alternative to using wear or wear rate as the measure of wear behavior, the life of the coating in the test is often used to rank and compare coatings. The life can be identified by examination of the surface for presence of the coating after tests of different exposure or duration. Frequently, the life can also be determined by monitoring the friction during the test. With wear-through, there is often a change in the coefficient friction that can be detected by such measurements. Additional information on unique aspects of wear tests with coatings and examples of wear tests for coatings can be found in Refs. 73 and 74.

9.3. OPERATIONAL WEAR TESTS

The examples of phenomenological wear tests discussed in the prior section illustrate some of the main attributes of that category of wear tests. One is that the phenomenological tests tend to address major or generic wear situations with the result that actual test configurations are noticeably different than the practical devices or configurations. Another is that the focus tends to be on the ranking or the determining of appropriate wear coefficients or parameters of materials and material pairs. As will be seen, operational wear tests tend to focus on the wear situation associated with individual devices or applications. While they are also used to rank and select materials, operational tests frequently allow the effect of other parameters associated with the application to be evaluated. In addition, with operational wear tests, there may be several potential wear sites and situations. As a result, the wear mechanisms involved might change as the conditions of the test change. These aspects, as well as some of the more general aspects of wear testing, will be illustrated by the consideration of several examples of these types of tests.

9.3.1. Jaw Crusher Gouging Abrasion Test

This test utilizes a jaw crusher to evaluate the wear resistance to what is termed gouging abrasion. This is a coarse form of abrasion in which macro-gouges and -grooves are produced in a single action. Figure 9.68 shows an example of this type of wear. Fracture of the abrasives is also a common feature for this type of abrasion. Jaw crushers, which are used for the crushing of ore and stone in mining operations, are examples of applications in which this type of wear occurs. The test is a replica of this type of application and can be done with a jaw crusher. This is the primary reason for its classification as an operational test. The test also has a phenomenological aspect as well. For example, rankings from this type of test have been applied to earth moving equipment, which experience similar wear but under different conditions (75). Because of the similarity of the wearing action, this type of apparatus and test has been used successfully to rank coatings for icebreaker hulls (76). A standard method (ASTM G81) has been developed for this type of apparatus when used to crush rock.

The test configuration is shown in Fig. 9.69. Basically, a jaw crusher consists of a pair of jaws, one stationary and the other articulated. The material to be crushed is fed between these two jaws and is squeezed by the action of the movable jaw against the stationary one. In the wear test, two pairs of wear plates are mounted on these jaws; one member of the pair is a reference material and the other is the material to be tested. They are mounted in such a manner that the reference and test specimens oppose one another, as shown in Fig. 9.70. The test basically consists of crushing a minimum amount of rock in a series of steps. At the end of the series, the wear plates are removed and the wear

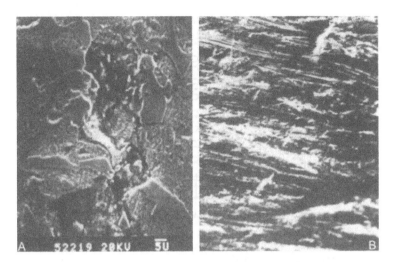

Figure 9.68 Example of gouging abrasion. "A" shows the morphology of the worn surface of a Hadfield steel from a jaw crusher application. "B" shows the morphology of the wear surface of cast iron bucket teeth. (From Ref. 124, reprinted with permission from ASME.)

determined by mass loss, which is then converted to volume loss. A wear ratio is established for both pairs, dividing the volume of wear of the test specimen by that of the reference specimen. The two values of this ratio (i.e., the one for the movable and the one for the stationary jaws) are averaged. It is this average wear ratio that is used to rank materials against a reference material. A value of less than 1 means improved performance over the reference material, whereas a value of greater than 1 means poorer performance.

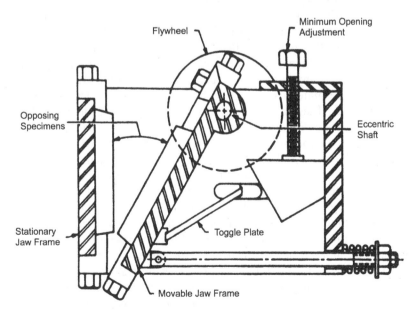

Figure 9.69 Schematic of a jaw crusher apparatus. (From Ref. 125, reprinted with permission from ASTM.)

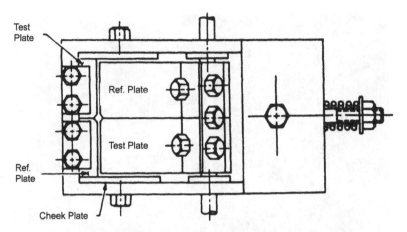

Figure 9.70 Position of reference and test specimens in the ASTM jaw crusher test. (From Ref. 125, reprinted with permission from ASTM.)

A single apparatus is not specified in the standard method. However, key dimensions and tolerance are given, as well as guidelines to monitor performance. One key dimension is the minimum jaw opening, which is specified to be 3.2 mm and the jaws are to be re-adjusted to this value after crushing 225 kg of rock. The minimum amount of rock to be crushed is 900 kg and twice that amount is recommended for the evaluation of very wear resistant materials. Because of wear, this dimension is re-adjusted several times during the course of a test. As a means of control and calibration, it is also suggested that three tests be run sequentially with all the test plates being reference material. If the average wear ratios for the last two tests of this sequence vary by more than 3%, the apparatus should be examined for signs of deterioration and lack of conformance to the specifications. It is also recommended that a single test of this type be performed after every six or so normal tests to monitor performance of the apparatus. Finally, it is recommended that the crushed size of the rock (i.e., after it has gone through the jaw crusher) be monitored. If the size changes, then state of the apparatus and the consistency of the rock used should be examined and any variations or degradations be addressed.

Specific reference materials or rock to be crushed are not identified or used in the standard test procedure but the significant attributes of both are identified. For example, the reference material should have uniform and consistent properties. The size of the rock to be crushed is a key factor and is specified. It must be precrushed to a particle size 25–50 mm and it should be hard and tough. A morainal rock of a specific composition is given as an example of an appropriate material to be used in the test. The test method requires that the rock and reference material used should be reported, along with an adequate description of their properties, when reporting the rankings obtained with this test. However, while absolute amounts of wear vary with the type of rock crushed, it has been found that similar rankings are obtained with different types of rock. This is a consequence of using an index based on relative performance to a reference material, which tends to reduce the effect of different rock properties on results.

The standard test method also addresses technique, specimen preparation and cleaning procedures. Because of the gross nature of this wear process, there is less concern with cleaning, surface preparation, and surface control than with many other wear tests. Cleaning is part of the test method mainly to insure accurate mass change data.

For the same reason, surface preparation is required to insure that initial rust or slag layers are removed.

When the test is used as an operational test for a jaw crusher, the rankings can be used directly in establishing absolute performance. This requires knowing the life of the reference material in that application. The lives of other materials in that application can be determined by multiplying the life of the reference material by the reciprocal of the index. This illustrates an advantage that many operational tests can provide (i.e., direct scaling of field performance).

However, the test does not directly provide a means of determining absolute wear performance in any application involving gouging abrasion. It simply provides a ranking for gouging abrasion resistance with respect to an arbitrary reference material, based on an operational type of test (e.g., crushing of a particular amount of rock). It does not provide the value of a wear coefficient that can provide a relationship between wear and parameters as load, type of rock crushed, or the amount of rock crushed. Such a type of relationship is needed to provide a means of quantitatively relating the test result to an application. Values of these test parameters can be factors affecting the absolute difference in the relative performance of materials found in the test and in the application. Consequently, the correlation of the test rankings to relative performance in an application other needs to be addressed on a case-by-case basis. An alternate way of stating this is that the sensitivity of the test with respect to an application needs to be established on a case-by-case basis. For example, large test differences might correspond to negligible differences in an application. Alternatively with some applications, a small difference in test results might correspond to a very large difference in the application. However, reproducibility of this type of test is good. When this test is performed in the manner described in ASTM G81, variation should be within 5%.

9.3.2. Cylindrical Abrasivity Test

This test is one of several that have been used to determine the abrasivity of magnetic tapes. An acronym for this particular test is SCAT, which stands for SpinPhysics Cylindrical Abrasivity Test (77,78). This test is quite similar in concept to a rod wear test developed in the 1960s (79). The basic configuration of the test is shown in Fig. 9.71. The test

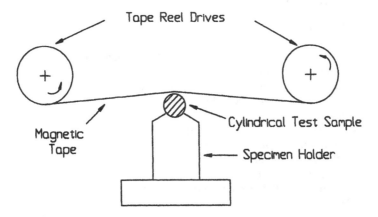

Figure 9.71   The basic configuration of the rod wear test and the SCAT wear tests.

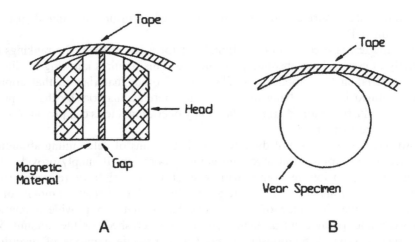

Figure 9.72 "A" illustrates the contact configuration between the head and the tape in a typical application. "B" shows the contact configuration between a worn cylinder and the tape in the SCAT and rod tests.

utilizes a commercially available tape drive, replacing the normal magnetic head with a cylindrical wear specimen. The similarity of the contact between the tape and a recording head and the tape and a cylindrical specimen with some wear is shown in Fig. 9.72. The basic test consists of running the tape across the surface of the cylinder for 10 hr and measuring the depth of wear produced during that period. The depth rate of wear is used as a measure of the abrasivity of the tape. This rate is used as a figure of merit for tape abrasivity, rather than as an absolute measure. This is because it is determined for only one representative condition and the methodology of the test does not address the effects that differences in various operational parameters, such as tension, speed, etc., can have on actual wear performance.

A summary of the parameters used in the test is given in Table 9.4. The load between the tape and the cylinder is determined by the tension in the tape and wrap

Table 9.4 Parameters in the SCAT Test

Wear bar	Material	Spinalloy
	Dimensions	
	Shape	Cylinder 1.5 in. long, 0.250 in. diameter
	Finish	1 µin average
Support	Accuracy	Azimuth and tilts ± 1 min of arc
Test transport	Drive	Honeywell Model 7600
	Tape	
	Wrap	5° each side of test bar
	Tension	8 oz./in. of tape width
	Speed	60 in./sec
	Width	0.5 or 1 in.
Environment	Moisture	Relative humidity to be controlled within ±2% of the desired value
Duration	Test time	10 hr

Source: Ref. 77.

angle of the tape around the cylinder. The size, roughness, and composition of the cylindrical wear specimen were selected to be representative of magnetic heads. The test allows the use of standard reels of either 1/2″- or 1″-wide tapes. All the mechanical and material parameters have to be controlled for reproducibility. In addition, since moisture can have an effect on the wear between heads and tapes, a tolerance is placed on the relative humidity under which the test is conducted. The actual temperature and humidity of the test should be given with the figure of merit that is determined in the test.

The test utilizes a novel way of measuring wear depth, involving a form of break-in. The basic cylinder of Spinalloy, a head material made by SpinPhysics, is installed in the apparatus. With a sample of the tape, a small flat spot or window is developed on the cylinder. This window is the wear region for the test. The cylinder is then removed and a series of micro-hardness indentations are placed in this window, using a diamond pyramid indenter. The cylinder is then coated with a sputtered coating of ceramic to fill the indentations. This helps maintain the edges of the indentations during wear and enables small amounts of wear to be determined. The cylinder is remounted in the apparatus and a sample of the tape to be tested is used to remove the ceramic coating in the window area. The cylinder is removed and the diagonal of the diamond indentations is measured and recorded as the initial values. The cylinder is then replaced, an unused tape sample is mounted, and the test performed. The cylinder is removed at the end of the test and the diagonals of the indentations are measured again. The depth of wear, d, is determined by the following equation:

$$d = 0.1428 \times (D_i - D_f) \tag{9.22}$$

where D_i and D_f are the initial and final diagonal measurements. Eq. (9.22) is based on the geometry of the diamond micro-hardness indenter used. This technique is illustrated in Fig. 9.73.

The SCAT test provides an opportunity to compare directly an operational test and a phenomenological test used for the same purpose. A phenomenological test has also been used to rank magnetic tapes in terms of their abrasivity (79). The test utilizes a ball–plane contact situation (Fig. 9.74). A sphere is used to press the moving magnetic tape against a flat wear specimen. This produces a spherical-shaped wear scar in the flat wear specimen. The volume of this wear scar can be determined from profilometer measurements and the use of geometrical relationships, as has been described with other test methods (see Secs. 9.2.6, 9.2.8, 9.2.11, and 9.2.12). The test provides a

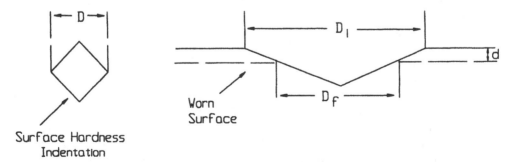

Figure 9.73 The use of microhardness indentations in determining wear depth in the SCAT test.

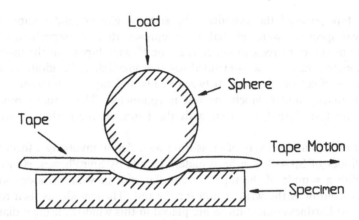

Figure 9.74 The configuration of a phenomenological test used for magnetic head applications.

great deal of flexibility in the load, sphere radius, speed, and materials used, as well as with the conditions surrounding the test (e.g., lubrication, temperature, and humidity). To a large degree, these parameters can be adjusted to simulate magnetic recording applications. However, the basic geometry of the contact situation and stress system are significantly different than in typical applications (e.g., those shown in Fig. 9.72). The method used to rank magnetic tapes in terms of their abrasivity is similar to that used for printer ribbons, namely, to determine a wear coefficient. This wear coefficient is the volume of wear divided by the product of the normal load, speed, and time of the test. The higher the value of this coefficient, the more abrasive the tape is.

The controllable parameters and the use of a wear coefficient in this test provide a more general description of the wear situation and makes the results more suitable to general models and theories for wear. In the SCAT test, many of these parameters are indirectly controlled and not specifically identified. This illustrates some of the differences typically associated with the two categories of tests. In general, the phenomenological test tends to have a potential for greater applicability or generalization than the operational test.

With appropriate values of parameters, the spherical test can rank tapes of significantly different abrasiveness that is consistent with field data. However, the wear rates in the test are typically several orders of magnitude higher than observed in practice (e.g., 10^4 times higher). While this is the case, the wear rates from the SCAT test are typical of those observed in practice. This difference between the two tests is the result of poor simulation of the contact situation in the spherical test and good simulation in the SCAT test. As this implies, the spherical test provides a high degree of acceleration and allows tests times to be reduced from hours, required with the SCAT test, to minutes. At the same time such a difference introduces more concern regarding correlation with the field and the ability of the test to provide accurate rankings. This illustrates another typical difference between phenomenological and operational tests. With an operational test, there is usually less concern with correlation, since the degree of simulation is high, and a minimal amount of effort is needed to establish correlation. With a phenomenological test, there is generally more concern in this area and more effort is required to establish correlation. In this particular case, there is a higher degree of confidence in the rankings provided by the SCAT test than those provided by the spherical test, because of the higher degree of simulation in the SCAT test.

9.3.3. Coin Wear Test

One type of test that has been used to simulate the wear of coins is a tumbling test. Test coins are placed inside a suitably designed drum and tumbled (80,81). The basic concept of the test is to simulate the rubbing that coins experience against one another in pockets, change drawers, etc. This is achieved by placing coins is a plastic drum, which is lined with a rubber-backed cloth. The interior of the drum has an axial ridge, which provides agitation to the coins as the drum is rotated (Fig. 9.75). To simulate possible chemical effects associated with handling of coins, the coins can be coated or dipped with artificial perspiration. Basically the test method consists of tumbling the coins for a period of time and determining mass loss. This is done at intervals so that a wear curve can be developed, which is used to establish an average wear rate (i.e., wear per unit time). In addition to directly using the mass loss as a measure of wear, thickness reduction is also used particularly when comparisons of different materials are involved. This is done by dividing the mass loss by the product of density and the nominal surface area of the coin (including the side area).

This test was investigated and used to evaluate the wear rates of different coin materials (80). As part of the investigation of this test and its correlation to the field, the influence of several of the test parameters were investigated to select optimum and desirable values for these parameters. The goal was to select values which result in producing similar wear characteristics to that found in the field, to minimize test time, and to reduce scatter in the data. Included in these were the effects of the number of coins in the drum, cylinder size, drum rotational speed, and the influence of chemical agents. It was concluded that the number of coins in the drum, the size of the drum, and total linear distance of rotation influence the wear. Total linear distance is the product of the inner circumference, revolutions per unit time, and time. It was found that different sizes of coins would wear at different rates, depending on the number of coins tumbled, up to a total of 12 coins after which the same wear rate was obtained. Thus, 12 coins are used in the standard test. For a given drum size it was found that the wear curve was linear and that the amount of wear was simply dependent on total linear distance. However it was found that the wear rate was higher with smaller diameter drums; hence a smaller drum and higher rpm was

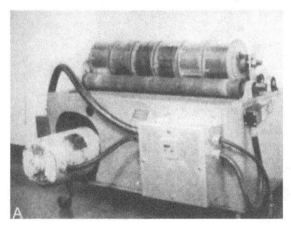
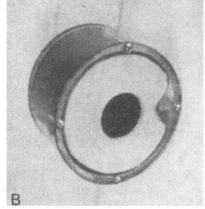

Figure 9.75 "A" shows the overall apparatus used to study coin wear. "B" shows an individual drum that is used to tumble the coins. (From Ref. 126, reprinted with permission from ASME.)

chosen as the standard test. It was also found that the chemical agents and the manner of their application influence wear behavior, which lead to the well-controlled use of artificial perspiration as a standard part of the test procedure.

In addition to these sensitivity evaluations during the initial portion of the development of the test, the morphology of tested coins was also compared to that of field-worn coins to insure that it was representative. As illustrated in Fig. 9.76, the appearance of tested coins was similar to used coins. In addition since field-worn coins tended to show increases in hardness with time in use, tested coins were also examined for an increase in hardness. A similar behavior was found with tested coins, supporting the case for simulation. Furthermore, this increase in hardness suggests that a major factor in the wear is coin/coin interactions. This is consistent with the results of the sensitivity studies. The increased wear rate observed with larger number of coins and small drum size indicate that wear by the drum liner is minor. The standard test conditions selected tend to enhance this type of interaction.

This is an accelerated test in terms of time or use. Comparison of test wear rates and field wear rates indicated acceleration factors in the range of 100–1000 times. The actual value is dependent on the conditions of use (e.g., degree of coin-usage in the society and the general environmental conditions associated with that society). For example, comparison with Canadian coins indicates a value of 600, while a similar comparison with El Salvadorian coins indicates a value of 300. A partial reason for this difference is probably associated with the difference in coin-usage in the Canadian and El Salvadorian societies (i.e., more coin-usage is considered likely in El Salvador). Also, the environmental conditions of the two countries are significantly different.

Figure 9.76 Examples of coins worn in the tumbling apparatus. (From Ref. 126, reprinted with permission from ASME.)

El Salvador has a hotter, more humid, marine-type atmosphere than does Canada. The laboratory tests indicate that such environmental conditions will tend to increase wear rate.

While this test does not provide a characterization of wear behavior in terms of more fundamental parameters, such as load, speed, or sliding distance, it not only provides a ranking of materials but it also provides a quantitative assessment of field performance through an acceleration factor. These are frequently features of operational tests. With these attributes, such tests are very useful for specific applications. On the other hand, it is difficult to apply the results of such tests to other applications or to general wear behavior.

9.3.4. Test for Rolling with Misalignment

This test was developed to address problems associated with a linear stepper motor used in a robot (Fig. 9.77). In this motor, the stator acts as a rail upon which the armature moves back and forth. Conventional ball bearings are used as wheels for the armature, with the outer race being the wheel surface, which engage the stator. In this type of actuator, the wheels serve to support the weight of the armature and any other weight that is being transported and to provide guidance with a minimum of friction. The magnetic coupling between the stator provides accelerations and decelerations parallel to the surface of the rail and the armature. Some degree of misalignment between the ball bearings and the surfaces of the stator was likely to be present in most assemblies and considered to be a significant factor in wear life. To obtain adequate life, there was a need to select and optimize several design parameters in terms of their influence on wear. This included the selection of materials and the evaluation of lubricants. In addition, it was also necessary to select bearing size and the contour of their outer race, evaluate the effect of load, and determine tolerances needed to control alignment. All of these can affect wear and are interrelated tribosystem parameters.

The wear situation may be described as rolling wear with some slip. Based on the general trend, it is expected that the wear rate will increase with the amount of slip

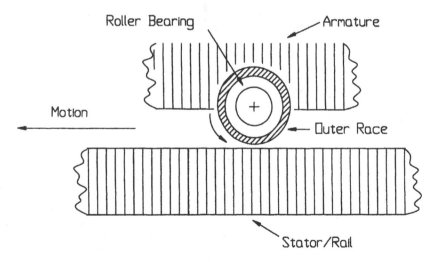

Figure 9.77 Diagram of the rolling contact situation occurring in a linear stepper motor application.

occurring in the contact and that this will be a major factor in overall wear behavior. There are two potential sources of slip: one is misalignment, with the amount of slip increasing with the degree of misalignment present and the other is slip during starting and stopping of the motor. Since the rollers do not provide traction for starting or stopping, the latter source was considered unlikely and actual motor tests with different ratios of start/stops to total distance traveled confirmed this. Consequently, it was concluded that it was not necessary to simulate start/stop behavior in a wear test used to evaluate materials and design parameters.

A test apparatus and technique was developed to investigate the influence of the various design parameters in this wear situation and to select materials (82) (Fig. 9.78). It consisted of a driven central cylindrical wear specimen, which rotates about its axis. This cylinder simulates the rail or stator. Pressed against the wear specimen are three ball bearings located 120° apart at three different locations along the axis of the wear specimen. These can be varied and simulate the rollers in the application. The apparatus was designed so that the normal load, rotational speed, and roller alignment could be varied as well. Hence, the contact situation in this apparatus was very similar to that in the application. There were some differences, which were not considered significant. One apparent difference is that in the test apparatus the contact is between two cylinders, while in the application it is between a cylinder and a flat. This difference is minor since the general nature, i.e., a line contact, and stress distributions of the contacts are the same according to Hertz contact theory. Another difference was in terms of the relative wearing action that the two members experience. The bearing surfaces in the tester experience about 2.5 times more wear action than the wear specimen. In the application, the bearings

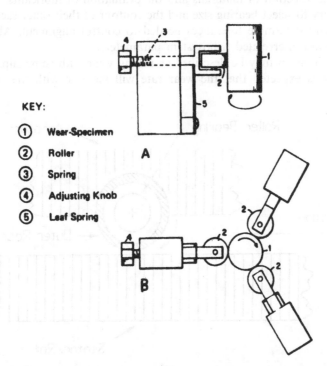

KEY:
① Wear-Specimen
② Roller
③ Spring
④ Adjusting Knob
⑤ Leaf Spring

Figure 9.78 Configuration of the tester used to investigate the wear between the roller and the stator in a linear stepper motor application. (From Ref. 127, reprinted with permission from ASME.)

experienced a varying amount, depending on the movement associated with the robotic action and the ratio can vary from something less than 1 to more than 10, with higher values being more typical. This was not considered to be significant since the ratio in the test was in the range of that for the application. The use of this type of wear apparatus was considered to be more advantageous than motor tests, because it provided adequate simulation with a greater degree of control of key design parameters of load, alignment, speed, ball bearing geometry, materials, and lubrication.

The rollers were positioned in the test apparatus such that they did not interfere with one another. Their wear tracks are separate and, since 120° separate them, wear debris from one track would not contaminate another. As a result, it is possible to simultaneously conduct tests at three different alignments at the same time. This has an advantage in that it facilitates the assessment of misalignment effects as a function of the other design parameters. For example, the developers of the test utilized three standard angles for much of their evaluations, namely 0°, 0.117° and 0.235° of axial misalignment. Examples of some of the differences seen in this manner are shown in Fig. 9.79.

The test method consisted of characterizing and measuring the amount of wear as a function of number of revolutions. This was done for different combinations of design parameters. For the wear specimen, the maximum depth of the scar was used as a measure of wear. This was determined by means of profilometer traces through the wear scar, which gives a natural reference surface for this type of measurement. Since the outer races of the ball bearings are usually significantly harder than the surfaces of the rail materials, little wear is produced on the roller surfaces. However, surface modifications did occur. These were characterized by optical and SEM micrographs, EDX, and roughness measurements. Features such as the occurrence of transfer, oxidation, polishing, and smoothing were noted and used in the overall wear assessment of the system. These techniques were used to examine the wear specimen as well, and similar notations were made.

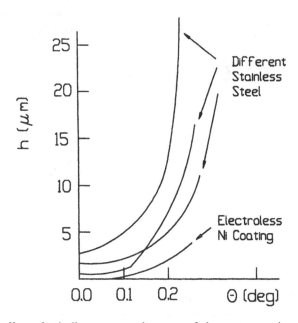

Figure 9.79 The effect of misalignment on the wear of the center specimen in the rolling test. (From Ref. 127.)

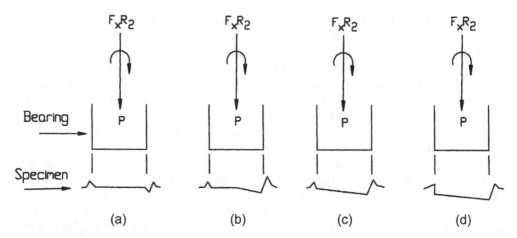

Figure 9.80 Illustration of the effects of misalignment on the loading between the roller bearing and the center specimen and the profile of the wear scars on the center specimen. (From Ref. 127.)

In addition to the measurements and characterization of the wear scars, wear debris was collected and characterized in terms of size, morphology, and composition. These observations were then integrated into an overall wear assessment of the system.

The maximum depth was used since the scar profile is typically tapered because of the misalignment. As is pointed out in Ref. 82, misalignment produces a moment at the interface, which results in nonsymmetric loading of the interface. As a result, a nonuniform wear scar tends to occur. A typical profile for a misaligned contact situation, along with a loading diagram, is shown in Fig. 9.80.

A wide range of wear behavior was observed in this test. Depending on the amount of misalignment, loads, and materials involved mild or severe wear resulted. Wear scars, which have a fretting corrosion morphology, were obtained, as were ones with a morphology similar to gross sliding. Figure 9.81 shows some of the conditions observed in the test.

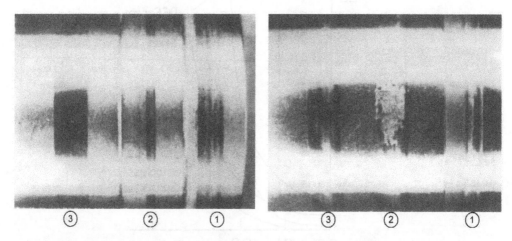

Figure 9.81 Examples of the wear scars occurring on plated center specimen in the rolling test with misalignment. The amount of misalignment in the test increases from left to right, 0°, 0.117°, and 0.235°. Flaking of the plating is evident in the center scar of "B". (From Ref. 127, reprinted with permission from ASME.)

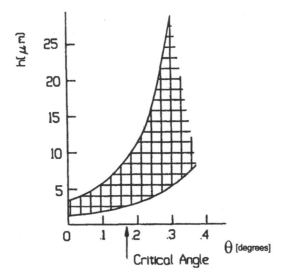

Figure 9.82 The general relationship between wear depth and misalignment observed in the rolling tests. (From Ref. 127, reprinted with permission from ASME.)

Similar wear behavior, including the occurrence of tapered wear scars, was found in motor tests, providing verification of the simulative aspects of the test. Of the many parameters investigated with this test, alignment was the overriding factor. Misalignment was found to affect both the magnitude and nature of the wear. The magnitude of wear increased with increasing amounts of misalignment. Transitions not only in the type of wear but from mild to severe wear were also found to be a function of misalignment. The experimental data indicated a critical angle for misalignment, above which a rapid increase in wear rate was observed (Fig. 9.82).

In addition to the experimental aspects of the study, a theoretical model for such behavior was developed and used to explain the general trends observed (82). In this model, a critical angle is identified. Misalignment above this value results in slip over the entire contact region. Below this value, the compliance of the surfaces is sufficient to limit the slip to regions within the contact zone. The following equation for this critical angle identifies the parameters involved:

$$\theta_c = \frac{3.625\mu P}{wbE}\left[c_1(v) + c_2(v)^{10} \times \log\left(\frac{w}{b}\right)\right] \tag{9.23}$$

where μ is the coefficient of friction; P is the normal load; w is the length of the contact; b is the width of the contact; E is the reduced modulus; v is Poisson's ratio. Values for c_1 and c_2 are given in Table 9.5. This relationship is illustrated graphically in Fig. 9.83.

This test was used to investigate a number of design parameters and aspects as well as the evaluation of materials and lubricants. The test was used to determine the allowed range of misalignment for the design, the size of the rollers, whether or not crowned rollers should be used, and to evaluate the effect of edge conditions. Since the testing was done with speeds and loads appropriate to the application, selections made on the basis of this test were directly applicable to the application and consistent with its load capacity and life requirements. The high degree of simulation allowed the estimation of wear life by the extrapolation of the data.

Table 9.5 Values for C_1 and C_2 of Eq. (9.23) for
the Critical Slip Angle

	Poisson's ratio	
v	C_1	C_2
0	0.1177	0.2022
0.1	0.1198	0.2224
0.2	0.1202	0.2426
0.3	0.1188	0.2628
0.4	0.1156	0.2831
0.5	0.1107	0.3033

Source: Ref. 126.

A phenomenological rolling wear test, which could be conducted with slip (Sec.
3.2.10), could probably have been used to rank materials in terms of their wear resistance
under combined conditions of rolling and sliding. However, that test cannot provide all
the information that the operational test did. The various effects of misalignment that
were found to occur in the application and significant in terms of life cannot be studied
in the phenomenological test (e.g., critical angle and loading alterations), since these
depended on the unique conditions of this application. The slip in the rolling wear test
is the result of different rotational speeds for the rollers, not from misalignment. In the
test performed the effect on wear results from the combined influence of misalignment,
loading, and material properties. As a consequence, the phenomenological test only provides rankings and cannot be used directly in addressing load capacity and life requirements. This illustrates a typical difference between these two types of tests.

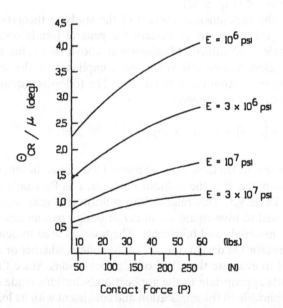

Figure 9.83 The relationship between critical angle of misalignment, friction, modulus, and load in
rolling. (From Ref. 127.)

9.3.5. Bearing Tests

The wear life of a bearing can be influenced by a variety of factors, which are difficult to address in phenomenological tests. Also, there can be several wear points and aspects of the wear that potentially determine life and that are influenced by different factors and interactions. For example, in the case of roller or ball bearings the mixture of rolling and sliding that takes place at various locations in the bearing is a determining factor in the life. In these types of bearings, there are two general wear situations which potentially determine life. One is the contact between the cage and the balls or rollers; another is the contact between the balls or rollers and the races. Which one determines life and the life itself is significantly affected by the general loading conditions (e.g., mixture of axial and radial loads, nature of time varying loads), type of motion (e.g., oscillatory or rotational), and preload. Other factors are also involve such as geometrical tolerances, lubrication, and thermal expansion. Frequently, these elements interact in a complex way to determine the nature and the location of the contacts among the various elements of the bearing (83–86). With journal bearings, the clearance between shaft and bearing can be a factor in wear behavior, as well as alignment, type of loading, and motion (e.g., frequent or rare stop/starts). In journal bearings, these elements can directly influence the wear in terms of contact pressure or location of the contact and indirectly influence wear by their influence on the type of lubrication which occurs (e.g., boundary or elastohydrodynamic (EHD)). Again, the effects of these parameters can be convoluted and interactive. Consider the situation with respect to clearance; temperature effects clearance and clearance influences heat dissipation and therefore temperature. Clearance, per se, also can effect load and pressure distributions and the formation of tribofilms. These factors in turn can influence friction, which influences temperature, etc. (87–89). These complex interactions and their influence on the nature and location of the wear are difficult to simulate in phenomenological tests. As a result, bearing wear tests are used to investigate these complex aspects, determine design information, and verify performance.

A wide variety of bearing testers, differing in complexity, instrumentation, and size have been used for this purpose. Several of these are illustrated in Figs. 9.84 and 9.85. The basic element of these testers is a bearing configuration representative of an application or type of application. Figure 9.86 shows this generic representation. Many of the bearing tests contain instrumentation to measure both friction and temperature. A variety of wear measures are used in these tests, depending on a large degree on the nature of the bearing and the failure criteria. For journal bearings common ones are various forms of measuring increases in clearance between shaft and bearing (e.g., such as end-play, elongation of hole, etc.), dimensional and roughness changes on shaft and bearings, and volumetric and mass changes; the last of these provides a more fundamental characterization. Sometimes increases in friction level and operating temperatures have been used as failure criteria in these types of evaluations. With ball and roller bearings, the measures tend to be indirect. Frequently, vibration characteristics are used to monitor bearing performance, since vibrations tend to increase as wear takes place. Also, play or slop in the bearing can be used as a measure. More basic wear measurements on the various components of the bearing can be used as well (e.g., depth of wear scar on race, mass loss of cage). However, in many cases, the useful life of the bearing is associated with very small geometrical changes on these components. For example, when increased vibration levels associated with the end of life are detected, the wear on the rollers and races may only manifest themselves as increases in roughness, such as shown in Fig. 9.87. Increases in temperature and friction in these bearings are also used as measures of life in some cases.

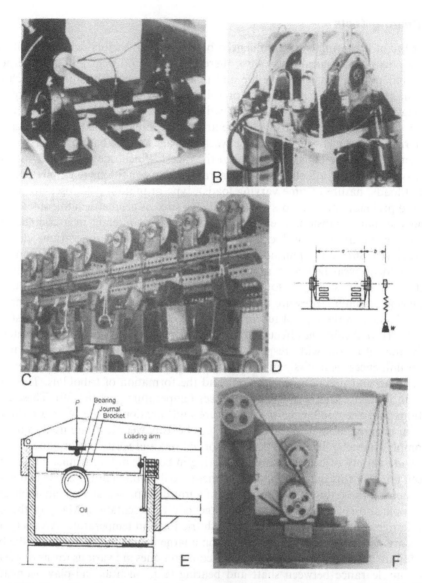

Figure 9.84 Examples of testers used in journal bearing wear tests. ("A" from Ref. 89; "B" from Ref. 129; "C" and "D" from Ref. 130; "E" and "F" from Ref. 131; reprinted with permission from Butterworth Heinemann Ltd.)

The general goal of these bearing tests is to correlate design and application parameters with life. This is typically done by adjusting and controlling the bearing and test conditions to those of interest, including such things as the materials, dimensions, and lubrication system of the bearings. Also, the loading conditions of the applications are simulated, as well as the nature of the motion, stop/start conditions, and the environment. The general approach is to monitor the wear measure during the test. In the case of ball and roller bearings, the duration of the test is usually until failure occurs. This is often the case with journal bearing tests as well, but these tests may be carried out only to the point needed to establish a stable wear rate. These stable wear rates are then used to project life.

These tests provide a value for the life of the bearing for a set of operational conditions and, if the simulation is complete enough, this can be a direct assessment of field life. However, there are more sophisticated approaches and uses of the data beyond this direct and simple use of the test results. These methods of using the data vary with the type of bearing. For ball and roller bearings, the tests are frequently used to establish the parameters in the load-life relationships and adjustment factors used for these types of bearings (86,90–93). For journal bearings, a common approach is to use a P-V concept to interpret the data (94). In this case, the concept is either to identify acceptable

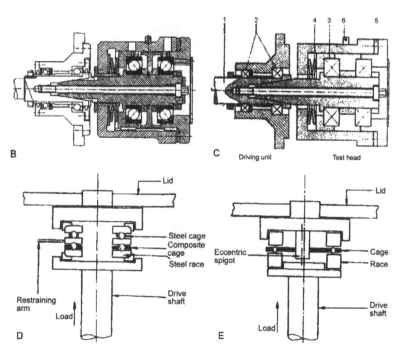

Figure 9.85 Examples of testers used in roller and ball bearing wear tests. ("A", "B", and "C" from Ref. 131, reprinted with permission from FAG Bearing Corp.; "D", "E", and "F" from Ref. 132, reprinted with permission from Butterworth Heinemann Ltd.)

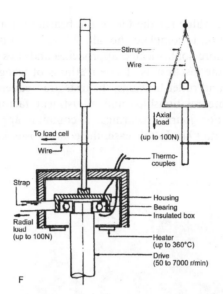

Figure 9.85 (*continued*)

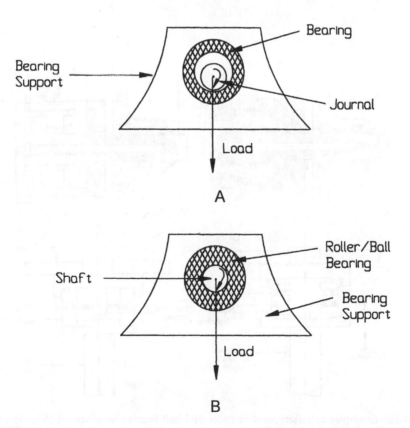

Figure 9.86 General configuration of bearing wear tests: "A", journal bearings; "B", ball and roller bearings.

Figure 9.87 "A" shows an example of the appearances of the wear scars on the races of bearings during the initial stages of wear. "B", in the intermediate stages. "C", in the final stages. Which stage represents "end-of-life" depends on the application. For most applications "B" would be more representative than "A", and "C" is generally unacceptable. (From Ref. 93. Original source SKF Industries, reprinted with permission from Texaco's magazine *Lubrication*.)

combinations of pressure and velocity as a function of all the other variables or to determine the wear rate for conditions where the product of pressure and velocity are constant as a function of the other parameters. These methods are covered in Secs. 2.7 and 2.8 (EDW2E).

Bearing tests of the type discussed are generally quite long, since they are similar to applications in terms of operating parameters. Because of this and the complex nature of these devices, these type of tests are generally combined with phenomenological tests. Phenomenological tests provide initial screening to select candidate material and conditions for the bearing tests. They also provide the opportunity to investigate in more detail specific aspects that are involved in overall bearing performance (56,87,95). A variety of sliding and rolling wear tests is used in this fashion. Pin-on-disk and crossed cylinder tests (Secs. 9.2.8 and 9.2.7, respectively) have been used to simulate sliding wear aspects in journal bearings under boundary lubrication conditions and between balls and cages in ball bearings. The rolling wear test discussed previously in the phenomenological section (Sec. 9.2.10) and ball rolling tests have been used to simulate race and roller wear in ball and roller bearings.

9.3.6. Brake Material Wear Tests

Dynamometers of various types are typically used to evaluate the wear and friction behavior of brake material systems (96–98). Such tests are used to evaluate both the friction (brake) material and the counterface (rotor) material. Two illustrations of typical dynamometer configurations are shown in Figs. 9.88 and 9.89. With this type of apparatus, the wear behavior is evaluated under braking conditions which simulate application conditions. This typically results in a complex testing procedure or sequence to provide adequate simulation. A typical example of a test sequence used for automotive brakes is shown in Table 9.6. These sequences simulate the synergistic effects of break-in, high speed stops, low speed stops, pulsed and continuous braking, etc. These same test sequences are also used in conjunction with different ambient conditions to

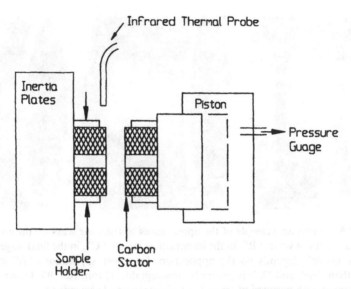

Figure 9.88 Basic configuration of a research dynamometer used for wear and friction evaluations of brake materials. (From Ref. 96.)

simulate use under different climatic conditions and at elevated temperatures. Typically, mass loss or thickness measurements are used to determine wear of the brake material. These techniques are usually not appropriate for the rotor material since their wear is typically quite small in these tests. As a result, the degree or severity of rotor wear that

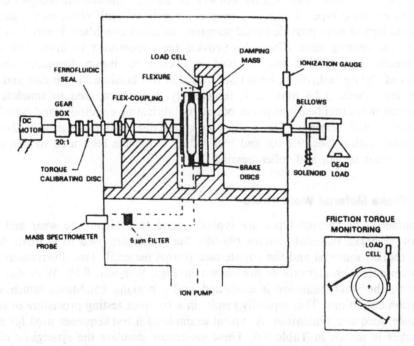

Figure 9.89 Configuration of a high-vacuum dynamometer used for wear and friction evaluations of brake materials. (From Ref. 133, reprinted with permission from ASME.)

Table 9.6 Test Sequence Used in a Chase Sample Dynamometer Wear and Friction vs Temperature Procedure

Initial burnishing	Twenty min drag at 312 rpm with 100 lb load with a maximum temperature of 200°F
Speed of drum	325 rpm
Load	350 in.-lbs
Test sequences,	
Drum temperatures (°F)	250, 350, 450, 550, 650, 780, 250, 350, 450
Applications	Forty of 20 sec duration at each temperature
Wear measurements	Sample weighted and thickness measured after each test at a different temperature

Source: Ref. 99.

occurs is usually characterized in terms of roughness changes. These wear measurements are taken at various points in the test sequence.

Since the typical use of these dynamometer tests is to provide a basis for selection of materials and design parameters for more extensive and costly field evaluations, these tests are designed to provide a relative ranking rather than an absolute determination of wear behavior or wear coefficients. One approach is to simulate the contact conditions with a standard configuration and a small sample of the brake material. In this case, the dynamometers are smaller and less complex, and the tests are easier to implement. An example of this approach is the use of the Chase dynamometer for these evaluations (99). While this approach is attractive from an implementation standpoint, the reduction in scale reduces simulation with the result that there is often poor correlation with field performance (99). Another approach that is used is to evaluate full-sized brake systems. This generally results in the need for larger, more complex, and expensive dynamometers and more complex tests. However, this approach has generally been found to correlate well with field performance (99). Because of this, full-sized testing is the recommended method for establishing material rankings. The smaller scale tests are used for more general purposes, such as investigating the effect of vacuum on performance and determining general trends (100), but care must be taken in extrapolating the results to specific applications.

The rankings in dynamometer tests are determined directly by the amount of wear generated in the test sequence. The best performer is the one that has the least amount of wear at the end of the test. While there is generally good correlation with test performed with full-sized brakes, the tests do not provide universal rankings since the rankings are for specific applications. Different rankings can be obtained with other tests. This is because the test sequence is selected to simulate a specific application and, when a full-size brake is used, the wear performance is relative to that design. Table 9.7 shows the results of tests on four different materials using two different test apparatuses and test sequences. In these tests, the rankings are based on thickness change of the brake material. It can be seen that differences in rankings are obtained with these two tests.

The complex and interactive nature of these full-scale dynamometer tests provides an effective means of assessing wear performance in terms of application parameters. For example, these tests provide a means of determining the effect of relative humidity, pulsing, or rotor roughness on brake performance. At the same time, this same nature inhibits the determination of basic wear parameters or coefficients of fundamental wear relationships. As an example, they do not provide a means of determining the coefficients of a

Table 9.7 Wear Data Comparison as a Function of Fade and
Recovery Performance Sequences

	Total wear (in)	
Friction material	Chase (Schedule J661A)	Inertial dynamometer (Schedule 111)
A	0.009	0.006
B	0.011	0.019
C	0.022	0.027
D	0.017	0.031

Source: Ref. 99.

fundamental equation proposed for wear of brake materials, Eq. (9.24), or verifying the applicability of the equation (101).

$$W = KP^a v^b T^c \tag{9.24}$$

In Eq. (9.24), W is the wear volume; P, the normal load; v, velocity; T, temperature. The situation is too complex and interactive in these full-scale dynamometer tests to isolate the effect of these parameters and to allow the determination of the coefficients, a, b, and c associated with them. This is a common feature of operational wear tests. The effect of these parameters can be isolated and the coefficients, including K, determined with phenomenological tests, which are generally simpler but less simulative.

9.3.7. Engine Wear Tests

Many of the wear situations encountered in engines are complex and are difficult to simulate. One aspect is the simulation of the local environment, for example, the combination of combustion products and temperature that exists in various locations within the engine. Another aspect is associated with the complex lubrication phenomena that take place in engines. Lubricants contain active agents which, along with combustion products, can react with the surfaces and form a variety of surface layers which can influence wear behavior in the boundary lubrication regime. The high speeds, amount of lubricant involved, and the conforming nature of the components can result in hydrodynamic lubrication as well. At many points within an engine, the occurrence of wear is associated with the presence or absence of this type of lubrication. These different aspects are interrelated as well. For example, temperature can influence both boundary and hydrodynamic lubrication, which in turn influence friction, which in turn influences temperature. Wear also influences the conformity between surfaces, which influences hydrodynamic and boundary lubrication, which influence wear. With this type of complexity to simulate laboratory engine tests provide a "natural" simulation and are often used as tests to address engine wear concerns. Additional advantages with this type of test are that naturally occurring wear points can be identified, and many wear points can be evaluated simultaneously.

An example of this type of test is in a study of wear between cylinder liners and piston rings (102). In this study, three identical engines were laboratory tested under different conditions of service. One engine was operated at high speed, another was operated at high load, and the third was operated under mixed conditions. At the end of the test sequence,

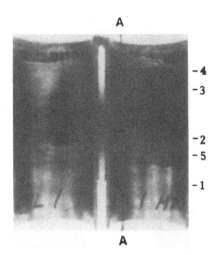

Figure 9.90 View of a cut cylinder liner from an engine test, indicating regions of different wear behavior. (From Ref. 102, reprinted with permission from ASME.)

the cylinder liners and piston were removed and examined for wear. Several wear regions could be observed on these parts, each with different wear characteristics. Figure 9.90 shows a cut-away section of a worn liner from these tests. The general nature of this wear (e.g., small with irregular outlines), along with the size and shape of the parts, makes it difficult to measure wear in terms of such measurements as volume, mass loss or dimensional change. This is a common situation with these types of tests and therefore as a result a variety of examination techniques and measurement are used to assess the severity of the wear and to provide a basis for comparison. In the liner study SEM, optical microscopy, X-ray fluorescence spectroscopy (XFS), energy dispersive X-ray spectroscopy (EDS), and metallographic techniques were used for this purpose. Examples of some of the results of these analysis techniques are shown in Fig. 9.91. Wear rates were estimated using the amount of original surface roughness remaining. In other engine-wear studies radioactive doping techniques have been used to quantify the wear (103) as have such measures as the average width of a wear scar or amount of edge rounding. Mass loss is usually not possible because of the large mass of the parts and the tendency for combustion products to coat the parts, as well as multiple wear locations.

The states of wear found for the three different operating conditions were very similar and were not considered as significantly different. A fourth test, which was a field test, was also done as a part of this study. The liners and pistons from the field test were examined in the same manner as those from the laboratory tests. Little difference was observed between the field and laboratory tests. This illustrates the good simulation that laboratory tests of this type (e.g., full scale) can provide. It should be recognized that at the same time it is difficult to quantify the results of these tests in terms of the more fundamental measures of wear.

9.3.8. Tests for Glazing Coatings on Plastics

The purpose of these coatings is to protect plastic surfaces from optical degradation, either in the form of reduced luster or transparency. These coatings are used extensively in the mass transit industry to coat surfaces of such components as windows, windshields, lights, and transparent panels. These surfaces can be worn or damaged by a variety of mechanisms

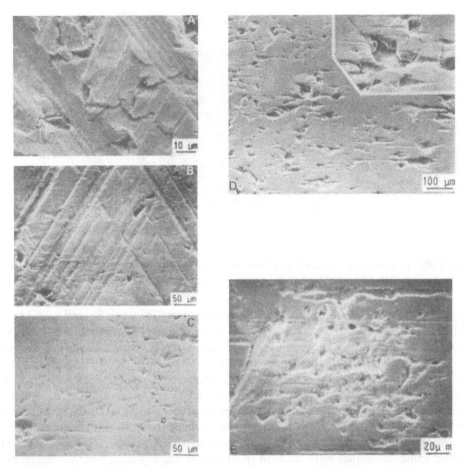

Figure 9.91 Wear morphology at the locations indicated in Fig. 9.90. "A", unworn surface, A1; "B", A2; "C", A3; "D", A4; "E", A5. (From Ref. 102, reprinted with permission from ASME.)

and as a result a variety of tests have been developed to address these different situations (73). One of the modes encountered is the wear that is associated with the use of large, rotating brushes in the washing of trains and buses; another is the wear associated with the action of a windshield wiper. In both of these situations, sand or other abrasive particles can be drawn across the surface of the plastic by the action of the brush and the wiper, causing wear. Operational tests have been developed to simulate both of these situations (104). The similarity of the tests to the wear situations is apparent in Figs. 9.92 and 9.93. Controlled slurries, selected to be representative of those encountered in practice, are used to provide the abrasive action in both tests. Loads and speeds representative of these applications are used in these tests.

Wear in these tests is quantified by measuring the haze of the surface, which is defined in ASTM D1003. These measurements are used in different manners in the two tests. In the Brush Abrasion Test, the change in the haze value from the initial value is used as a measure of wear for comparing and ranking coatings. In the Wiper Test, the amount of haze produced in the test is the measure since the specimens are initially transparent. This type of functional wear measure is commonly used for the evaluation of wear on optical surfaces.

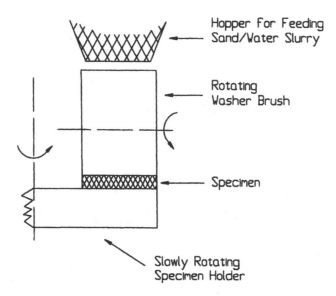

Figure 9.92 Diagram of test configuration used to simulate the wear caused by brushes during the cleaning of vehicles.

Both of these tests correlate well with field experience. This is to be expected as a result of their high degree of simulation provided by these tests. However less simulative, more phenomenological tests can provide good correlation as well, provided they simulate the basic wear mechanisms involved in the application. For example, tests with

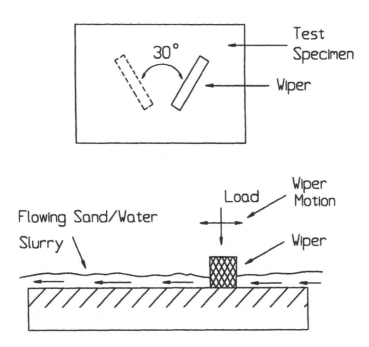

Figure 9.93 Diagrams of the test configuration used to simulate the wear between wiper blades and windshields.

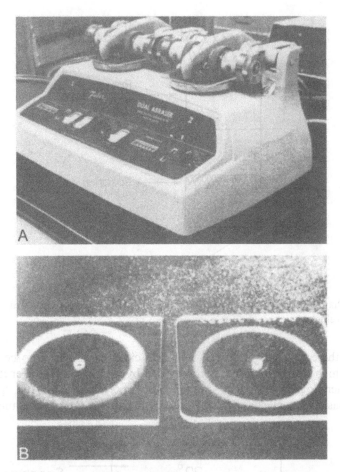

Figure 9.94 A commercial version of a Taber Abraser is shown in "A". "B" shows examples of the wear scar produced in the test. (From Ref. 104, reprinted with permission from ASTM.)

a Taber Abraser (ASTM D1044) (Fig. 9.94) correlate well with the Brush Test (105). The test with the Taber Abraser provides a dry, two-body abrasive wear condition. The common or basic element in these two tests is a mild abrasive wear, which appears to be the controlling factor in this type of application. This is to be expected since corrosive wear should not be a significant element in the wear of these coating materials in these applications, based on their properties.

9.3.9. Drill Wear Tests

Controlled drilling is a frequently used approach to address drill wear concerns (106,107). These tests consist of sequential hole drilling in a controlled work piece at a controlled speed, depth, and feed rate. The number of holes that can be drilled before failure determines the wear life of the drill. Several criteria for failure are used, depending somewhat on the type of drilling being done. One general type of criteria is the quality of the hole being drilled including hole dimensions, hole appearance, smear, and roughness. Periodic measurements or inspection of the holes are performed with these type of criteria. Another type of criteria is associated with various attributes of the drilling

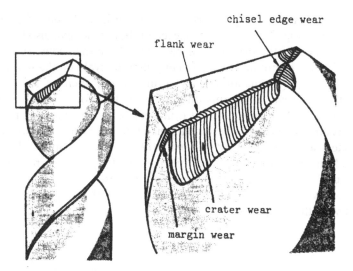

Figure 9.95 Twist drill geometry and location of wear zones. (From Ref. 107, reprinted with permission from ASME.)

processes. For example, the occurrence of squealing might be used to determine drill life. Other criteria could be the occurrence of chatter, increased torque, temperature or drill fracture. With all these criteria, drilling is continued until the event occurs and the number of operations is used as a measure of performance. Frequently, several of these criteria might be used with the nature of the failure changing with the drilling conditions or drill.

Tests of this nature are very operational in character, defining wear life in terms of the number of holes that can be successfully drilled rather than directly in terms of wear. In fact, there are several wear points or zones associated with drills. The wear conditions can be different at these points, and the wear at these points can have a different effect on

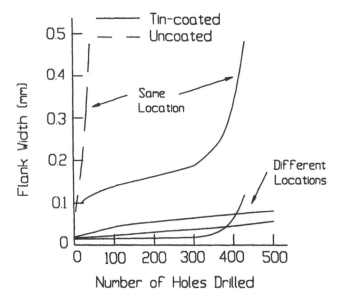

Figure 9.96 Wear curves for flank wear of twist drills. (From Ref. 105.)

performance. For example, the geometry of a twist drill, with the different wear zones identified, is shown in Fig. 9.95. The wear in each of these zones can also be addressed with these tests, in addition to the overall wear behavior. Since the geometry of these regions is complex, qualitative techniques are frequently used to evaluate the wear, such as SEM and optical microscopy. Sometimes a linear dimension, such as scar width, can be used to quantify the wear. Some quantitative wear data for flank wear are shown in Fig. 9.96, and some SEM micrographs of drill wear scars are presented in Fig. 9.97.

These tests are used not only to compare materials but also to investigate a wide variety of parameters associated with drills and drilling. For example drill speed, various dimensions and angles of the drill, hole thickness, and work piece properties are parameters, which can be studied and evaluated in these tests. Some data from these types of studies are shown in Figs. 9.98 and 9.99, illustrating the way the data are analyzed and presented with these types of tests.

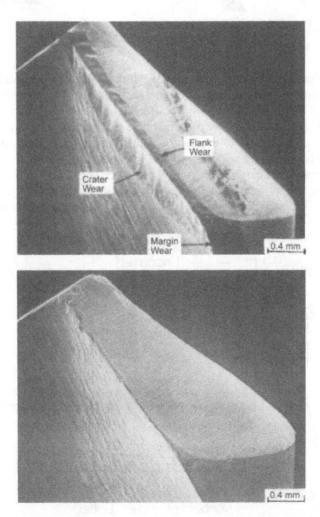

Figure 9.97 Examples of the wear of twist drills. (From Ref. 106, reprinted with permission from ASME.)

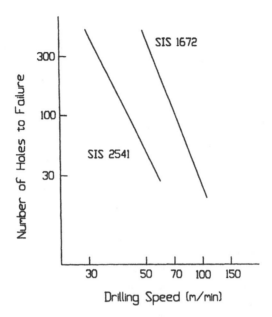

Figure 9.98 The effect of drill speed on drill life when drilling two different materials. (From Ref. 107.)

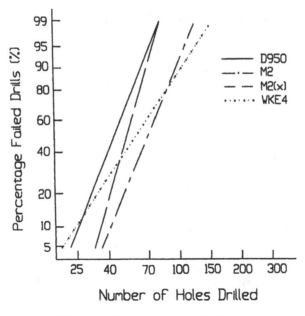

Figure 9.99 The effect of different drill material on drill life. A statistical measure is used to quantify performance in this case. (From Ref. 107.)

9.3.10. Seal Wear Tests

There are several aspects of the wear of seals that are of significance. One is directly related to materials loss (e.g., changes in dimensions and clearances). Another is changes in surface roughness, and the third can be the formation of transfer or third body films. All three can individually and jointly influence sealing and cause functional failure, namely leakage. All of these aspects can be influenced by the operational conditions associated with the application. Because of this complex relationship between wear and function, it is desirable to evaluate seal wear under highly simulative conditions in an operational type test (108). An example of this type of test is one that was used to investigate seals for Stirling engines (109).

A cross-section of the test apparatus is shown in Fig. 9.100. It replicates the sealing conditions of the engine rod, with the exception that inert gas at a fixed pressure is used to simulate the conditions in the combustion chamber. The fixed pressure is selected to be representative of the average pressure in the chamber under some generic operating conditions (e.g., highway or urban travel conditions). The other parameters, such as rpm, stroke length, speed, and temperature, are selected to be representative of the application and can be varied to represent different conditions of operation. The apparatus can accommodate different materials and design parameters. It was also designed so that leakage measurements, as well as seal temperature measurements, could be made. Thus, this apparatus can be used in the evaluation of a wide range of material, design, and use factors in terms of their effects on seal wear performance.

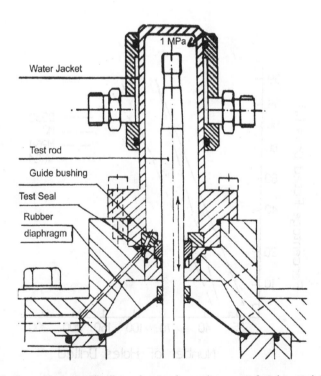

Figure 9.100 Diagram of an apparatus used to evaluate the wear of seal material. (From Ref. 134, reprinted with permission from ASME.)

Table 9.8 Data Obtained in Seal wear Test Used to Simulate the Stirling Engine Application

Rod surface	Time (hr)	Rod temp. (°C)	N_2 leakage (1/hr)	Initial roughness (R_a μm)	Final roughness (R_a μm)	Hardness (HV 1)	Wear rate (mg/h)	Comments
Nitrided Steel	78	55	1.3	0.06	0.03	1127	0.105	
	88	71	2.8	0.25	0.09	1162	0.132	
	70	47	7.3	0.04	0.14[a]	1132	0.250	Wavy[a]
Plasma sprayed molybdenum	71	64	2.1	0.22	0.18	530	0.107	
	70	67	2.9	0.21	0.31	589	0.110	
	70	74	1.7	0.24	0.18	620	0.117	
Plasma sprayed aluminium oxide	71	47	3.1	0.26	0.09	708[a]	0.086	Uncertain[a]
	70	51	4.9	0.43	0.26	680	0.033	
Plasma sprayed chromium oxide	70	58	13.6	0.65	0.55	617	0.343	
Nedox	70	56	14.9	0.22	[a]	739	0.409	Very Wavy[a]
Hard chromium on zinc	70	56	13.1	0.31	0.01	1039	0.554	

Seal material: Rulon LD.
Source: Ref. 134.

The basic methodology of the test is to run the apparatus for a specific length of time and to monitor leakage and temperature during that period. At the end of the complete test duration, the samples are removed and examined for wear. This included profilometer measurements of roughness, mass loss of the seals, and other characterizations of the rubbing surfaces, such as optical and SEM examinations. The type of quantitative data generated in this test is shown in Table 9.8. Leakage is considered to be the primary criteria for wear. The other measurements and observations aid in identifying the particular failure mode and differences in wear behavior. For example, in this study, it was concluded that the plasma-sprayed Mo coating performed the best, primarily because it maintained its roughness and allowed the formation of a very stable polytetrafluoroethene (PTFE) transfer film.

While this test provides good simulation of most of the aspects and allows wear phenomena to be correlated with performance (i.e., leakage), it does not provide complete simulation. Combustion does not occur, only the mean pressure of the duty cycle is simulated. As a consequence, this test is used to select material and design parameters for evaluation in a more simulative test in a laboratory engine. At the same time, more phenomenological type wear tests are used to identify promising candidates for this intermediate seal test (110). This is an example of a staged or multilevel testing program, with each level serving as a screen for selecting materials for the next and more simulative test.

9.3.11. Wear Test for a Magnetic Sensor

In this application, pieces of encoded magnetic tape are attached to or mounted on a variety of items, such as boxes, packages, or routing documents. To read the information, a

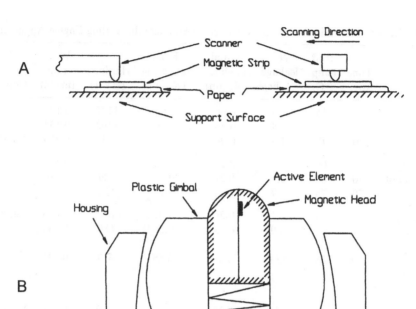

Figure 9.101 "A" illustrates the use of a magnetic hand scanner in reading stored information on magnetic strips located on labels, cards, and documents. "B" shows the design of the sensor in the region of the magnetic head.

magnetic head or sensor is pressed up against the surface of the tape section with a hand-held scanner and moved across the surface of the tape. A use for such a scanner is at the checkout station of a retail store, serving a similar purpose to the bar code readers found in many supermarkets. However, the magnetic tape may contain more information than provided by the bar code. Another use is to monitor the flow of materials through a manufacturing plant. In this case, the magnetic code not only provides information regarding identification but also about processing steps, etc. The scanning situation in these types of applications is illustrated in Fig. 9.101, along with a diagram of the magnetic head that was used.

The basic wear situation is similar to but significantly different than that encountered with more typical uses of magnetic tape (e.g., tape recorder and memory tape drives). Both are concerned with the wear of the head by the magnetic tape, but there are differences in speed and relative usage. Surface speeds are higher in magnetic recording. In both the head surface experiences the greater amount of sliding but the tape surface experiences much less use in this application than in recording and data storage applications. However, the major difference is that in tape recorders and drives, the tape surface is relatively clean. This is not the case in this sensor application since the tape surfaces are exposed to a wide variety of environments, some of which are quite dirty and contain very abrasive materials. Examples of these types of environments would be that of an open manufacturing or machining area, or receiving and shipping bays. Sand, iron oxide particles, aluminum oxide grit, and other abrasive particles are common in these environments (111). The abrasive action of these particles collected by the tape surfaces can become the primary

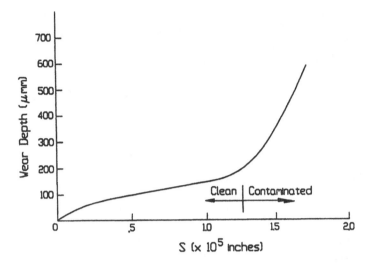

Figure 9.102 The effect of abrasive contamination on the wear of the magnetic head. (From Ref. 135.)

wear mode, masking the effect of the tape itself. Figure 9.102 shows the effect that contamination can have on the wear of these sensors. To address this wear situation, a wear test was developed that utilizes the drum configuration discussed previously in the section on phenomenological tests (Sec. 9.2.12) (42).

The drum test configuration was selected for the same reason that it was developed to address wear by paper and ribbon, namely that the tape surface wears more readily than the heads. In this use of the device, the magnetic tape is wrapped around the surface of the drum, and the magnetic head replaces the normal spherical wear specimen. To simulate the action of the abrasive contamination of the tape surface in the application, the surface of the tape is coated with different types and amounts of abrasive particles. This simulation was verified early in the development of the test by comparing wear scars produced in the field to those produced in the test for a variety of abrasive dust coatings on the surface of the tape. Figure 9.103 shows such a comparison. It was concluded that several methods of coating could be used to provide simulation, and that the primary reason for selecting one or the other was control and ease of application. A spray coating of AC Fine Test Dust, which is mainly sand (SiO_2), and gelatin were selected for this purpose. This technique provided a uniform coating of abrasive particles protruding above the surface

Figure 9.103 "A" shows an example of the wear morphology found on heads worn in field tests. "B" and "C" show examples of the morphology obtained in laboratory tests on tape surfaces contaminated with abrasives. (From Ref. 135, reprinted with permission from Elsevier Sequoia S.A.)

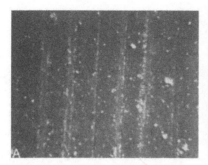

Figure 9.104 Examples of contaminated tape surfaces after a wear test. In "A", the tape was coated with loose AC Fine Test Dust. In "B", the tape was coated with a mixture of AC Fine Test Dust and gelatin. (From Ref. 135, reprinted with permission from Elsevier Sequoia S.A.)

of the gelatin (Fig. 9.104). Tests were done to investigate the possible influence that the gelatin might have on the wear. This was done by comparing wear behavior with this type of coating to those without the gelatin, that is dust coated tapes. No difference was found. Wear was found to depend solely on the amount, size, and nature of the abrasives on the surface. This is probably because the gelatin layer is weak, allowing easy movement of the particles, and thin enough not to mask or bury the particles. The gelatin layer is less than 2 μm, while approximately 80% of the sand particles are greater than 2 μm. To insure consistency of these coatings a short test, which utilized only a small portion of the tape, was conducted with a 52100 steel ball on each coated tape. If the amount of wear in this test fell outside an acceptable range, the tape was not used. Within the acceptable range, these same results provided a means to scale the individual tests to improve resolution.

As was described in Sec. 9.2.12, the wear specimen moves across the surface of the drum in an axial direction while the drum rotates, producing a helical path on the surface of the drum. A high enough axial speed for the specimen can insure that the specimen is always sliding on a new or fresh tape surface. Such a condition was used for most of the evaluations done with this test. This eliminated the complexities introduced by changes in the abrasive characteristics of the tape surface with wear. It also provided a worst case situation, since the abrasivity of both the uncoated and coated tape surfaces tend to decrease with wear. A standard drum rotational speed was also selected with two concerns in mind. One was to maintain simulation. The other was to reduce test time. Similar wear behavior was observed for surface speeds up to 300 cm/sec; however, testing with abrasive coatings at the higher speeds produced temperature increases which were beyond those found in the application. As a result, a speed more representative of the application and one which did not produce a significant temperature rise was selected. The drum rotational speed selected was 36 cm/sec and the specimen speed was 0.25 mm/rev or 0.02 mm/sec.

A gimbal-spring loaded mounting was used for the magnetic head in the application. This insured that even though the sensor was hand-held a consistent load and proper orientation would occur at the head/tape interface. A limited amount of testing was done to characterize wear behavior as a function of load. However, the majority of the tests, particularly those done to evaluate materials and design options, were done using the load in the application, 50 gm.

The relative wear performance of the heads was measured in terms of the amount of sliding under standard test conditions that was required to produce a 5-ohm change in the resistance of the magnetic element, located beneath the surface (see Fig. 9.101). The 5-ohm change was a functional criteria for the device. This was accomplished by monitoring this resistance during the test and stopping the test when a change equal to or greater than 5 ohms occurred. An automatic system was developed for this. In these tests, as well as in the field, a 5-ohm change could be caused by several different phenomena which could occur in the abrasive environment. This included gradual wear of the surfaces down to the magnetic element, micro- and macro-fracture down to the element, and penetration by abrasive particles of the seam between the two halves. These conditions are illustrated in Fig. 9.105. In most cases a combination of gradual wear, micro-fracture of the edges, and penetration determined life. The relative contribution of each was assessed at the end of the test by optical and SEM microscopy, and profilometer measurements. The profilometer measurements were also used to estimate wear volume, which was normalized with load and distance of sliding to provide a wear coefficient for the material and abrasive condition. With these types of tests, the evaluations of a wide variety of design and manufacturing aspects were addressed. These included radius, load, magnetic element location, material selection, adhesive thickness, deposition techniques for the magnetic element, grinding and polishing of the head, environmental effects, and orientation of the seam with sliding direction. The test was also used to precondition heads in a controlled manner for use in corrosion evaluations.

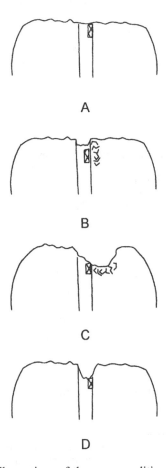

Figure 9.105 Illustrations of the wear conditions found on heads.

With the head replaced by a spherical-ended wear specimen, material studies were performed. In this type of test, the volume of wear after a fixed amount of sliding was determined in the same manner as in the test for ribbon abrasivity (Sec. 9.2.12) (112). A wear coefficient was developed from these data by dividing the volume by the load and distance of sliding. This was used to compare and rank materials.

In addition to the development of rankings and the identification of dependencies, this test was also used to determine coefficients of a wear model that was developed to describe wear by contaminated tape (42) The basic equation of this model is

$$V = [(1 - \alpha)K' + \alpha K'']PS = KPS \qquad (9.25)$$

where V is the volume of wear; P, the load, S, the distance of sliding; K', the wear coefficient for a clean tape surface; K'', the wear coefficient of a tape surface saturated by abrasives; α is the fraction of the tape surface covered by the abrasive. K is the effective wear coefficient for that condition of contamination. K' is the wear coefficient determined with a drum test on clean, uncoated tape. K'' is related to the wear coefficient determined by tests on the tape coated with the gelatin mixture, which was found to be equivalent to a saturated amount of abrasives. (The details of the model are presented in a case study in Sec. 5.12 EDW2E.) Some values of K' and K'' for different conditions are shown in Table 9.9.

Based on the studies that were done to investigate correlation with field performance, it was concluded that the test method provided good simulation of the field and could be used to predict field performance when used in conjunction with the model, that is, Eq. 9.25. Scatter in the test data was typically in the range of 10%.

Table 9.9 Abrasive Wear Coefficients for 52100 Steel Specimens for Various Abrasive Conditions

	Abrasive condition			Abrasive coefficient	
Type	Average size (μm)	Coverage amount (μg/cm²)	Saturation coverage (μg/cm²)	K (μm³/g-m)	K'' (μm³/g-m)
Clean tape	0	0	–	≈58	–
SiO_2	4	4	67	118	1661
	6	69	100	1796	2584
	6	255	100	2707	2707
	8	33	100	664	1969
	8	11	100	317	2584
	8	8	100	121	1969
	8	36	100	926	2461
	8	120	100	3076	3076
	15	170	100	7874	7874
	40	170	100	10,458	10,458
Al_2O_3	50	170	100	24,483	24,483

Saturated coverage is the surface coverage above which there is no change in the abrasive coefficient K'' is the maximum abrasive coefficient or saturation abrasive coefficient for particles of a given composition and size.
Source: Ref. 135.

REFERENCES

1. H Avery. Classification and precision of abrasion tests. In: D Rigney, W Glaeser, eds. Source Book on Wear Control Technology. Metals Park, OH: ASM, 1978, pp 57–66.
2. P Swanson. Comparison of laboratory and field abrasion tests. Proc Intl Conf Wear Materials. ASME 519–525, 1985.
3. C Saltzman. Wet-sand rubber-wheel abrasion test for thin coatings. In: R Bayer, ed. Selection and Use of Wear Tests for Coatings. STP 769, West Conshohocken, PA: ASTM, 1982, p 71.
4. J Miller. The miller number-a New Slurry Rating Index. American Institute of Mining, Metallurgical, and Petroleum Engineers (AIME) Paper 73-B-300. Society for Mining, Metallurgy, and Exploration (SME) Meeting, Pittsburgh, 1973.
5. J Miller, JD Miller. The miller number - a review. In: A Ruff, R Bayer, eds. Tribology: Wear Test Selection for Design and Application, STP 1199. West Conshohocken, PA: ASTM, 1993, pp 100–112.
6. Standard Test Method for Determination of Slurry Abrasivity (Miller Number) and Slurry Abrasion Response of Materials (SAR Number). West Conshohocken, PA: ASTM, G75.
7. J Young, A Ruff. Particle erosion measurements on metals. J Eng Materials Tech Trans ASME 99(2):121–125, 1977.
8. J Hansen. Relative erosion resistance of several materials. In: W Alder, ed. Erosion: Prevention and Useful Applications, STP 664. West Conshohocken, PA: ASTM, 1979, pp 148–162.
9. F Wood. Erosion by solid-particle impacts: A testing update. J Testing Eval 14:23–27, 1986.
10. I Finnie, J Wolak, Y Kabil. Erosion of metals by solid particles. J Materials 2:682–700, 1967.
11. A Ninham, I. Hutchings. A computer model for particle velocity calculation in erosion testing. Proceedings of the 6th International Conference on Erosion by Liquid and Solid Impact. New York: Cambridge University Press, 1983, pp 50.1–50.7.
12. R Barkalow, J Goebel, F Pettit. Erosion-corrosion of coatings and superalloys in high-velocity hot gases. In: W Adler, ed. Erosion: Prevention and Useful Applications, STP 664. West Conshohocken, PA: ASTM, 1979, pp 163–192.
13. J Hobbs. Experience with a 20-kc cavitation erosion test. In: Erosion by Cavitation or Impingement, STP 408. West Conshohocken, PA: ASTM, 1967, pp 159–185.
14. Proceedings of the ASME Symposium on Cavitation Research Facilities and Techniques. New York, NY: ASME, 1964.
15. R Knapp, J Daily, F Hammitt. Cavitation. New York: McGraw-Hill, 1970.
16. Cavitation. In: Treatise on Material Science and Technology, Vol. 16 Erosion. C Preece, Ed. New York: Academic Press, 1979, pp 249–308.
17. C Preece, I Hansson. A Metallurgical Approach to Cavitation Erosion. Advances in the Mechanics and Physics of Surfaces I. Harwood, London: Academic Pub, 1981, pp 199–254.
18. A Karimi, J Martin. Cavitation erosion of materials. Intl Metals Rev 31(1):1–26, 1986.
19. H Wiegand, R Schulmeister. Investigations with vibratory apparatus on the effect of frequency, amplitude, pressure and temperature on cavitation erosion (in German). MTZ Motorechnische Zeitschrift 29(2):41–50, 1968.
20. M Matsumura. Influence of test parameters in vibratory cavitation erosion tests. In: W. Adler, ed. Erosion: Prevention and Useful Applications, STP 664. New York, NY: ASTM, 1979, pp 434–460.
21. I Hannson, K Morch. Guide vanes in the vibratory cavitation system to improve cavitation erosion testing. Proceedings of the 6th International Conference on Erosion by Liquid and Solid Impact. New York: Cambridge University Press, 1983, pp 9.1–9.9.
22. Standard Test Method for Calibration and Operation of the Falex Block-on-Ring Friction and Wear Testing Machine. West Conshohocken, PA: ASTM, D2714.
23. K Budinski. Wear of tool steels. Proc Intl Conf Wear Materials. ASME 100–109, 1977.

24. R Bayer. Predicting wear in a sliding system. Wear 11:319–332, 1968.
25. K Budinski. Incipient galling of metals. Proc Intl Conf Wear Materials. ASME 171–178, 1981.
26. W Schumacher. The galling resistance of silver, tin and chrome plated stainless steels. Proc Intl Conf Wear Materials. ASME 186–196, 1981.
27. R Morrison. Test date let you develop your own load/life curves for gear and cam materials. Machine Design. 8/68. Cleveland, OH: Penton Publishing Co., 1968, pp 102–108.
28. E Buckingham, G Talbourdet. Roll Tests on Endurance Limits of Materials. Metals Park, OH: ASME, 1950.
29. R Benzing, I Goldblatt, V Hopkins, W Jamison, L Mecklenburg, M Peterson. Friction and Wear Devices. Park Ridge, IL: ASLE, 1976.
30. A Wayson. A study of fretting of steel. Wear 7:435–450, 1964.
31. R Bayer, J Sirico. Wear of Electrical Contacts due to Small-amplitude Motion. IBM J R&D 15(2):103–107, 1971.
32. R Bayer. Wear of a C Ring seal. Wear 74:339–351, 1981–1982.
33. N Payne, R Bayer. Friction and wear tests for elastomers. Wear 130:67–77, 1991.
34. R Bayer, N Payne. Wear evaluation of molded plastics. Lub Eng May:290–293, 1985.
35. R Bayer, A Trived. Molybdenum disulfide conversion coating. Metal Finishing Nov: 47–50, 1977.
36. Standard Test Method for Wear Testing with a Pin-on-Disk Apparatus. West Conshohocken, PA: ASTM. G99.
37. R Bayer, T Ku. Handbook of Analytical Design for Wear. New York: Plenum Press, 1964.
38. D Roshon. Testing machine for evaluating wear by paper. Wear 30:93–103, 1974.
39. R Bayer. Wear by ribbon and paper. Wear 49:147–168, 1978.
40. D Roshon. Mechanism of wear by ribbon and paper. IBM J R&D 22(6):668–674, 1978.
41. R Bayer. Electroplated diamond-composite coating for abrasive wear resistance. In: R Denton, K Keshavan, eds. Wear and Friction of Elastomers, STP 1145. West Conshohocken, PA: ASTM, 1992, pp 681–686.
42. R Bayer. Tribological approaches for elastomer applications in computer peripherals. Wear 70:114–126, 1981.
43. R Bayer. A model for wear in an abrasive environment as applied to a magnetic sensor. Wear 92:93–117, 1983.
44. R Bayer. Aspects of paper abrasivity. Wear 100:517–532, 1984.
45. R Bayer, D Baker, T Ku. Abrasive wear by paper. Wear 12:277–288, 1968.
46. P Engel, R Bayer. Abrasive impact wear of type. J Lub Tech 98:330–334, 1976.
47. R Lewis. Paper No. 69AM 5C-2. Proceedings of 24th ASLE Annual Meeting, Philadelphia, 1969.
48. J Theberge. A Guide to the Design of Plastic Gears and Bearings. Machine Design. 215/70. Cleveland, OH: Phenton Publishing Co., 1970, pp 114–120.
49. R Gates, J Yellets, D Deckman, S Hsu. The development of a wear test methodology. In: C Yust, R Bayer, eds. Selection and Use of Wear Tests for Ceramics, STP 1010. West Conshohocken, PA: ASTM, 1988, pp 1–23.
50. W Wei, K Beaty, S Vinyard, J Lankford. Friction and wear testing of ion beam modified ceramics for high temperature low heat rejection diesel engines. In: C Yust, R Bayer, eds. Selection and Use of Wear Tests for Ceramics, STP 1010. West Conshohocken, PA: ASTM, 1988, pp 74–87.
51. L Fiderer. Unique friction and wear tester for fundamental tribology research. In: C Yust, R Bayer, eds. Selection and Use of Wear Tests for Ceramics, STP 1010. West Conshohocken, PA: ASTM, 1988, pp 24–42.
52. C Yust, R Bayer, eds. Selection and Use of Wear Tests for Ceramics, STP 1010. West Conshohocken, PA: ASTM, 1988.
53. C Yust, J Leitnaker, C DeVore. Wear of alumina-silicon carbide whisker composite. Proc Intl Conf Wear Materials. ASME 277–284, 1987.

54. M Gee. Meeting on standardization of wear test methods for ceramics, cermets, and coatings. In: C Yust, R Bayer, eds. Selection and Use of Wear Tests for Ceramics, STP 1010. West Conshohocken, PA: ASTM, 1988, pp 88–92.

55. G Tennenhouse, F. Runkle. Pin-on-disk wear tests for evaluating ceramic cutting tool materials. In: C Yust, R Bayer, eds. Selection and Use of Wear Tests for Ceramics, STP 1010. West Conshohocken, PA: ASTM, 1988, pp 43–57.

56. L Wedeven, R Pallini, C Hingle. Systematic testing of ceramic rolling bearing elements. In: C Yust, R Bayer, eds. Selection and Use of Wear Tests for Ceramics, STP 1010. West Conshohocken, PA: ASTM, 1988, pp 58–73.

57. K Dufrane, W Glaeser. Wear of ceramics in advanced heat engine applications. Proc Intl Conf Wear Materials. ASME 285–292, 1987.

58. P Engel. Impact Wear of Materials. Tribology Series. New York: Elsevier Science Publishing Co., 1978.

59. W Glaeser. Failure mechanisms of reed valves in refrigeration compressors. Wear 225–229:918–924, 1999.

60. P Engel, Q Yang. Impact wear of multiplated electrical contact. Wear 181–183:730–742, 1995.

61. R Anderson, T Adler, J Hawk. Scale of microstructure effects on the impact resistance of Al_2O_3. Wear 162–164:1073–1080, 1993.

62. P Engel. Impact wear. In: P Blau, ed. Lubrication, and Wear Technology. 18. Materials Park, OH: ASM Handbook, 1992, pp 263–270.

63. E Iturbi, I Greenfield, T Chou. Surface layer hardening of polycrystalline copper by multiple impact. J Materials Sci 15:2331–2334, 1980.

64. P Engel. Impact Wear of Materials. Tribology Series. New York: Elsevier Science Publishing Co., 1978, pp 180–243, 264–290.

65. R Bayer, ed. Effects of Mechanical Stiffness and Vibration on Wear, STP 1247. West Conshohocken, PA: ASTM, 1995.

66. P Engel. Impact Wear of Materials. Tribology Series. New York: Elsevier Science Publishing Co., 1978, pp 228–235.

67. R Bayer, P Engel, E Sacher. Impact wear phenomena in thin polymer films. Wear 32: 181–194, 1975.

68. A Ramamurthy, G Buresh, M Nagy, M Howell. Novel instrumentation for evaluating stone impact wear of automotive paint systems. Wear 225–229:936–948, 1999.

69. K Rutherford, R Tresona, A Ramamurthy, I Hutchings. The abrasive and erosive wear of polymeric paint films. Wear 203–204:325–334, 1997.

70. A Ramamurthy, J Charest, M Lilly, D Mihora, J Freese. Friction induced paint damage - A novel method for objective assessment of painted engineering plastics. Wear 203–204: 350–361, 1997.

71. S Jacobsson. Scratch testing. In: P Blau, ed. Lubrication, and Wear Technology. 18. Materials Park, OH: ASM Handbook, 1992, pp 430–437.

72. In: P Blau, ed. Lubrication, and Wear Technology. 18. Materials Park, OH: ASM Handbook, 1992, pp 433–436.

73. R Bayer, ed. Selection and Use of Wear Tests for Coatings, STP 769. West Conshohocken, PA: ASTM, 1982.

74. B Kushner, E Novinski. Thermal spray coatings. In: P Blau, ed. Lubrication, and Wear Technology. 18. Materials Park, OH: ASM Handbook, 1992, pp 829–833.

75. F Borik, D Sponseller. Gouging Abrasion Test for Materials used in Ore and Rock Crushing: Part I-Description of the Test; Part II-Effect of Metallurgical Variables on Gouging Wear. J Materials 6(3):576–589 and 590–605, 1971.

76. S Calabrese, S Murray. Methods of evaluating materials for Icebreaker hull coatings. In: R Bayer, ed. Selection and Use of Wear Tests for Coatings, STP 769. West Conshohocken, PA: ASTM, 1982, pp 157–173.

77. Test developed by SpinPhysics Div. Eastman Kodak Co., 3099 Science Park Road, San Diego, CA, 92121.

78. J Carroll, R Gotham. The measurement of abrasiveness of magnetic tape. IEEE Trans Magnetics MAG-2(1):6–13, 1966.

79. A van Groenou. The sphere-on-tape: A quick test on wear of materials used in magnetic recording. Proc Intl Conf Wear Materials. ASME 212–217, 1983.

80. M Ruscoe. A predictive test for coin wear in circulation. Proc Intl Conf Wear Materials. ASME 1–12, 1987.

81. P Hammer. Schmierungstechnik 4(6):165, 1973.

82. P Engel, C Adams. Rolling wear study of misaligned cylindrical contacts. Proc Intl Conf Wear Materials. ASME 181–191, 1987.

83. S Carver. Rolling Bearings-I. Lubrication 60(Jul–Sept). Beacon, NY: Texaco, Inc., 1974.

84. S Carver. Rolling Bearings-II. Lubrication 60(Oct–Dec). Beacon, NY: Texaco, Inc., 1974.

85. E Bamberger, T Harris, W Kacmarsky, C Moyer, R Parker, J Sherlock, E Zaetsky. Life Adjustment Factors for Ball and Roller Bearings. Engineering Design Guide. New York, NY: ASME, 1971.

86. E Zaretsky, cd. Life Factors for Rolling Bearings, SP-34. Park Ridge, IL: STLE, 1999.

87. McGrew. The design and wear of sliding bearings. Stand News 2(9):22–28, 56, 1974.

88. Bearings. In: M Neale, ed. Tribology Handbook. New York: John Wiley and Sons, 1973.

89. F Barwell, ed. Bearing systems. New York: Oxford University Press, 1979.

90. E Bamberger, T Harris, W Kacmarsky, C Moyer, R Parker, J Sherlock, E Zaetsky. Life Adjustment Factors for Ball and Roller Bearings. Engineering Design Guide. New York, NY: ASME, 1971, pp 2–24.

91. S Carver. Rolling Bearings-I. Lubrication 60(Jul–Sept). Beacon, NY: Texaco, Inc., 1974, pp 54–60.

92. S Carver. Rolling Bearings-I. Lubrication 60(Oct–Dec). Beacon, NY: Texaco, Inc., 1974, pp 67–70.

93. S Carver. Rolling Bearings-III. Lubrication 61(Jan–Mar). Beacon, NY: Texaco, Inc., 1975.

94. J Lancaster, ed. Special Issue on Dry Bearings. Tribology 6(6), 1973.

95. A Begelinger, A de Gee. Wear in lubricated journal bearings. Proc Intl Conf Wear Materials. ASME 298–305, 1977.

96. H Chang. Correlation of wear with oxidation of carbon-carbon composites. Proc Intl Conf Wear Materials. ASME 544–547, 1981.

97. H Hawthorne. Wear debris induced friction anomalics of organic brake materials in vacuo. Proc Intl Conf Wear Materials. ASME 381–388, 1987.

98. M Jacko, P Tsang, S Rhee. Automotive friction materials evolution during the past decade. Wear 100:503–515, 1984.

99. P Tsang, M Jacko, S Rhee. Comparsion of chase and inertial brake dynamometer testing of automotive friction materials. Proc Intl Conf Wear Materials 129–137, 1985.

100. H Hawthome. Wear debris induced friction anomalies of organic brake materials in Vacuo. Proc Intl Conf Wear Materials. ASME 381–388, 1987.

101. S Rhee. Wear 18:471, 1971.

102. S Hogmark, J Alander. Wear of cylinder liners and piston rings. Proc Intl Conf Wear Materials. ASME 38–44, 1983.

103. W Pike, R Pywell, S Rudston. Wear Phenomenon of Chromium Plated Rings as Revealed by Radioactive Traces. SAE paper No. 699773, 2504–2512, 1969.

104. L Haluska. Wear Testing of abrasion-resistant coated plastics. In: R Bayer, ed. Selection and Use of Wear Tests for Coatings, STP 769. West Conshohocken, PA: ASTM, 1982, pp 16–27.

105. R Bayer, ed. Selection and Use of Wear Tests for Coatings, STP 769. ASTM, 1982, p 26.

106. C Young, S Rhee. Wear Process of TiN coated drills. Proc Intl Conf Wear Materials. ASME 543–550, 1987.

107. S Soderber. Performance and Failure of High Speed Steel Drills Related to Wear. In: O Vingsbo, M Nissle, eds. Proc Intl Conf Wear Materials. ASME, 1981, pp 456–467.

108. J Dray. Friction and Wear of Seals. In: P Blau, ed. Lubrication, and Wear Technology. 18. Materials Park, OH: ASM Handbook, 1992, pp 546–552.

109. G Lundholm. Comparison of seal materials for use in stirling engines. Proc Intl Conf Wear Materials. ASME, 1983, pp 250–255.

110. B Bhushan, D Wilcock. MTI Report No. 80ASE 141ER9. Latham, NY: Mechanical Technology, Inc, Aug 1980.

111. R Bayer, R Ginsburg, R Lasky. Settable and airborne particles in industrial environments. Proceedings of the 35th Meeting of the IEEE Holm Conference on Electrical Contacts. IEEE, 1989, pp 155–166.

112. Standard Test Method for Abrasiveness of Ink-Impregnated Fabric Printer Ribbons. West Conshohocken, PA: ASTM, G56.

113. P Swanson, R Klann. Abrasive wear studies using the wet sand and dry sand rubber wheel tests. Proc Intl Conf Wear Materials. ASME 379–389, 1981.

114. Standard Test Method for Measuring Abrasion Using the Dry Sand/Rubber Wheel Apparatus. West Conshohocken, PA: ASTM, G65.

115. L Ives, A Ruff. Electron microscopy study of erosion damage in copper. In: W Adler, ed. Erosion: Prevention and Useful Applications, STP 664. ASTM, 1979, pp 5–35.

116. Standard Test Method for Conducting Erosion Tests by Solid Particle Impingement Using Gas Jet. West Conshohocken, PA: ASTM, G76.

117. Standard Test Method for Cavitation Erosion Using Vibratory Apparatus. West Conshohocken, PA: ASTM, G32.

118. Standard Test Method for Ranking Resistance of Materials to Sliding Wear Using Block-on-Ring Wear Test. West Conshohocken, PA: ASTM, G77.

119. R Tucker, A Miller. Low stress abrasive and adhesive wear testing. In: R Bayer, ed. Selection and Use of Wear Tests for Metals. West Conshohocken, PA: ASTM, 1977, pp 68–90.

120. R Bayer. Mechanism of Wear by Ribbon and Paper. IBM J R&D 26:668–674, 1978.

121. A Ajayi, K Ludema. Surface damage of structural ceramics: Implications for wear modeling. Proc Intl Conf Wear Materials. ASME 349–360, 1987.

122. T Fischer, M Anderson, S Jahanmir, R Salher. Friction and wear of tough and brittle zirconia in nitrogen, air, water, hexadecane and hexadecane containing stearic acid. Proc Intl Conf Wear Materials. ASME 257–266, 1987.

123. Standard Practice for Liquid Impingement Erosion Testing. West Conshohocken, PA: ASTM, G73.

124. Y-J Liu, N-P Chen, Z-R Zhang, C-Q Yang. Wear behavior of two parts subjected to 'Gouging' abrasion. Proc Intl Conf Wear Materials. ASME 410–415, 1985.

125. Standard Test Method for Jaw Crusher Gouging Abrasion Test. West Conshohocken, PA: ASTM, G81.

126. M Ruscoe. A Predictive Test for Coin Wear in circulation. Proc Intl Conf Wear Materials. ASME 1–12, 1987.

127. P Engel, C Adams. Rolling wear study of misaligned cylindrical contacts. Proc Intl Conf Wear Materials. ASME 181–191, 1979.

128. J Lancaster. Dry Bearings: A survey of materials and factors affecting their performance. Trib Intl 6:219–252, 1973.

129. A Kearse, J Rudman. Radial elastomeric bearings: testing and indications of failure. Trib Intl 15(5):249–254, 1982.

130. B Ginty. Bearing materials for small electric motors. Trib Intl 15(2):85–88, 1982.

131. E Kleinicin. The testing of rolling bearing lubricants. Ball Roller Bearing Eng 13:29–33, 1974.

132. K Stevens, M Todd. Parametric study of solid-lubricant composites as ball-bearing cages. Trib Intl 15(5):293–302, 1982.

133. H Hawthorne. Wear debris induced friction anomalies of organic brake materials in vacuo. Proc Intl Conf Wear Materials. ASME 381–388, 1987.

134. G Lundholm. Comparison of seal materials for use in stirling engines. Proc Intl Conf Wear Materials. ASME 250–255, 1983.

135. R Bayer. A model for wear in an abrasive environment as applied to a magnetic sensor. Wear 70:93–117, 1981.

136. R Bayer, J Sirico. Some observations concerning the friction and wear characteristics of sliding systems involving cast ceramic. Wear 16:421–430, 1970.

137. R Bayer. Abrasiveness of Electroerosion Papers. 1984 International Printing and Graphic Arts/Testing Conference, 56–60.

138. G Massouros. Model of wear in a plain bearing under boundary lubrication. Trib Intl 15(4):193–198, 1982.

10
Friction Tests

Friction tests are tests to determine the coefficient of friction of material pairs or rank material pairs in terms of their friction. The amount of motion in these tests is limited and generally small, being just sufficient to allow the measurement to be made. The methodology elements are the same with friction tests as they are with wear tests with the exception of acceleration (1,2). There is no need to consider this aspect with friction tests since these tests are typical short enough so that there is no practical need for acceleration. As with wear tests simulation is the key element in friction tests. The same parameters need to be considered for simulation and control, though their significance in a friction test can be different than in a wear test. One trend is that initial surface films tend to be more significant in friction tests than in many wear tests. This is particularly true for unlubricated conditions and material pairs that have high coefficients of friction. This is because such films can provide significant lubrication but are usually quickly worn away, so that their effect on long-term wear is often negligible. However, friction tests are generally limited to initial and not long-term conditions. For example, residual oil films on metal surfaces can reduce the coefficient of friction from near 1 to near 0.3. As sliding progresses, a gradual increase in friction from this low value to the higher value will occur, as a result of removal of this film from the contact area. This behavior is illustrated in Fig. 10.1. When determining the coefficient of friction for a particular application, the general nature of the contact geometry in the application should be replicated in the test. For example, a flat-on-flat test configuration should be used for an application where the contact is a flat on a flat. Similarly, if the application involves a web material wrapped around a cylinder or similar curved surface, the test configuration should also have the web material wrapped over a cylinder as illustrated in Fig. 10.2 (3). For nominal values of the coefficient, this degree of simulation is generally not necessary.

There are a large number of configurations and test methods used for measuring and characterizing friction. Many of these test methods evolved from the nature of different applications where friction is important. There are a large number of friction tests for which standard test methods have been developed. This is illustrated in Table 10.1, where sketches of the configuration and methods of measurements used in these different tests are shown. In addition, there is an ASTM standard, G115, which provides guidelines for measuring and reporting the coefficient of friction. Most methods result in a value for the coefficient of friction. With some only the static coefficient of friction, μ_s, is measured. In others, the kinetic coefficient of friction, μ_k, is determined and with some both the static and kinetic coefficient can be measured. However, not all friction tests result in a coefficient of friction. With some of these standard methods, differences in friction are used to characterize friction behavior of materials. Still others use the energy dissipated to characterize friction. Tests using pendulums are of this type. In these, the height that the

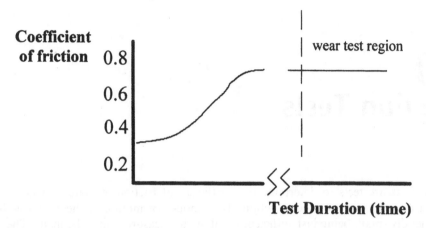

Figure 10.1 Nominal behavior of the coefficient of friction between unlubricated steel specimens when organic films contaminate the surfaces. The effect on long-term wear behavior is generally negligible.

pendulum reaches after contact is used as a measure. Generally, friction tests that provide values for the coefficient of friction are more relevant to wear behavior than those that do not.

There are three general relationships that are used in tests for determining the coefficient of friction. For tests in which one body slides across another, the coefficient of friction is given by

$$\mu = \frac{F}{N} \tag{10.1}$$

F is the friction force and N is the normal force. For friction tests using an inclined plane the following is generally used:

$$\mu = \tan \phi \tag{10.2}$$

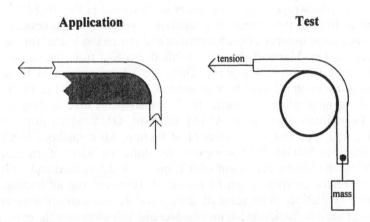

Figure 10.2 Example of a friction test (ASTM G143) used to simulate tape or other web material sliding over a guide.

Table 10.1 Standard ASTM Test Methods for Friction

Standard (committee)	Title	Materials (parameters measured)	Test configuration
B460 (B-9 on metals)	Dynamic Coefficient of Friction and Wear of Sintered Metal Friction Materials Under Dry Conditions	Friction materials vs. metals (μ_k vs. temperature)	
B461 (B-9 on metals)	Frictional Characteristics of Sintered Metal Friction Materials Run in Lubrication	Friction materials vs. metals (μ_k vs. number of engagements) (μ_k vs. velocity)	
B526 (B-9 on metal powders)	Coefficient of Friction and Wear of Sintered Metal	Friction materials vs. gray cast iron (μ_s and μ_k)	
D1894 (D-20 on plastics)	Static and Kinetic Coefficients of Friction of Plastic Films and Sheeting	Plastic film vs. stiff or other solids (μ_s and μ_k)	Nylon 200g Speed: 2 – 16 mm/s 50% RH

Table 10.1 (*Continued*)

Standard (committee)	Title	Materials (parameters measured)	Test configuration
D2047 (D-21 on polishes)	Static Coefficient of Friction of Polish Coated Floor Surfaces as Measured by the James Machine	Flooring materials vs. shoe heels and soles (μ_s and μ_k)	
D2394 (D-7 on wood)	Simulated Service Testing of Wood and Wood-base Finish Flooring	Wood and wood-base flooring vs. sole leather (μ_s and μ_k)	
D2714 (D-2 on lubricants)	Calibration and Operation of Alpha Model LFW-1 Friction and Wear Testing Machine	Steel ring vs. block lubricated with standard oil (μ_k)	

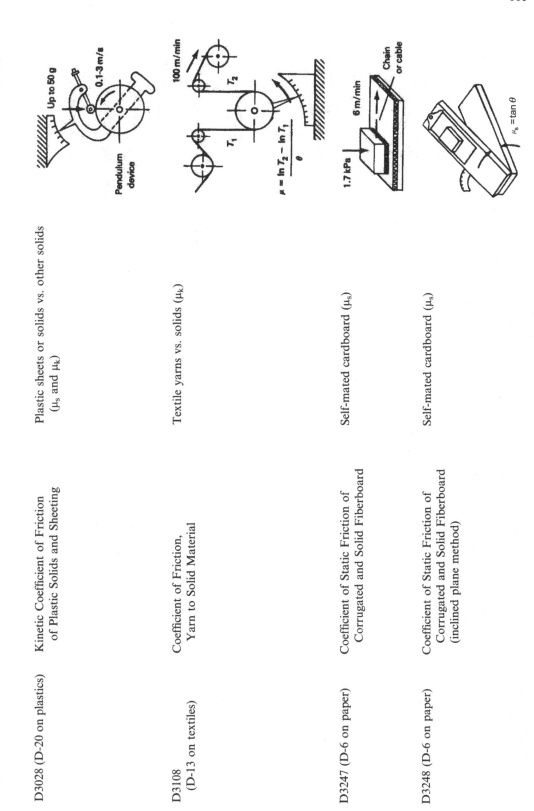

D3028 (D-20 on plastics) Kinetic Coefficient of Friction of Plastic Solids and Sheeting Plastic sheets or solids vs. other solids (μ_s and μ_k)

D3108 (D-13 on textiles) Coefficient of Friction, Yarn to Solid Material Textile yarns vs. solids (μ_k)

D3247 (D-6 on paper) Coefficient of Static Friction of Corrugated and Solid Fiberboard Self-mated cardboard (μ_s)

D3248 (D-6 on paper) Coefficient of Static Friction of Corrugated and Solid Fiberboard (inclined plane method) Self-mated cardboard (μ_s)

Table 10.1 (*Continued*)

Standard (committee)	Title	Materials (parameters measured)	Test configuration
D3334 (D-13 on textiles)	Testing of Fabrics Woven from Polyolefin Monofilaments	Self-mated woven fabric (μ_s)	$\mu_s = \tan \theta$
D3412 (D-13 on textiles)	Coefficient of Friction, Yarn to Yarn	Continuous filament and spun yarns self-mated (μ_s and μ_k)	$\mu = [\ln(T_2/T_1)]/\theta$ where: μ = coefficient of friction θ = wrap angle, rad T_1 = input tension, gf (or mN) T_2 = output tension, gf (or mN)

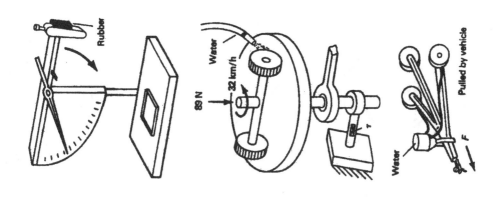

E303 (E-17 on traveled surfaces)	Measuring Surface Frictional Properties Using the British Pendulum Tester	Rubber vs. pavement (BPN, British Pendulum Number)
E510 (E-17 on traveled surfaces)	Determining Pavement Surface Frictional and Polishing Characteristics Using a Small Torque Device	Rubber vs. pavement (TN, torque number)
E670 (E-17 on traveled surfaces)	Side Force Friction on Paved Surfaces Using the Mu-Meter	Tires vs. pavement (μ) ($F_{\text{fry}} - F_{\text{wet}}$)

Table 10.1 (*Continued*)

Standard (committee)	Title	Materials (parameters measured)	Test configuration
E707 (E-17 on traveled surfaces)	Skid Resistance of Paved Surfaces Using the North Carolina State University Variable Speed Friction Tester	Rubber tires vs. pavement (VSN, variable speed number)	Locked tire / Water
F489 (F-13 on footwear)	Rating of Static Coefficient of Shoe Sole and Heel Materials as Measured by the James Machine	Leather and rubber sole and heel material vs. walking surfaces (μ_s)	6.9–90 kPa / 38–152 mm/s
F609 (F-13 on footwear)	Test Method for Static Slip Resistance of Footwear, Sole, Heel or Related Materials by Horizontal Pull Slipmeter (HPS)	Footwear materials vs. walking surfaces (μ_s)	6.9–90 kPa / 38–152 mm/s

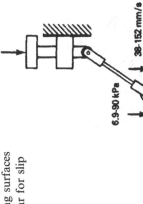

F695 (F-13 on footwear) Evaluation of Test Data Obtained by Using the Horizontal Slipmeter or the James Machine for Measurement of Static Slip Resistance of Footwear, Sole, Heel, or Related Materials Footwear materials vs. walking surfaces (reliable ranking of footwear for slip resistance)(μ_s)

F732 (F-4 on medical and surgical materials) Reciprocating Pin-on-Flat Evaluation of Friction and Wear Properties of Polymeric Materials for Use in Total Joint Prostheses Materials for human joints (μ_k)

Table 10.1 (*Continued*)

Standard (committee)	Title	Materials (parameters measured)	Test configuration
G143 (G-2 on wear and erosion)	Measurement of Web/Roller Friction Characteristics	Plastic films and other flexible web materials vs. roller surfaces (μ_s and μ_k)	$\mu = [\ln (T/W)]/\theta$ where: μ = coefficient of friction θ = wrap angle, rad T = input tension, gf (or mN) W = output tension, gf (or mN)
G164 (G-2 on wear and erosion)	Determination of Surface Lubrication on Flexible Webs	Lubricated flexible web materials vs. steel (detection of lubricant film on surface) (μ_s)	paper clip $\mu_s = \tan \theta$

Source: Refs. 1–4.

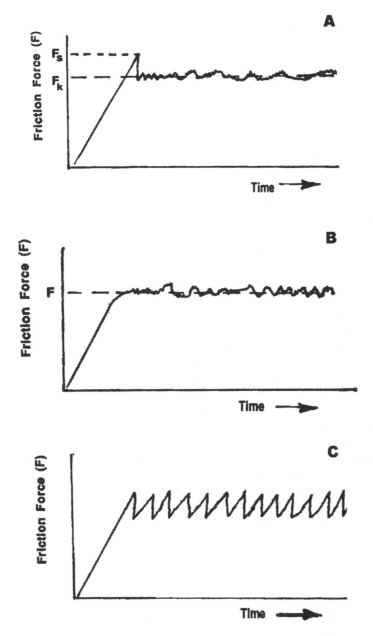

Figure 10.3 Examples of friction behavior in sliding tests. "A" illustrates a friction curve where the force required to initiate sliding, F_s, is higher than the average force needed to sustain motion, F_k. The static and dynamic coefficients are generally based on these two values, respectively. "B" illustrates a tribosystem which does not exhibit a higher breakaway force. This system has a single value for the coefficient of friction, based on F, the average force to sustain motion. Stick-slip behavior is illustrated in "C". In this case, the peak value is used to determine the static coefficient of friction. (From Ref. 1. Reprinted with permission from ASM International.)

is the tilt angle. The tilt angle at which sliding starts gives the static coefficient of friction. The minimum angle, which is needed to just sustain motion, gives the kinetic coefficient. For tests involving sliding over a cylindrical mandrel, the following relationship is used:

$$\mu = \frac{\ln(T_1/T_2)}{\theta} \tag{10.3}$$

In this equation, T_1 and T_2 are the tensions on each side of the mandrel ($T_1 > T_2$) is the difference in tension across the mandrel and is wrap angle in radians.

In friction tests where Eq. (10.1) is used, the friction force is usually measured as a function of time during the test. Figure 10.3 shows two general forms of these data when plotted as a function of time or sliding distance. Curve A illustrates a situation where the static coefficient of friction is different than the kinetic coefficient of friction. In this case, the static coefficient is based on the initial peak value of the force. The kinetic coefficient is generally based on a time-average of the force after slip takes place. Some initial stick-slip behavior is also shown in this curve. Curve B is a general form where there is no difference between the static coefficient and the dynamic, as well as no stick-slip. In either case, these two graphs illustrate the fact that the friction force is not usually constant but fluctuates in these tests. The ability to resolve such fluctuations, including stick-slip behavior, curve C, and a difference between static and kinetic coefficients, depends on the resolution capability of the measuring system. In addition, the stiffness of the apparatus can be a factor in stick-slip behavior and differences between the static and kinetic coefficients. It is possible that both types of curves could be observed for the same materials with different instrumentation and apparatus stiffness. When the normal force is not constant, instantaneous values of the coefficient of friction can be defined by simultaneously measuring the friction and normal. Again the ability to resolve short-term fluctuations depends on the instrumentation.

An illustration of stick-slip behavior is also shown in Fig. 10.3, curve C. When this occurs, only the static coefficient of friction can be determined, using the peak values of the friction force. It is possible that variations in the peak force might be observed over time, as a result of changing surface conditions.

As with wear test, it is generally recommended that details of the test be reported, along with the results.

REFERENCES

1. K Budinski. In: P Blau, ed. Friction, Lubrication, and Wear Technology, ASM Handbook. 18. Materials Park, OH: ASM International, 1992, pp 45–58.
2. Standard Guide for Measuring and Reporting Friction Coefficients. West Conshohocken, PA: ASTM G115.
3. Standard Test Method for Measurement of Web/Roller Friction Characteristics. West Conshohocken, PA: ASTM G143.
4. Standard Test Method for Determination of Surface Lubrication on Flexible Webs. West Conshohocken, PA: ASTM G164.

Glossary of Wear Mechanisms, Related Terms, and Phenomena

Abrasion: A process in which hard particles or protuberances are forced against and moving along a solid surface. (See abrasive wear.)

Abrasion–corrosion: A synergistic process involving both abrasive wear and corrosion, in which each of these processes is affected by the simultaneous action of the other.

Abrasion erosion: Erosive wear caused by the relative motion of solid particles, which are entrained in a fluid, moving nearly parallel to a solid surface.

Abrasive wear: Wear by displacement of material caused by hard particles or hard protuberances or wear due to hard particles or protuberances forced against and moving along a solid surface.

Adhesive wear: Wear by transference of material from one surface to another during relative motion due to a process of solid-phase welding or wear due to localized bonding between contacting solid surfaces leading to material transfer between two surfaces or loss from either surface. (Note: Sometimes used as a synonym for dry (unlubricated) sliding wear.)

Anti-wear Number (AWN): The base-10 log of the inverse of the wear coefficient.

Asperity: A protuberance in the small-scale topographical irregularities of a solid surface.

Atomic wear: Wear between two contacting surfaces in relative motion attributed to migration of individual atoms from one surface to the other.

Beilby layer: An altered surface layer formed on a surface as a result of wear.

Boundary lubricant: A lubricant that provides boundary lubrication.

Boundary lubrication: A condition of lubrication in which the friction and wear behavior are determined by the properties of the surfaces and by the properties of fluid lubricants other than their bulk viscosity.

Break-in: See run-in.

Brinelling: Indentation of the surface of a solid body by repeated local impact or impacts, or static overload or damage to a solid bearing surface characterized by one or more plastically formed indentations brought about by overload.

Brittle erosion behavior: Erosion behavior having characteristic properties that can be associated with brittle fracture of the exposed surface, such as little or no plastic flow and the formation of intersecting cracks that create erosion fragments.

Brittle fracture: A form of wear in rolling, sliding, and impact contacts, characterized by the formation of tensile cracks in a single loading cycle.

Burnish (ing): To alter the original manufactured surface of a sliding or rolling surface to a more polished condition or to apply a substance to a surface by rubbing.

Catastrophic wear: Rapidly occurring or accelerating surface damage, deterioration, or change of shape caused by wear to such a degree that the service life of a part is appreciably shortened or its function destroyed.

Cavitation erosion: Progressive loss of original material from a solid surface due to continued exposure to cavitation or wear of a solid body moving relative to a liquid in a region of collapsing vapor bubbles that cause local high-impact pressures or temperatures.

Checking: See craze cracking.

Chemical wear: See corrosive wear.

Coefficient of friction: Ratio of the force required to initiate or maintain motion between to bodies, F, and the force pressing these bodies together, N, F/N.

Compound impact wear: Impact wear when there is a component of relative velocity parallel to the interface between the impacting bodies.

Coulomb friction: A term used to indicate that the frictional force is proportional to the normal load.

Corrosive wear: A wear process in which chemical or electrochemical reaction with the environment predominates. (Also called chemical wear.)

Craze cracking: Irregular surface cracking associated with thermal cycling. (Also called checking.)

Deformation wear: Sliding wear involving plastic deformation of the wearing surface or in impact wear of elastomers, the initial stage of wear not involving material loss but progressive deformation, generally approaching an asymptotic limit.

Delamination wear: A wear process in which thin layers of material are formed and removed from the wear surface or a wear process involving the nucleation and propagation of cracks so as to form lamellar wear particles.

Diffuse wear: Wear processes involving diffusion of elements from one body into the other, such as those often occurring in high-speed cutting tool wear, generally requires high temperatures.

Diamond film: A carbon-composed crystalline film that has the characteristics of diamond.

Diamondlike Film: A hard, non-crystalline carbon film.

DLC: Diamondlike carbon coatings.

Droplet erosion: Erosive wear caused by the impingement of liquid droplets on a solid surface.

Dry-film lubrication: Lubrication resulting from the application of a thin film of a solid to a surface.

Dry sliding wear: Sliding wear in which there is no intentional lubricant or moisture introduced into the contact area.

Dynamic friction: See kinetic friction.

Ductile erosion behavior: Erosion behavior having characteristic properties that can be associated with ductile fracture of the exposed solid surface, such as considerable plastic deformation preceding or accompanying material loss from the surface which can occur by gouging or tearing or by eventual embrittlement through work hardening that leads to crack formation.

Electrical discharge wear: Material removal as a result of electrical discharge.

Electrical pitting: The formation of surface cavities by removal of metal as a result of an electrical discharge across an interface.

Elastohydrodynamic lubrication: A form of fluid lubrication in which the friction and film thickness is a function of the deformation of the surfaces and the viscous properties of the fluid lubricant.

EP lubricant: See extreme-pressure lubricant.

Erosion: Loss of material from a solid surface due to relative motion in contact with a fluid that contains solid particles or progressive loss of original material from a solid surface due to mechanical interaction between that surface and a fluid, multi-component fluid, and impinging liquid, or solid particles.

Erosive wear: See erosion.

Erosion–corrosion: A conjoint action involving corrosion and erosion in the presence of a corrosive substance.

Extreme-pressure lubricant (EP lubricant): A lubricant that imparts increased load-carrying capacity to a rubbing surface under severe operating conditions.

False Brinelling: Damage to a solid bearing surface characterized by indentations not caused by plastic deformation resulting from overload, but thought to be due to other causes such as fretting or fretting corrosion or local spots appearing when the protective film on a metal is broken continually by repeated impacts.

Fatigue wear: Removal of particles detached by fatigue arising from cyclic stress variations or wear of a solid surface caused by fracture arising from material fatigue.

Ferrography: Characterization of magnetic wear debris from oil samples.

Flash temperature: The maximum local temperature generated at some point in a sliding contact.

Flow cavitation: Cavitation caused by a decrease in static pressure induced by changes in the velocity of a flowing liquid.

Fluid erosion: See liquid impingement erosion.

Fluid friction: Frictional resistance due to the viscous or rheological flow of fluids.

Fluid lubrication: A form of lubrication with a fluid in which the friction and thickness of the film is a function of the viscosity of the fluid. (See elastohydrodynamic and hydrodynamic lubrication.)

Fracture: See brittle fracture.

Fretting: Wear phenomena occurring between two surfaces having oscillatory relative motion of small amplitude. (Note: Term also can mean small-amplitude oscillatory motion.)

Fretting corrosion: A form of fretting in which chemical reaction predominates.

Fretting fatigue: The progressive damage to a solid surface that arises from fretting and leads to the formation of fatigue cracks.

Fretting wear: See fretting.

Friction: The tangential force between two bodies that opposes relative motion between these bodies.

Friction coefficient: See coefficient of friction.

Friction polymer: An organic deposit that is produced when certain metals are rubbed together in the presence of organic liquids or gases.

Full-film lubrication: Fluid lubrication when the surfaces are completely separated by the fluid film.

Galling: A severe form of scuffing associated with gross damage to the surface or failure or a form of surface damage arising between sliding solids, distinguished by macroscopic, usually localized roughening and creation of protrusions above the original surface, often includes plastic flow or material transfer or both or a severe form of adhesive wear.

Gouging abrasion: A form of high-stress abrasion in which easily observable grooves or gouges are created on the surface.

Heat checking: A process in which fine cracks are formed on the surface of a body in sliding contact due to the buildup of excessive frictional heat.

High-stress abrasion: A form of abrasion in which relatively large cutting force is imposed on the particles or protuberance causing the abrasion, and that produces significant cutting and deformation in the wearing surface.

Hydrodynamic lubrication: A form of fluid lubrication in which the friction and film thickness is a function of the viscous properties of the fluid lubricant.

Impact wear: Wear of a solid surface resulting from repeated collisions between that surface and another solid body.

Impingement corrosion: A form of erosion–corrosion generally associated with the impingement of a high-velocity flowing liquid containing air bubbles against a solid surface.

Incubation period: An initial amount of wearing action that is needed for the occurrence of some wear mechanisms or for these mechanisms to become detectable.

IRG Transition Diagram: See transition diagram.

Kinetic friction: Friction associated with sustained motion.

Lapping: A surface finishing process involving motion against an abrasive embedded in a soft metal or rubbing two surfaces together with or without abrasives, for the purpose of obtaining extreme dimensional accuracy or superior surface finish.

Limiting PV: The value of the PV Factor above which severe wear results. Typically use to characterize the wear behavior of plastics and other bearing materials.

Liquid impact erosion: See erosion.

Liquid impingement erosion: See erosion.

Low-stress abrasion: A form of abrasion in which relatively low contact pressure on the abrading particles or protuberances cause only fine scratches and microscopic cutting chips to be produced.

Lubricant: Any substance interposed between two surfaces in relative motion for the purpose of reducing the friction or wear between them.

Lubrication: The reduction of wear or friction by the use of a lubricant.

Lubricated impact wear: Impact wear with lubrication.

Lubricated rolling wear: Rolling wear with lubrication.

Lubricated sliding wear: Sliding wear with lubrication.

Measurable-wear: In the context of the Zero and Measurable Wear Models for sliding and impact, it describes a state of wear in which the wear exceeds the magnitude of the surface roughness.

Mechanical wear: Removal of material due to mechanical processes under conditions of sliding, rolling, or repeated impacts; includes adhesive wear, abrasive wear, and fatigue wear but not corrosive wear and thermal wear.

Metallic wear: Typically, wear due to rubbing or sliding contact between metallic materials that exhibits the characteristics of severe wear, such as, significant plastic deformation, material transfer, and indications that cold welding of asperities possibly has taken place as part of the wear process.

Mild wear: A form of wear characterized by the removal of material in very small fragments.

Mixed lubrication: A condition of lubrication in which the friction and wear behavior are determined by the properties of the surfaces and by the viscous and non-viscous properties of fluid lubricants.

Oxidative wear: A corrosive wear process in which chemical reaction with oxygen or oxidizing environments predominates.

Peening wear: Removal of material from a solid surface caused by repeated impacts on very small areas.

Pitting: A form of wear involving the separation of particles from a surface in the form of flakes, resulting from repeated stress cycling, generally less extensive than spalling.

Plowing: The formation of grooves by plastic deformation of the softer of two surfaces in relative motion.

Polishing wear: An extremely mild form of wear, which may involve extremely fine-scale abrasion, plastics smearing of micro-asperities, and/or tribochemical material removal.

PV Factor: Product of pressure and velocity. (See Limiting PV.)

Quasi-hydrodynamic lubrication: See mixed lubrication.

Ratcheting: A sliding wear process involving progressive deformation, ultimately leading to the formation of loose fragments.

Rehbinder Effect: Modification of the mechanical properties at or near the surface of a solid, attributed to interaction with a surface-active substance or surfactant.

Repeated-cycle deformation wear: Mechanical wear mechanisms requiring repeated cycles of mechanical deformation or engagement.

Ridging (wear): A deep form of scratching in parallel ridges usually caused by plastic flow.

Rolling-contact fatigue: Wear to a solid surface that results from the repeated stressing of a solid surface due to rolling contact between that surface and another solid surface or surfaces, generally resulting in the formation of sub-surface cracks, material pitting, and spallation. (Note: Often used as a synonym for rolling-contact wear.) See rolling-contact wear.

Rolling-contact wear: Wear due to the relative motion between non-conforming solid bodies whose surface velocities in the nominal contact location are identical in magnitude, direction, and sense; most common form is rolling-contact fatigue.

Run-in: An initial transition process occurring in newly established wearing contacts. (As a verb, run in, refers to an initial operation designed to improve wear and friction performance of a device.)

Scoring: The formation of severe scratches in the direction of sliding or a severe form of wear characterized by the formation of extensive grooves and scratches in the direction of sliding. (Note: Sometimes also called scuffing in USA.) See scuffing and scratching.

Scouring abrasion: See abrasion.

Scratch: A groove produced in a solid surface by the cutting and plowing action of a sharp particle or protuberance moving along that surface.

Scratching: The formation of fine scratches in the direction of sliding; a mild form of scoring.

Scuffing: Localized damage caused by the occurrence of solid-phase welding between sliding surfaces, without local surface melting or a mild degree of galling that results from the welding of asperities due to frictional heat or a form of wear occurring in inadequately lubricated tribosystems which is characterized by macroscopically observable changes in surface texture, with features related to the direction of relative motion. (Note: Sometimes also called scoring in USA.) See scoring.

Selective transfer: A wear process involving the transfer and attachment of a specific species from one surface to the mating surface during sliding.

Self-lubricating material: Any solid material that shows low friction without application of a lubricant.

Severe wear: A form of wear characterized by removal of material in relatively large fragments.

Shelling: A term used in railway engineering to describe an advanced phase of spalling.

Single-cycle deformation wear: Mechanical wear mechanisms requiring only a single cycle of contact or engagement.

Sliding wear: Wear due to relative sliding between two bodies in contact.

Slurry abrasion: Three-body abrasive wear involving a slurry.

Slurry erosion: Erosion produced by the movement of a slurry past a solid surface.

Smearing: Mechanical removal of material from a surface, usually involving plastic shear deformation, and redeposition of the material as a thin layer on one or both surfaces.

Solid impingement erosion: Progressive loss of original material from a solid surface due to continued exposure to impacts by solid particles.

Solid lubricant: Any solid used as a powder or thin film on a surface to reduce friction and wear.

Solid particle erosion: See solid impingement erosion.

Sommerfeld Number: A dimensionless number that is used to characterize the state of lubrication of a bearing. (See Stribeck curve.)

Spalling: A form of wear involving the separation of particles from a surface in the form of flakes as a result of repeated stressing, generally more extensive than pitting or a form of wear involving the separation of macroscopic particles from a surface in the form of flakes or chips, usually associated with rolling but may also result from impact.

Specific wear rate: Wear volume divided by load and distance of sliding. (See wear factor.)

Static friction: Friction associated with the initiation of motion.

Stick-slip: A relaxation oscillation in friction, which is generally characterized by a sharp decrease, followed by a more gradual increase in the force of friction. It generally causes jerky-type motion and squeaking.

Stiction: Term used to signify the condition in which the frictional resistance is sufficient to prevent macroscopic sliding.

Stress fracture: See brittle fracture.

Stribeck curve: A graph showing the relationship between the coefficient of friction for a journal bearing and the dimensionless Sommerfeld Number. There is a general correlation between this number and the different forms of lubrication with a liquid, boundary, mixed, and fluid. (See.)

Surface distress: In bearings and gears damage to the contacting surfaces that occurs through intermittent solid contact involving some degree of sliding and surface fatigue.

Traction: The transmission of tangential stress across an interface.

Traction coefficient: Ratio of the traction force to the normal force pressing the surfaces together.

Traction Force: The tangential force transmitted across an interface.

Transfer film: A tribofilm composed wear debris from the counterface.

Tribochemistry: Chemistry dealing with interacting surfaces in relative motion.

Tribofilm wear: Wear processes that are controlled by the formation of tribofilms, such as transfer and third-body films.

Tribosystem: All those elements that affect friction and wear behavior.

Thermal wear: Removal of material due to softening, melting, or evaporation during sliding, rolling, or impact.

Thermoelastic instability: Sharp variation of local surface temperatures with passing of asperities leading to stationary or slowly moving hot spots of significant magnitude, resulting in local expansion and elevation of the surface.

Third body: A solid interposed between two contacting surfaces.

Third-body film: A tribofilm containing wear debris from the surface, generally a mixture of wear debris from both surfaces.

Three-body abrasion: Abrasive wear when the abrasive particles are free to move.

Two-body abrasion: Abrasive wear from protuberances or attached abrasive particles.

Transfer: The process by which material from one sliding surface becomes attached to another surface, possibly as the result of interfacial adhesion.

Transition diagram: A form of wear map, involving to or more experimental or operating parameters, which is used to indicate boundaries between different regimes of wear or surface damage or effectiveness of lubrication, such as the *IRG Transition Diagrams*.

Vibratory cavitation: Cavitation caused by the pressure fluctuations within a liquid, induced by the vibrations of a solid surface immersed in the liquid.

Wear: Damage to a solid surface, generally involving or leading to progressive loss of material, that is due to the relative motion between that surface and a contacting substance or substances.

Wear coefficient: Normally defined as the non-dimensional coefficient, k, in the following equation, $V = KPS/3p$, where V is the volume of wear, P is the load, S is the distance of sliding, and p is hardness. Less specific, it is the dimensionless form of a wear factor obtained dividing it by the hardness of the wearing material.

Wear curve: Plot of wear as a function of usage, e.g., wear depth vs. sliding distance and wear volume vs. time.

Wear factor: Constant in a linear wear equation, $V = KPS$, where V is the volume of wear, P is the load, and S is the sliding distance. Wear volume divided by load and sliding distance. (Note: An alternative definition is based on the differential form of this equation, $\Delta V = KP\Delta S$, where ΔV is the incremental increase in wear volume over an incremental amount of sliding, ΔS.)

Wear-in: (See run-in.)

Wear map: A graphical characterization of wear behavior in terms of independent operational parameters of the tribosystem, such as speed and load. Various forms of wear maps are used to identify ranges and combinations of operational parameters for different wear mechanisms, same wear rates, and acceptable operating conditions. (See transition diagram.)

Wedge formation: In sliding metals, the formation of a wedge or wedges of plastically sheared metal in local regions of interaction between sliding surfaces.

Zero-wear: In the context of the Zero and Measurable Wear Models for sliding and impact, it describes a state of wear in which the wear or damage is less than the magnitude of the surface roughness.

Appendix

GALLING THRESHOLD STRESS

Material Pair		Threshold Stress	
		MPa	Kpsi
Silicon bronze (Rb 94)	Silicon bronze (Rb 94)	28	4
	304 SS (Rb 77)	300	44
1020 (Rb 90)	440C SS (Rc 58)	14	2
1034 (Rc 45)	1034 (Rc 45)	14	2
	Nitronic 32 (Rb 94)	14	2
4337 (Rc 48)	Nitronic 60 (Rb 94)	>350	>51
4337 (Rc 51)	4337 (Rc 45)	14	2
	Nitronic 60 (Rb 94)	>350	>51
201 SS (Rb 94)	201 SS (Rb 94)	105	15
	304 SS (Rb 77)	14	2
	630 SS [17-4Ph] (Rc 41)	14	2
	S24100 (Rc 21)	284	41
	Nitronic 32 (Rb 98)	250	36
	Waukesha 88 (Rb 77)	>350	>51
301 SS (Rb 87)	416 SS (Rc 37)	21	3
	440C (Rc 56)	21	3
	Nitronic 60 (Rb 94)	>350	>51
303 SS (Rb 81)	303 SS (Rb 81)	14	2
	304 SS (Rb 77)	14	2
	316 SS (Rb 81)	21	3
	410 SS (Rc 38)	28	4
	416 SS (Rc 37)	60	8
	430 SS (Rb 84)	14	2
	440C SS (Rc 56)	35	5
	630 SS [17-4Ph] (Rc 45)	21	3
	Nitronic 32 (Rb 99)	>350	>51
	Nitronic 60 (Rb 94)	>350	>51
303 SS (Rb 85)	303 SS (Rb 85)	138	20
303 SS (Rb 89)	Waukesha 88 (Rb 77)	>350	>51
304 SS (Rb 77)	Silicon Bronze (Rb 94)	300	44
	201 SS (Rb 94)	14	2
	303 SS(Rb 81)	14	2

(Continued)

Appendix (Continued)

Material Pair		Threshold Stress	
		MPa	Kpsi
	304 SS (Rb 77)	14	2
	316 SS (Rb 81)	14	2
	410 SS (Rc 38)	21	3
	416 SS (Rc 37)	165	24
	430 SS (Rb 84)	14	2
	440C SS (Rc 56)	21	2
	630 SS [17-4Ph] (Rc 33)	14	2
	630 SS [17-4Ph] (Rc 45)	14	2
	630 SS [17-4Ph] (Rc 47)	14	2
	Nitronic 32 (Rb 99)	210	30
	Nitronic 32 (Rc 43)	90	13
	Nitronic 50 (Rb 94)	28	4
	Nitronic 60 (Rb 94)	>350	>51
304 SS (Rb 86)	304 SS (Rb 86)	55	8
	304 SS (Rc 27)	28	4
	440C (Rc 55)	28	4
	S20910 (Rb 97)	69	10
	S2410 (Rc 23)	>104	>15
	Custom 450 (Rc 43)	21	3
	Custom 455(Rc 48)	124	18
	Nitronic 30 (Rb 96)	41	6
304 SS (Rc 27)	304 SS (Rb 86)	28	4
	304 SS (Rc 27)	17	2.5
316 SS (Rb 77)	Stellite 6B (Rc 45)	240	35
316 SS (Rb 81)	303 SS (Rb 81)	21	3
	304 SS (Rb 77)	14	2
	316 SS (Rb 81)	14	2
	316 SS (Rc 27)	55	8
	410 SS (Rc 38)	14	2
	416 SS (Rc 37)	290	42
	430 SS (Rb 84)	14	2
	440C SS (Rc 56)	14–255	2–37
	630 SS [17-4Ph] (Rc 45)	14	2
	Nitronic 32 (Rb 99)	21	3
	Nitronic 60 (Rb 94)	260	38
316 SS (Rb 83)	316 SS (Rb 83)	48	7
316 SS (Rb 90)	440C SS (Rc 58)	7	1
	Nitronic 60 (Rb 92)	35	5
316 SS (Rb 94)	Waukesha 88 (Rb 77)	>350	>51
316 SS (Rc 27)	316 SS (Rc 27)	35	5
	329 SS (Rc 27)	14	2
329 SS (Rc 25)	316 SS (Rc 27)	14	2
	329 SS (Rc 25)	7	1
410 SS (Rb 87)	410 SS (Rb 87)	7	1
410 SS (Rc 32)	416 SS (Rc 38)	28	4
	420 SS (Rc 50)	21	3

(*Continued*)

Appendix   (Continued)

		Threshold Stress	
Material Pair		MPa	Kpsi
410 SS (Rc 34)	416 SS (Rc 40)	28	4
	420 SS (Rc 48)	21	3
410 SS (Rc 38)	303 SS (Rb 81)	28	4
	304 SS (Rb 77)	14	2
	316 SS (Rb 81)	14	2
	410 SS (Rc 38)	21	3
	416 SS (Rc 37)	28	4
	430 SS (Rb 84)	21	3
	440C SS (Rc 56)	21	3
	630 SS [17-4Ph] (Rc 45)	21	3
	Nitronic 32 (Rb 99)	320	46
	Nitronic 60 (Rb 94)	>350	>51
410 SS (Rc 43)	410 SS (Rc 43)	21	3
	440C (Rc 55)	35	5
416 SS (Rb 83)	416 SS (Rc 32)	76	11
416 SS (Rb 95)	416 SS (Rb 95)	21	3
416 SS (Rc 32)	416 SS (Rb 83)	76	11
	416 SS (Rc 32)	42	6
416 SS (Rc 34)	430 SS (Rb 90)	21	3
416 SS (Rc 37)	301 SS (Rb 87)	21	3
	303 SS (Rb 81)	60	9
	304 SS (Rb 77)	165	24
	316 SS (Rb 81)	290	42
	410 SS (Rc 38)	28	4
	416 SS (Rc 37)	90	13
	430 SS (Rb 84)	21	3
	440 SS (Rc 55)	159	23
	440C SS (Rc 56)	145	21
	630 SS [17-4Ph] (Rc 45)	14	2
	20Cr-80Ni (Rb 89)	50	7
	Nitronic 32 (Rb 99)	310	45
	Nitronic 60 (Rb 94)	>350	>51
416 SS (Rc 37)	410 SS (Rc 32)	28	4
	416 SS (Rc 37)	62	9
416 SS (Rc 40)	410 SS (Rc 34)	28	4
420 SS (Rc 49)	410 SS (Rc 32)	21	3
	420 SS (Rc 49)	55	8
	Nitraonic 60 (Rb 96)	>345	>50
420 SS (Rc 55)	420 SS (Rc 55)	125	18
430 SS (Rb 84)	303 SS (Rb 81)	14	2
	304 SS (Rb 77)	14	2
	316 SS (Rb 81)	14	2
	410 SS (Rc 38)	21	3
	416 SS (Rc 37)	21	3
	430 SS (Rb 84)	14	2
	440C SS(Rc 56)	14	2
	630 SS [17-4Ph] (Rc 45)	21	3

(Continued)

Appendix   (Continued)

Material Pair		Threshold Stress	
		MPa	Kpsi
	Nitronic 32 (Rb 99)	21	3
	Nitronic 60 (Rb 94)	250	36
430 SS (Rb 90)	416 SS (Rc 34)	21	3
430 SS (Rb 98)	430 SS (Rb 98)	10	1.5
430F SS (Rb 92)	430F SS (Rb 92)	14	2
430C SS (Rc 56)	301 SS (Rb 87)	21	3
	303 SS (Rb 81)	35	5
	304 SS (Rb 77)	21	3
	304 SS (Rb 86)	28	4
	316 SS (Rb 81)	250	36
	410 SS (Rc 38)	21	3
	410 SS (Rc 43)	35	5
	416 SS (Rc 37)	145	21
	430 SS (Rb 84)	14	2
	440C SS (Rc 56)	75	11
	440C SS (Rc 59)	80	12
	630 SS [17-4Ph] (Rc 45)	21–76	3–11
	Nitronic 32 (Rb 99)	>350	>51
	Nitronic 60 (Rb 94)	>350	>51
440C SS (Rc 58)	1020 (Rb 90)	14	2
	316 SS (Rb 90)	7	1
	440C SS (Rc 58)	35–75	5–11
	17-5PH (Rc 43)	7	1
	17-5PH nitrided (Rc 70)	>500	>73
	S30430 (Rb 74)	83	12
	Nitronic 60 (Rb 92)	14	2
	Stellite 6 (Rc 42)	55	8
	Tribaloy 400 (Rc 48)	140	20
630 SS [17-4Ph] (Rc 33)	304 SS (Rb 77)	14	2
	630 SS [17-4Ph] (Rc 34)	35	5
	Nitronic 60 (Rb 96)	>350	>51
630 SS [17-4Ph] (Rc 36)	Nitronic 60 (Rb 94)	>350	>51
630 SS [17-4Ph] (Rc 38)	S13800 (Rc 46)	14	2
	S24100 (Rc 23)	76	11
	Custom 455 (Rc 48)	55	8
630 SS [17-4Ph] (Rc 41)	201 SS (Rb 94)	14	2
	S17700 (Rc 41)	21	3
	Nitronic 32 (Rb 94)	75	11
	Nitronic 32 (Rb 43)	21	3
630 SS [17-4Ph] (Rc 43)	631 SS (Rc 43)	21	3
630 SS [17-4Ph] (Rc 44)	Waukesha 88 (Rb 77)	>350	>51
630 SS [17-4Ph] (Rc 45)	303 SS (Rb 81)	14	2
	304 SS (Rb 77)	14	2
	316 SS (Rb 81)	14	2
	410 SS (Rc 38)	21	3
	416 SS (Rc 37)	14	2

(Continued)

Appendix (Continued)

Material Pair		Threshold Stress	
		MPa	Kpsi
	430 SS (Rb 84)	21	3
	440C SS (Rc 56)	21	3
	630 SS [17-4Ph] (Rc 45)	14	2
	S13800 (Rc 46)	62	9
	S17700 (Rc 44)	14	2
	Custom 450 (Rc 48)	76	11
	Nitronic 32 (Rb 99)	>350	>51
	Nitraonic 60 (Rb 94)	>350	>51
630 SS [17-4Ph] (Rc 47)	304 SS (Rb 77)	14	2
	630 SS [17-4Ph] (Rc 47)	69	10
	631 SS (Rc 47)	14	2
631 SS (Rc 43)	630 SS [17-4Ph] (Rc 43)	21	3
17-5Ph (Rc 43)	440C SS (Rc 58)	7	1
	Stellite 6 (Rc 42)	35	5
	Tribaloy 400 (Rc 48)	48	7
17-5Ph nitrided (Rc 70)	440C SS (Rc 58)	>500	>73
660 SS (Rc 28)	660 SS (Rc 28)	21	3
	Nitronic 60 (Rb 94)	>350	>51
N08020 (Rb 87)	N08020 (Rb 87)	14	2
	Nitronic 60 (Rb 92)	48	7
S13800 (Rc 46)	630 SS [17-4Ph] (Rc 38)	14	2
	630 SS [17-4Ph] (Rc 45)	62	9
	S13800 (Rc 46)	21	3
	Nitronic 60 (Rb 96)	>345	>50
S17700 (Rc 41)	630 SS [17-4Ph] (Rc 41)	21	3
S17700 (Rc 44)	630 SS [17-4Ph] (Rc 44)	14	2
S18200 (Rb 98)	S1820 (Rb 98)	35	5
S20900 (Rb 96)	S20900 (Rb 96)	48	7
S20910 (Rb 95)	S24100 (Rc 43)	90	13
	Nitronic 60 (Rb 96)	>345	>50
S20910 (Rb 97)	S20910 (Rb 97)	35	5
	304 SS (Rb 86)	69	10
S20910 (Rc 34)	S24100 (Rc 23)	55	8
S24100 (Rc 21)	201 SS (Rb 94)	284	36
S24100 (Rc 23)	304 SS (Rb 86)	>104	>15
	630 SS [17-4Ph] (Rc 39)	76	11
	S20910 (Rc 34)	55	8
	S24100 (Rb 23)	97	14
S24100 (Rc 43)	S20910 (Rb 95)	90	13
S30430 (Rb 74)	S30430 (Rb 74)	35	5
	440C SS (Rc 55)	83	12
S66286 (Rc 30)	S66286 (Rc 30)	14	2
20 Cr-80Ni (Rb 89)	416 SS (Rc 37)	50	7
	Nitronic 60 (Rb 94)	250	36
	Waukesha 88 (Rb 77)	>350	>51
Ti-6Al-4V (Rc 36)	Nitronic 60 (Rb 94)	>350	>51

(Continued)

Appendix   (Continued)

Material Pair		Threshold Stress	
		MPa	Kpsi
Custom 450 (Rc 29)	Custom 450 (Rc 29)	69	11
	Nitronic 30 (Rb 96)	55	88
Custom 450 (Rc 33)	Custom 450 (Rc 33)	14	2
Custom 450 (Rc 38)	Custom 450 (Rc 38)	17	2.5
Custom 450 (Rc 43)	304 SS (Rb 86)	21	3
	Custom 450 (Rc 43)	55	8
	Nitronic 30 (Rb 96)	62	9
Custom 450 (Rc 48)	630 SS [17-4Ph] (Rc 45)	76	11
Custom 455 (Rc 36)	Custom 455 (Rc 36)	28	4
Custom 455 (Rc 43)	Custom 455 (Rc 43)	59	8.5
Custom 455 (Rc 48)	304 SS (Rb 86)	124	18
	630 SS [17-4Ph] (Rc 38)	55	8
	Custom 455 (Rc-48)	90	13
Gall Tough (Rb 95)	Gall Tough (Rb 95)	104	15
Hard Anodized	Hard Anodized	>270	>39
Nitronic 30 (Rb 96)	304 SS (Rb 86)	41	6
	Custom 450 (Rc 29)	55	8
	Custom 450 (Rc 43)	62	9
	Nitronic 30 (Rb 96)	166	24
	Nitronic 30 (Rc 35)	104	15
Nitronic 30 (Rc 35)	Nitronic 30 (Rb 96)	104	15
	Nitronic 30 (Rc 35)	62	9
Nitronic 32 (Rb 94)	1034 (Rb 94)	14	2
	630 SS [17-4Ph] (Rc 41)	75	11
	Nitronic 32 (Rc 43)	235	34
	304 SS (Rb 77)	50	7
Nitronic 32 (Rb 98)	201 SS (Rb 94)	250	36
	303 SS (Rb 81)	>350	>51
	304 SS (Rb 77)	210	30
	316 SS (Rb 81)	21	3
	410 SS (Rc 38)	315	46
	416 SS (Rc 37)	310	45
	430 SS (Rb 84)	55	8
	440C SS (Rc 56)	>350	>51
	630 SS [17-4Ph] (Rc 45)	>350	>51
	Nitronic 32 (Rb 99)	210	30
	Nitronic 50 (Rc 34)	55	8
	Nitronic 60 (Rb 94)	>350	>51
Nitronic 32 (Rc 43)	304 SS (Rb 77)	90	13
	630 SS [17-4Ph] (Rc 41)	21	3
	Nitronic 32 (Rb 43)	235	34
	Nitronic 50 (Rb 94)	90	13
Nitronic 50 (Rb 94)	304 SS (Rb 77)	28	4
	Nitronic 32 (Rc 43)	90	13
	Nitronic 50 (Rb 94)	14	2
	Nitronic 60 (Rb 94)	>350	>51

(Continued)

Appendix (Continued)

Material Pair		Threshold Stress	
		MPa	Kpsi
Nitronic 50 (Rc 34)	Nitronic 32 (Rb 99)	55	8
Nitronic 60 (Rb 92)	316 SS (Rb 90)	35	5
	440C (Rc 58)	14	2
	N08020 (Rb 87)	48	7
	Nitronic 60 (Rb 92)	14–104	2–15
Nitronic 60 (Rb 94)	4337 (Rc 48)	>350	>51
	4337 (Rc 51)	>350	>51
	301 SS (169)	>350	>51
	303 SS (Rb 81)	>350	>51
	304 SS (Rb 77)	>350	>51
	316 SS (Rb 81)	260	38
	410 SS (Rc 38)	>350	>51
	416 SS (Rc 37)	>350	>51
	420 SS (Rc 48)	>350	>51
	430 SS (Rb 84)	250	36
	440C SS (Rc 56)	>350	>51
	630 SS [17-4Ph] (Rc 36)	>350	>51
	630 SS [17-4Ph] (Rc 45)	>350	>51
	660 SS (Rc 28)	>350	>51
	20Cr-80Ni (Rb 89)	250	36
	Ti-6Al-4V (Rc 36)	>350	>51
	Nitronic 32 (Rb 99)	>350	>51
	Nitronic 50 (Rb 94)	>350	>51
	Nitronic 60 (Rb 94)	>350	>51
	Stellite 6B (Rc 45)	>350	>51
Nitronic 60 (Rb 96)	301 SS (Rb 87)	>345	>50
	420 SS (Rc 50)	>345	>50
	630 SS [17-4Ph] (Rc 33)	>350	>51
	S13800 (Rc 44)	>345	>50
	S20910 (Rb 95)	>345	>50
Stellite 6 (Rc 42)	440C SS (Rc 58)	55	8
	15-5Ph (Rc 43)	35	5
Stellite 6B (Rc 45)	304 SS (Rb 77)	240	35
	316 SS (Rb 77)	25	3.5
	Nitronic 60 (Rb 94)	>350	>51
	Stellite 6B (Rc 45)	>350	>51
	Tribaloy 400 (Rc 54)	>350	>51
	Tribaloy 700 (Rc 47)	>350	>51
Tribaloy 400 (Rc 48)	440C SS (Rc 58)	140	20
	15-5 Ph (Rc 43)	48	7
Tribaloy 400 (Rc 54)	Tribaloy 400 (Rc 54)	>350	>51
Tribaloy 700 (Rc 47)	Tribaloy 700 (Rc 47)	185	27
Waukesha 88 (Rb 77)	201 SS (Rb 94)	>350	>51
	303 SS (Rb 89)	>350	>51
	316 SS (Rb 94)	>350	>51
	630 SS [17-4Ph] (Rc 44)	>350	>51
	20Cr-80Ni (Rb 89)	>350	>51

Sources:
1. J Magee. Wear of Stainless Steels, In: P Blau, Ed., Friction, Lubrication and Wear Technology, ASM Handbook. Vol. 18. Materials Park, OH: ASM International, 1992, 710–724.
2. B Bhushan, B Gupta. Handbook of Tribology. Chapter 4. New York: McGraw-Hill, 1991.
3. K Budinski. Proc Intl Conf Wear Materials ASME 171, 1981.

Index